When Silver Was K[illegible]g
Arizona's Famous 1880[illegible] Mine

By Jack San Felice

First Edition

2006

When Silver Was King
Arizona's Famous 1880s Silver King Mine

Published by: Millsite Canyon Publishing — Mesa, Arizona 85215

First Printing 2006
Photographs not otherwise credited are the author's.
Front cover photo of King's Crown Mountain by Author
Inset photo of Silver King Town courtesy Velma Bowen Tucker
Back cover photo of Headframe courtesy Dick Bynum
Inset photo of author courtesy Greg Davis
Cover and book design by Tony Felice

Publishers Cataloging-in-Publishing Data
San Felice, Jack
When Silver Was King—Arizona's Famous 1880s Silver King Mine, by Jack San Felice 1st. Edition
p: photos, illustrations.
Bibliography
Includes name index

1. History--2. Silver King Mine (AZ)--3. Silver Mining --4. Ore Processing--5. Cornish Miners--6. Arizona Pioneers--7. Ghost Towns--8. Pinal Mountains (AZ)--(9) Stoneman Grade--10. Central Arizona Freighters--11.Pinal City (AZ)

Library of Congress Catalog Card Number 99-070028
ISBN 1-890216-06-2
ISBN 9781890216061

10 9 8 7 6 5 4 3 2 1

Silver King miners with child cir. 1880s —Bowen Family Collection

ACKNOWLEDGEMENTS

Special Recognition is given to Jack Carlson, Gregory E. Davis, Sam Michael, Gladys Walker, Ronald R. Deen Sr. family and the Velma Bowen Tucker family.

Acknowledgement is given to the following people and agencies: The staffs of the; Arizona Historical Society at the University of Arizona, Tucson; Historical Foundation at Arizona State University, Tempe; Arizona State Archives at the Capitol, Phoenix; the late Mason Coggins and the staff of the Arizona Bureau of Mines and Mining Museum, Bancroft Library, Berkely, California; William Haak and the staff of the Gila County Historical Society; Pinal County Recorder's Office; Superstition Mountain Historical Society; Superior Historical Society; Connie Stotts, Debbie Jennings, and Shawn Stouffer, of Superstition Springs OfficeMax, Inc.; Charles G. Taylor of the BHP staff; E. S. Crosswhite of Boyce Thompson Arboretum; Richard Bynum and George Sites; Claressa Pell and Dee Wardwell, (the decendants of William Wardwell); Ken and Nancy Barkley McCollough; John Swearengin; Pinal County Historical Society; Dennis Karkoz and Dave Lira, former Magma Mine employees; Bob Schoose of Goldfield Ghost Town; Ron, Jayne, Jesse and Josh Feldman of the OK Corral; Greg Beseth; Tom and Roberta Clary; Marcie Greenberg; Jesse Gage; James Herron family; Margaret Down and Judy Pintar (Phy and Gabriel decendants); *Arizona Silver Bell* newspaper staff, Globe; Dan Wright, Ann Knight Rose; Michael Papaiannni; Marshall Trimble; Laura Devine, Tim Williams; Laura Hubbard and Mindy Colwell of Costco, Superstition Springs.

Thanks to my readers Jack Carlson, Pat Harrison, Elizabeth Stewart, Bob Stambach and Wynne San Felice.

Thanks to my cover and book designer Tony Felice.

To my children, Tony, Rusty, Cheri (and like a daughter Stephanie) and my grandchildren Lauren, Tony, Michael, Rachael, Brandon, Gabby, Evan, Maia plus little Vincent, I hope this inspires one of them to be a historian or writer.

A special thanks is given to my wife Winnie for her patience during the long research and writing of this book.

In Memorium

My Father, Anthony San Felice 1915-2001

My Sister, Rose Marie San Felice Hazen 1938-2003

Dedication

To Freddie Joe Deen for his courage, and the late Velma Bowen Tucker for her love of history.

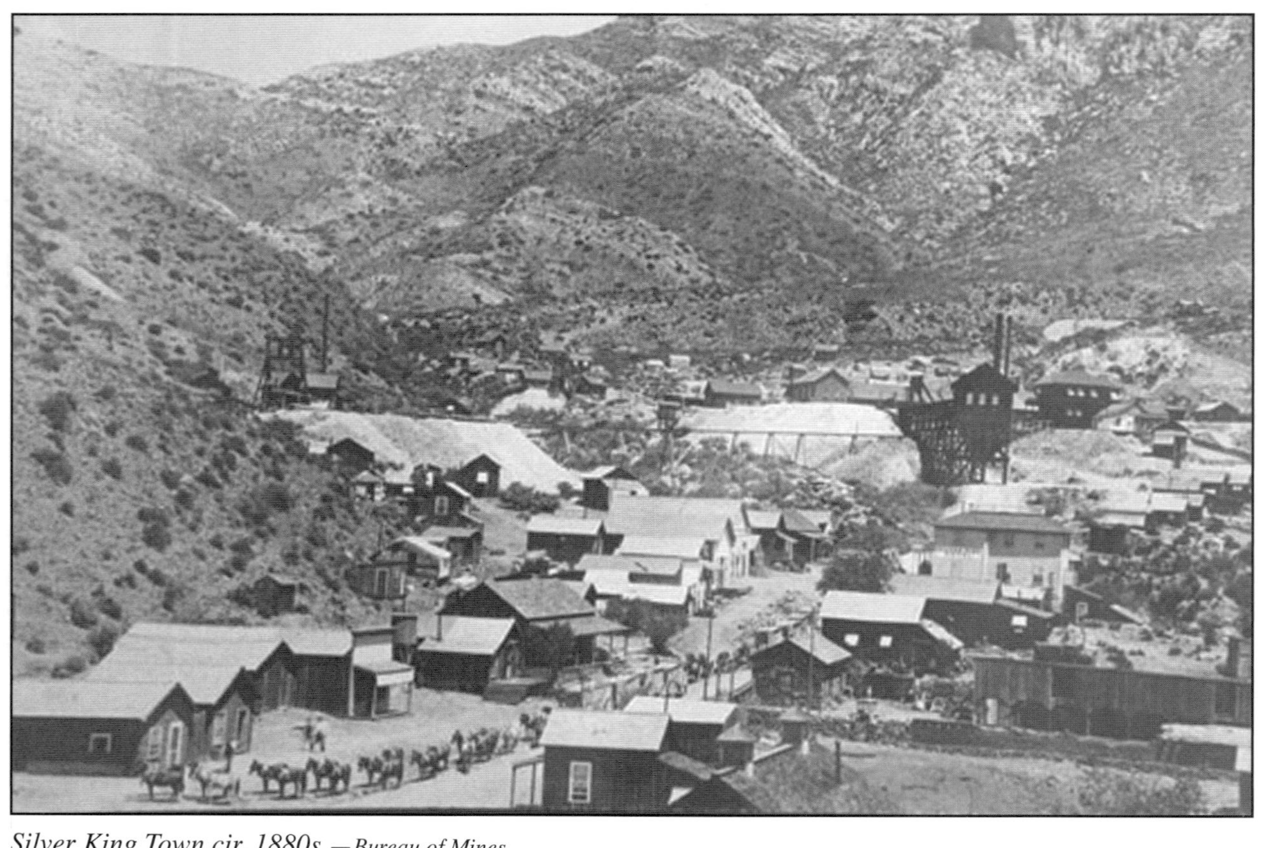

Silver King Town cir. 1880s —*Bureau of Mines*

WHEN SILVER WAS KING

Jack San Felice © 2006

TABLE OF CONTENTS

Acknowledgements
Introduction i

Chapters

1. Discovery of the Silver King Mine 1
2. When Silver Was King—1880s Silver King Mine 35
3. Development of the Silver King Mine 49
4. Processing Ore at the Silver Mills - Pinal City 61
5. The Mining Men of the Silver King Mine 73
6. Mule Skinners, Freighters & Cowboys 95
7. Life at Silver King Town and Pinal City 114
8. Pioneer Women of Silver King and Pinal 154
9. A Pioneer Family—The Bowens of Silver King 170
10. Stagecoaches, *Bullion * & Bullets * 182
11. Frontier Justice and the Outlaws 194
12. Tragic Tale of the Lost Soldiers Mine 223
13. Silver King Mine's Impact on Nearby Mining 245
14. Death and Re-births of the Silver King Mine 277
15. In Search of History—Camp Picket Post to Camp Pinal 297
16. 1996 Silver King Reopening by the Deens 308

Footnotes by Chapter 338
Bibliography 351
Glossary of Mining Terms 358
Index 362
About the Author 371

CAPITAL STOCK
$10.000.000
100.000 SHARES
$100.00 EACH.

LOCATION
PIONEER DISTRICT
PINAL CO.
ARIZONA

Nº

★25★

SILVER KING
MINING CO.

San Francisco, 188

This Certifies that is entitled

to Shares of the Capital Stock of the

SILVER KING MINING CO.

Transferable on the books of the Company by endorsement hereon and surrender of this Certificate

SECRETARY.

INCORPORATED
MAY 3D
1877.

PRESIDENT.

The Bancroft Company Lith S.F.

1877 Silver King Stock Certificate—Sam Michael Collection

INTRODUCTION

The Author's Involvement with the Silver King Mine

The Old Silver King Mine has always held a fascination for me, as it impacted so much on the early history of the Arizona Territory. While doing research on Arizona history over ten years ago, I came across several references to the old mine. This was also the location where several of the famous searches for the Lost Dutchman Mine were initiated, so I wanted to see it for myself. Also, I was avidly interested in stories of the Lost Dutchman Mine and the Lost Soldiers Mine.

After several visits to the Old Silver King Mine I began to do some research in order to be more informed, and to make a class I was teaching, called the Lore of the Superstitions, more interesting. Here was something real and tangible relating to the tale of the Lost Dutchman. There was so much written and much unwritten about the old mine, that I began to assemble quite a large collection of material. I felt that the Silver King Town and Pinal City stories had to be told in the context of those 1880s mining days, when people lived during the west's most turbulent time period. This was a very exciting era of the west and the time of Apache warfare, rugged prospectors, miners, rich ore strikes, stagecoach holdups, and the difficult yet interesting lifestyles of the people of the 1880s.

I began to really delve into the old files at the Mining and Mineral Museum in Phoenix. My research into the old files began with the assistance of the late Director Mason Coggins and his staff. I checked the files at the Historical Foundation at the Hayden Library in Tempe, the Arizona Historical Society's files at Tucson, as well as all the area's public libraries. I also contacted any person who had any information on the mine, and the old mining camps around where the mine was located. I received assistance from my friend Gregory Davis, the person in charge of Research and Acquisitions for the Superstition Mountain Museum who had numerous files on the old mine. I met Sam Michael, an antique dealer, who shared his experiences at Silver King and his files with me. Then, I was referred to Gladys Walker, a local historian at Superior, who had many old newspaper clippings about the old mine as well as some old photos.

Since I had been going to the old abandoned mine and townsite beginning in the early 1990s, I was quite surprised in October of 1996 to see that someone had actually begun to work in the old mine. It was then that I became acquainted with the Deen family. They were working on draining the water out, and had it drained down whereby the old 256 foot level was now open. I wanted to see for myself the underground workings and climbed down by way of ladders that went down 16 feet to each small platform, where the ladder then went down another 16 feet, etc. I climbed down into the "hole," where apparently no one had been since the old miners of the 1880s. As I reached the 256-foot level Ron Deen Jr., Dick Bynum and a miner named Roy Wheeler were shoring up the old timbers and doing some exploratory work. They were examining the old quartz pipe, and taking ore samples up to the surface to be assayed. At the 256-foot level, the tunnels were filled with old timbers and the work was slow going.

After speaking with Ron Deen Sr., Joe Deen, and Ron Deen Jr., and viewing the underground workings of the old mine, I decided to do some serious research on the old mining towns of Silver King and Pinal. The Deens gave me the right to do their story, and unlimited access to the property, as it was now closed and locked with a steel gate. In return for their access to the mine and sharing underground mining knowledge, I said that I would share any important documents that I located with them. So began this story.

I was able to locate several of the old mining reports of the 1880s, and some old maps that described in detail, the workings of the underground mining operations. In return, they began to explain the process of underground mining, processing the mine dumps and tailings, and the complicated process of milling the ore. A friendship was formed and I was able to learn about the old Silver King literally from the ground up as the saying goes, or in this case from the ground down.

On the subsequent weeks and months that followed I went to the mine on a regular basis to photograph the old tunnels and their operations. I started with the 114-foot level. It was a much safer tunnel to examine than the 256-foot level, as it was a hard rock tunnel, which did not require timbers to shore it up. I personally obtained several specimens of excellent silver ore from these levels. While the climb down was not too difficult, the climb up the almost vertical ladders from the 256 foot level was difficult.

Roy Wheeler was somewhat of a superstitious person, as many of the old miners were, and I always had a ghost story to tell him about the miners who had died in the hole. I used to revel and enjoy a good laugh with the Deens when I told these stories. On occasions I accompanied Roy to photograph the old tunnels, drifts and winzes. He was very knowledgeable on the workings of underground mining. Later, I accompanied Joe or Ron Deen Jr., sometimes with Dick Bynum, into the depths of the mine. Meanwhile, I was gathering first hand knowledge of mining, present and past.

During the late 1990s, Jack Carlson and I began our quest for the *elusive* Stoneman Grade, which is where the story of the Silver King Mine really begins. It starts with the building of the Stoneman Grade in 1870 in order to effectively pursue the renegade Apaches up the side of King's Crown Mountain. It took us two years to finally trace the old trail from Picket Post Mountain, the old site of General Stoneman's army camp, up past the Silver King Mine, and further up King's Crown Mountain, on to Pinal Ranch. During this time I was gathering more information from old news records and all the local historical societies. (Chapter 15 describes the search for the famous old trail.) In 2001 I contacted Velma Bowen Tucker (via Greg Davis), granddaughter of Robert Bowen (Silver King Superintendent), who lives in Mesa. She had done a marvelous job of collecting the Bowen family history, along with many original photographs taken at the Silver King Mine. Velma's story contained many personal glimpses of life at Silver King during the 1880s.

A soldier under General Stoneman's command, John Sullivan, originally located the silver lode in 1870, but he kept its location a secret. Later in 1875, four prospectors from Florence, Copeland, Mason, Long and Regan, found Sullivan's silver lode and founded the famous Silver King Mine. This is a story full of happiness, sadness and old west action. The central theme is the Silver King Mine, but it is the people and their human-interest vignettes that create the fabric of the book. It is a tale about prospectors, miners, Apaches, muleskinners, ranchers, cowboys, merchants, pioneer women, stagecoach robbers and greedy men. These notorious men would steal the riches of the hardworking; honest workers who helped settle the west. After all, this was the rip roaring, wild-west days of the 1870s and 1880s Arizona Territory, where men lived by their guns and their wits.

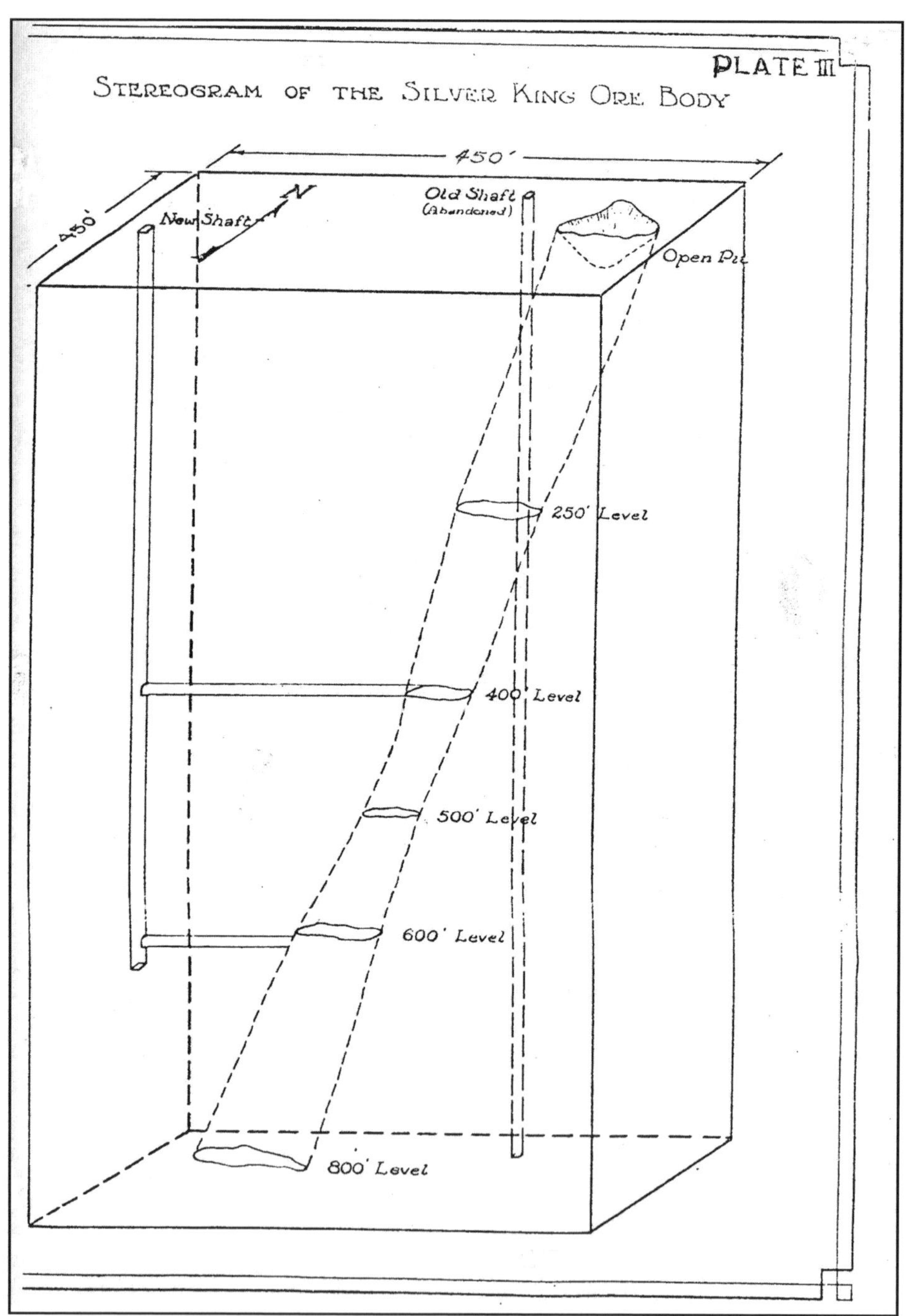

Stereogram of Silver King Ore Body —Author's Collection

Chapter One
Discovery of the Silver King Mine

The Silver King Mine was one of the great silver mines of the 19th Century. The Silver King's story speaks volumes of the great American dream. Four men eking out a living on the frontier, fighting fierce Apaches, the heat and difficult life of pioneers find a fortune in silver. This was not accomplished easily though, for Apaches killed a fifth member of their small prospecting party while they were heading home from yet another foray into the wilderness. The Silver King Mine goes on to become one of the most famous mines of Arizona and the west, for its riches provided prosperity for many of the territory's early pioneers, and it was the catalyst that enabled settlement of the Central Arizona Territory in the 1870s. The Silver King Mine produced between $7 and $17 million dollars according to official reports. Most of this production took place between 1875 and 1887, or a period of twelve years. Oh, they thought the silver would last forever, and a huge amount of high-grade silver was taken out of the "King," as it was often referred to. It had a great impact on the economy of Central Arizona Territory. Remember that the locating of the Silver King was at the height of the Apache Wars. I cannot tell this story without first telling of the brave prospectors, the soldiers and their adversaries, the Apaches.

After the 1849 California gold rush was over, and Colorado had been searched for rich metals, Arizona Territory was the place the prospectors came to. The American Civil War was over and the Army was now sent to the place referred to on some maps as Apacheria. It was called this because of the fierce Apaches who inhabited this new territory, which was created about 1865. Prospectors had requested the assistance of the Army to help protect them while they looked for the precious metals which could make their dreams come true. Let us then go back to those early days of toil and strife. It was the summer of 1870, and former Civil War calvary veteran Brevet Major General George Stoneman and his California Column had just made their appearance in war torn Arizona Territory. [1]

The following scenario depicts the Army and the Apaches during those sanguinary days. There were guns blazing as the troopers chased the dreaded Apaches up the yellow and red cliffs of the Pinal Mountains in Arizona Territory. The Apaches had just come back from a raid on the little settlement by the Gila River named Florence. As the Apaches crested the mountain, later called King's Crown, they raised their arms in defiance at the troopers. They had gotten away again. The calvary horses could not go up the rugged mountainside. There was no horse trail there and the Apaches knew it. They were safe in their stronghold again, high above the small military camp called Picket Post where the small detachment of soldiers was stationed. This scene sets the stage for the actual deployment of the troops and the confrontations that occurred in old Arizona Territory near what was to become the Silver King Mine.

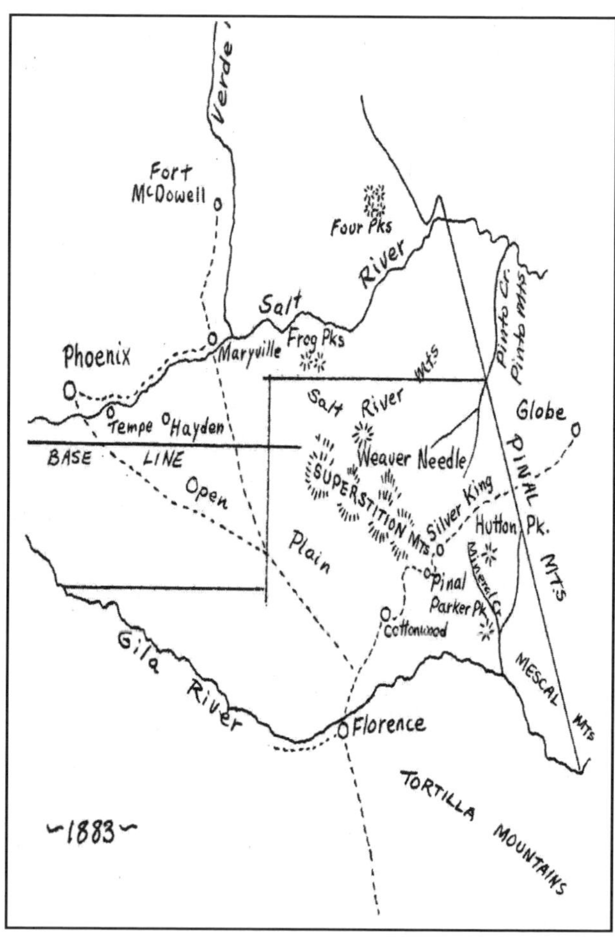

Silver King Area Map cir. 1883 —Authors Collection

The U.S. Calvary had been sent to Arizona Territory as part of the California Column. General George Stoneman led them to this wilderness area in order to protect the new settlements and prospectors who were making inroads towards civilizing a vast part of the empire won in the last war with Mexico. It was after the 49ers gold rush in California, and now prospectors were looking for new areas to search for riches. Many had heard of the tales of gold found around Prescott in the northern area of Arizona Territory. Now prospectors were looking for silver and gold where no white man had prospected before, in the Central Arizona Territory, the home of the fierce, warlike Apaches.

This is the setting for the soldiers who came to Arizona Territory to fight the Apaches and bring peace to this war torn area.

Importance of the Silver King Mine to opening up the Central Arizona Territory

The importance of the Silver King Mine in the history of mining in the United States, and particularly in the Territory of Arizona, lends more than the usual interest to the details of the discovery and location of the vein.

Silver King mine cir. 1878 —Bureau of Mines

The efforts to find the locality and to open up the mine date back to the period when the southwestern and central parts of Arizona were still in the possession of hostile Indians. The history of the Silver King Mine is closely connected with the development of the region and its settlement by Anglo-Americans.

In the 1870s the Silver King Mine was the leading factor in the reclamation of the Pinal Region of Arizona Territory from the Apaches. The Apaches dominated the whole region, and they made it almost inaccessible to the hardy and daring prospectors who pressed outward from the frontier settlements on the Gila River towards the Pinal mountains. According to William Middleton's wife, "We were over at Hayden's Ferry, when William (Middleton) was approached by a man named John Sullivan who pleaded with William to accompany him (Sullivan) to the Pinal Mountains where he knew of a rich silver deposit. He tried many others in the area, but no one would go due to the danger of marauding Apaches. The man (Sullivan) went on west."[3]

Apache scouts with soldier cir. 1870s —Gila County Historical Society

The establishment of large mining camps like the ones at Silver King and Picket Post (which later was called Pinal City) made the whole region safer from Indian attack and thus allowed for the development of many of thc smaller mining camps. Settlement by the miners allowed for the close following of merchants, freighters, cattlemen and a host of other business ventures, including stagecoach lines. The Silver King Mine opened up a whole new realm of activity, which paved the way for commerce and civilization locally from Phoenix to Tucson and eventually extended from coast to coast. The ore, then later the bullion, went to California, while New York investors bought up the mine itself. People flocked to see this very rich silver mine, and the second floor of the main office building at the Silver King was set up to accommodate visitors, who came from around the world, to see this magnificent mining venture. It is no doubt that the Silver King Mine opened up this part of the "wild-west" for civilization.

The Legend of Apache Leap

The little settlement on the Gila River named Florence was for a long time the scene of Apache troubles until a decisive attack on their stronghold (Apache) was made in 1870, when their power was broken in that region. General Stoneman was stationed, along with several companies of United States soldiers, at Camp Picket Post, below the old site of the Silver King

Mills, thirty miles north of Florence, in what was then called Tordillo Peak (Picket Post Butte) in the Pinal Mountains. The post was in a small valley on Queen Creek, overlooked from a high ledge of the mountains (known as sheer vertical mountain wall) now called Apache Leap, and all of Stoneman's movements were noted by the Apaches from their higher mountain fastness.

On top of the mountains above what would later become the Town of Superior was a rancheria (village) of Apaches. These Indians occasionally poured down some unknown pathway upon the settlers along the Gila Valley, stealing, burning and killing, and when pressed by troops, would vanish in the canyons. The location of the village was suspected from a solitary Indian now and then seen perched upon these peaks, who watched the proceedings at Picket Post. All attempts by Stoneman to get at them were fruitless. At length, emboldened by their successes, they raided a ranch near Florence and drove away a band of cattle. The Florentines armed themselves and began pursuit. Captain John D. Walker, of Company B of the Arizona Volunteers, led them, along with a large group of Pima Indians, and perhaps some Maricopa Indians as well.

Apache Leap near Superior —*Author's photo*

After several days of patient pursuit, they found the trail that led to the rancheria. The Apaches, doubtless feeling secure in this fastness, neglected to post videttes or lookouts, and thus the Florentines and their Indian allies were able to steal upon them by night. At daybreak they attacked the rancheria, which was situated only a few yards back from the bluff overlooking Picket Post Butte and the valley below. The Indians, seeing that they were surrounded, fired a few shots, then threw down their guns and went to meet the approaching Florentines and Indian comrades with their hands raised in token of surrender. The latter, seeing their advantage, and remembering the cruelty to the defenseless families on the Gila, continued firing upon them. When about two-thirds had fallen, and seeing no chance for quarter, the remainder of the Apaches panicked and retreated to the cliff's edge. One by one they hurled themselves out into space striking the rocks below. About 75 Apaches went over the bluff. The Florentines and Indian comrades could see their mangled remains from the place where they sprang over. Not a single warrior escaped, but the women and children were turned over to General Stoneman.

The place where this actually happened is just south of the present Town of Superior, and is still referred to as Apache Leap. The Apache women, who survived this attack, referred to the black obsidian nodules gathered at the foot of these cliffs as Apache Tears, for the tears they shed over their lost kin. Apache Tears are found in beds of Perlite, a form of ancient lava. An old timer from Superior told me that as late as the 1930s you could still see parts of skeletons in the crevices below Apache Leap.[12]

Captain John D. Walker —Indian Fighter/Medicine Man/Miner

There are many versions of this story, but according to historian James Barney, this event involved Company C of the Arizona Volunteers and was composed mainly of Pima Indians. Company C's commander was

J. D. Walker House cir. 1880s — John Swearengin Collection

Captain John D. Walker, a native of Nauvoo, Illinois, and was supposed to be part-Wyandot Indian. After the Civil War, he came to Arizona Territory with the Army's California Column and settled around Sacaton. He mastered the Pima and Maricopa Indian languages and soon became a big leader and chief medicine man for the Pima Tribe. Walker led many campaigns against the Apaches and other hostile Indians during the 1860s and documented his activities. Walker was said to have gone into battle wearing nothing but moccasins and a breech-clout. He whooped and yelled like an Indian warrior, in terrifying fashion. The Pimas, although peaceful to the white settlers, were traditional enemies of the Apaches. Walker, after leaving the army, worked for the Bichard brothers, who were traders in Sacaton.[13] In his pre-Pima days Walker had studied medicine under an army surgeon at nearby Ft. McDowell and was a self-educated man. He became the local physician at Sacaton, Arizona. He allegedly married a Pima woman called Churga under the Pima tribal laws and they had a daughter named Juana.

John D. Walker was supposedly taken to the rich silver deposits that would soon become the Vekol Mine by a Papago Indian (Tohono O'Odham) named Juan Gradello. Along with his brother Lucien Walker and Peter Brady, they founded the Vekol Silver Mine. (Vekol means grandmother.) The old mine workings are still located in the Vekol Mountains, south of Florence, Arizona. Walker became quite wealthy and was said to have used much of his earnings to assist the Pima Indians. Walker was one of the early founders of Florence, Arizona and later became a probate judge.

INCORPORATED UNDER THE LAWS OF
THE TERRITORY OF ARIZONA
No 527
Shares
VEKOL MINING COMPANY
CAPITAL $5,000,000.00
This Certifies That ... is the owner of ... Shares of the Capital Stock of
VEKOL MINING COMPANY FULLY PAID AND NON-ASSESSABLE
transferable only on the books of the Corporation by the holder hereof in person or by Attorney upon surrender of this Certificate properly endorsed.
In Witness Whereof, the said Corporation has caused this Certificate to be signed by its duly authorized officers and to be sealed with the Seal of the Corporation
SHARES $1.00

Vekol Mining Certificate —*Author's collection*

Later on in his life Walker's money caused family problems and, in a bizarre turn of events, his family committed him to an insane asylum in Napa, California. After a court hearing for his freedom, which he apparently lost, he was confined to the asylum and soon died while still in his fifties. Walker's Indian daughter Juana tried to inherit his wealth. But, she lost a lenthly court battle with the Walker family who did not recognize her as a legal heir.

J. D. Walker also figures prominently in the legends surrounding the Lost Dutchman Mine. Jacob Waltz's partner, Jacob Weiser, supposedly came to the Walker house in Florence after Indians wounded him in the Superstition Mountains. Weiser allegedly gave Walker a map to the Dutchman's gold mine after he believed Waltz dead. The map surfaced

Bob Stambach & Jack San Felice at Vekol Mine 2002
—Jack Carlson Photo

many years later in the hands of Tom Weeden of Florence. This famous map has become known as the "Weiser-Walker-Weeden" lost goldmine map. Robert Blair gives a lengthy description of J. D.Walker and this famous map, on page 75 of his book, *Tales of the Superstitions.* Also, Greg Davis lists a chronological history of the map in the *Superstition Mountain Journal Volume 18.*[14]

The Pinal Mountains and the Apaches

The Pinal Mountains rise to the eastward of the Gila and Salt Rivers and stretch northwest to the higher ranges of the White Mountains. They have many irregular areas that seem to leap out to meet the sky. This range of mountains has many canyons leading down to the Gila Valley which by reason of their extreme ruggedness and impassable nature, afforded an excellent refuge and secure hiding-place for the Apaches, after conducting marauding expeditions against early settlers or the destruction of trains of immigrants. One of the trails most traveled by the Apaches led up the Gila Valley to the little creek, now known as Queen Creek, and over the nearly vertical face of the mountain, within a stone's throw, almost, of the point where the Silver King Mine was afterward discovered.

This was the place where the Apaches came to hide, in the various caves and hidden canyons in the days of Apache warfare. This can be attested to, due to the many finds of broken pots and arrow-heads and old fire remains, that were found in the cliffs that overlooked that old Indian trail.

General Stoneman, the Apaches, and the Stoneman Grade

Colonel (later Brevet Major General) George Stoneman, an accomplished cavalry officer, became commander of the Military District of Arizona. He later became Governor of California. Stoneman decided to adopt more vigorous measures than had been made for the repression of Apache raids. He sought a way into "Apacheria," which that part of Arizona Territory between the Gila River and the Mongollon Range and White Mountains had been referred to. For the most part, people either walked or rode horses or mules across the trails. General Stoneman was determined to enter the Pinal Range and "Apacheria" from the west and establish a picket post or military camp from which to engage the hostile Apaches. He moved his command to the base of Signal Mountain in the Pinals and established Camp Pinal. Then he established Camp Picket Post at the base of current Picket Post Mountain (and Boyce-Thompson Arboretum). He then set up a set of heliograph stations. One was at Signal Mountain, near his headquarters, and the other at the top of Picket Post Mountain. He was then able to deploy his troops more effectively, according to the movements of the Apaches. He only lacked one thing, a good horse and mule trail up the side of the mountains.

General George Stoneman
—Gila County Hisorical Society

He needed a passable trail into the Apache areas in order to facilitate troop movements and supplies. After he could not find a satisfactory trail into the high country, he had his men construct a 5-mile mule trail upward from Camp Picket Post in a northward route that would traverse the steep mountain at an angle or passable grade. This grade is still called "Stoneman's Grade" today and its remnants can yet be found. This mountain would later be called King's Crown by the miners. Below the mountain's peak is a large rock formation that very much resembles a king's crown, hence the name.

Building of the Stoneman Grade—Sulivan Finds the Silver Nuggets

General Stoneman decided the best way to get the advantage over the Apaches, and to allow his horse soldiers to pursue them, was to build a road up the side of the mountains. He had assigned Captain Netterville the responsibility of building this road. It was another hot day in Arizona Territory. A group of his soldiers were working on this road about halfway up the mountain. The troops referred to this road as the Stoneman Grade. John Sullivan was one of these soldiers. He did not like road building and like many of the other troopers, thought it demeaning for horse soldiers to act as common laborers, for he had signed on to fight Apaches. He had not signed on to do pick and shovel work. That was the role of the miners and prospectors that he was sent to protect.

Sullivan had just sat down to take a break when he saw some strange black rocks. He picked up some and they felt unusually heavy. He tried breaking one of them open, but it would not break. It just flattened like lead. He tried another one and it did the same thing. His sergeant shouted: "Break time is over, get back to work." Sullivan said nothing to the other soldiers but put these black rocks in his pocket. They were silver sulphide nuggets, black nuggets of silver as the miners called them. Little did he know that he had located one of the richest silver strikes in all of Arizona Territory. He had found what the miners and prospectors would kill for, the *mother lode* of silver. Later when others asked him where he had found the nuggets he would just shrug and say "I found them on the Grade." They called that part of the Stoneman Trail just below King's Crown Mountain the Stoneman Grade for its steep upward climb.[2] By the time Sullivan had learned what the heavy black nuggets were he was no longer working on the Stoneman Grade and the troops had been removed from the area. Sullivan mustered out of the army and went to work for Charles G. Mason at his ranch on the Gila River near the little settlement of Florence. Sullivan tried several times to get prospectors to go with him to work the area and file a claim, but now the troops were gone. No one wanted to go back to the mountains and risk an Indian attack, not even for a very rich claim, as he claimed his was.

Sullivan Disappears

After unsuccessfully getting anyone to go back to his rich find, Sullivan disappeared and was not heard of or seen for years. It was supposed that he was killed by Apaches or died while returning to the place where he found the silver nuggets. Before Sullivan left Mason's ranch for California, he described the location for some cash, a pair of boots and other items. Sullivan told Mason to go to a prominent boulder located near Stoneman's Grade. "Within a stone's throw," he said, "one could he find his rich strike." Allegedly, Sullivan told Mason that if he found the strike to include his name on the claim. Several times Mason led prospecting parties and followed Sullivan's directions without success. They even made a location for a claim only a mile and a half from the present location of the Silver King and named it the Silver Queen. Later, the search party prospected over the Pinal Mountains. There is some controversy to who officially located the Silver Queen, however, as we shall soon see below.

Sullivan's find, it must be noted, was supposed to be known to some of the men of Captain Kerr's party (of the Miner Expedition) some of whom attempted unsuccessfully to re-locate it as they passed through the Stoneman Grade. Disappointed in their hope of finding rich ore, they located and registered a number of claims of less promise which they called collectively the Silver Queen Ledge.

The *Arizona Citizen* on November 18, 1871 published a letter by Captain Kerr to Governor Safford. It stated:

> "We were lucky in finding a ledge that I think will give all the boys interested a stake for life, if they are careful and do not throw it away. This is the Silver Queen Ledge; samples of the rock you have no doubt seen, as some has been tested in Tucson, and we have heard went over $6,000. Whether these assays are correct or not, I am unable to say. We have made some five locations in this (mining) District, all showing fair prospects, but none so rich as the Silver Queen."

The other mining districts located by the Kerr party were called the Kerr Lode, Holstead, Nevada and Yellow Jacket. The *Weekly Arizona Miner* on the following March 30, 1872, wrote an article describing the location and ores in the Silver Queen Ledge.

> "Two or three miles northeast from Picket post Mountain, rises stratified limestone without petrifaction's, and belonging to the primitive period too. Said limestone and partially quartzite bed of the Silver Queen Ledge. The minerals of the Silver Queen consist of sulphide of copper and sulphide of silver, oxide and carbonate of oxide of copper. The gangue is quartz and limestone. The richest silver ore was found at a place where on the northern wall the limestone changes into quartzite. Besides the silver Queen, eighteen ledges have been located in the neighborhood, containing mostly argentiferous galena."

Description of the Stoneman Grade

This pack-trail or mule trail, built by soldiers in 1870, goes diagonally up the face of King's Crown Mountain, leading over into the higher tablelands and valleys of the Pinal Range. This pack-trail became known as "the Stoneman Grade" (and passed right by what would become the Silver King Mine). Above the Stoneman Grade near a year-round spring the General set an intermediate camp, simply called Camp Supply, and had stone ovens built for baking bread. Camp Supply was in a meadow like area, next to what is now the Omya Marble Mine, at the top of the "Grade." It was described in the *Pinal Drill* on February 26, 1881: "There is first the beautiful view over the country to the west from the top of Pinal Mountain a wonderful site, (King's Crown Mountain). Then you pass old Nicks Ranch or Stoneman's Supply Camp, where there is splendid spring water."[4]

Stoneman grade from the top — Jack Carlson photo

The trail then drops down into and then out of Devil's Canyon. This is a place where the fantastic rock forms rise from the canyon floor. It was called Devil's Canyon, not for it's sinister look, but because the soldiers said it was a "devil of a trick" to take a mule train down the steep rocky trail to the canyon below. Thus, to this day, the old trail and grade is still called the Stoneman Grade on most maps. Ten miles further out from the Grade, below a mountain ridge called Signal Peak, General Stoneman set up his headquarters command base at a place he called Camp Pinal. This is not to be confused with Old Camp Pinal, at the base of what is now Picket Post Mountain. Stoneman's camp below Signal Peak is now know as the Pinal Ranch and is just off U. S. Route 60, near Oak Flats and an old village called Top of the World, above the Queen Creek Canyon Tunnel.

Col. James M. Barney, writing in *Arizona Municipalities,* in August 1940, states:

> "This pack trail was for years, the main traveled route into the mining districts of Globe and vicinity, and led up into the rocky defiles and lofty peaks of that range. This old trail, which was one of the most traveled by the Apaches, led up the valley of what is now known as Queen Creek and thence over the nearly vertical walls of mountain peaks to the broken fastness beyond. It commenced its tortuous windings almost within a stone's throw of the spot where the Silver King Mine was afterwards discovered."[5]

Stoneman's Camp Pinal

General Stoneman set up Camp Pinal at the site of what is now Pinal Ranch in what then was called Mason Valley in 1870. This was at the headwaters of the Mineral and Pinto Creeks, six miles west of the present town of Miami. He moved in about 400 troops and established his headquarters there. They were from the 21st Infantry, companies A.E.G.and I.[6] He then set up an outpost at Picket Post Mountain. Its purpose was to protect the miners in the area. He communicated from Camp Pinal's Signal Peak Mountain, which is 700 feet higher than Camp Pinal, by means of signals to Picket Post Mountain in order to keep apprised of Apache movements. He used heliograph mirrors to flash Morse Code signals across the mountains. Stoneman expected to maintain Camp Pinal as his permanent base and set up fortifications there.

However, General Stoneman's plans were not to last long. He was criticized of spending too much time on administrative duties, building roads and surveying the new Department of Arizona. In fact, the *Tucson Arizona Citizen* referred to Stoneman as "Economy Stoneman."[7] Stoneman was transferred to California and Camp Pinal moved most of the men to the base of Picket Post Mountain, establishing Camp Picket Post there. Thus Camp Pinal only lasted from November, 1870 to July, 1871.

Sullivan remained at Mason's ranch for some time and frequently showed the black ore to others. The miners and prospectors of this area knew this black ore as "nugget silver". It is reported that Sullivan tried to get Mason and Mason's friend Ben Regan, and others, to accompany him to the place where he found the black nuggets, which was later described by

Mason as black nuggets of silver sulfide. Although Mason was one of the early frontiersmen to brave the terror of Apache attacks, he thought it too dangerous to go into the Pinal Mountains at that time. Both Mason and Regan thought Sullivan had made a rich find, but neither would journey unprotected by the Army, or unless they were accompanied by a large group of prospectors, into the dangerous Apache country where the silver was supposed to be located.

General Stoneman Replaced by General Crook

General George Crook replaced General Stoneman as the military commander of Arizona Territory in mid 1871. Crook did not believe in small detachments assigned to scattered posts, so he closed Camp Picket Post and Camp Pinal. The soldiers were assigned to groups who were given the tasks of taking the war to the Apaches. If the Apaches did not come into settle on the reservations, then the troops were to wage a continuous war against them. This left the area of Central Arizona and its miners open to hostilities from the Apaches as the Army was always on the move. Prospectors and miners operated at their own risk and the only protection they had was the Anglo settlements on the Gila.

General George Crook
—National Archives photo

The location of a military camp encouraged many to speculate on new mines being opened in the eastern part of the Central Arizona Mountains, however, this was not to be the case. On June 4, 1871, when General Crook took over the Army's Department of Arizona, he closed the small, isolated military camps, and Camp Pinal, Camp Supply and Camp Picket Post were abandoned. The Stoneman Grade, however, proved a boon to prospectors and early settlers of that era as well as the soldiers of Fort McDowell and Camp San Carlos, who frequently chased the Apaches into the Pinal Mountains.

Picket Post Mountain

Picket Post Mountain was called by this name because the soldiers who were stationed there thought that the corners of the mountain resembled corner towers or *picket posts* of calvary forts, and the mountain is known as Picket Post to this day. The origin of the word *picket* comes from the

Picket Post Mountain —Tim Williams photo

calvary or horse soldiers. Literally, it means to stake a horse so it can feed without wandering off. This was done by means of connecting the halter rope from the horse to a *picket pin.* Also, on the western frontier the *picket* was a post set straight up in the ground. Several pickets would be placed together to form a *picket post* or stockade to resist attack from hostile Indians. Any of these explanations could have resulted in the mountain being named Picket Post. However, the shape of the mountain does resemble a frontier fort, if you use your imagination.[8]

Picket Post was first called Tordilla Mountain. Even after Generals Crook and Miles were in charge of the Department of Arizona's Army, Picket Post Mountain was an important station in the heliograph network that kept the army informed of the Apache movements. The telegraph did not extend to the Pinal Mountain range. The heliograph network allowed messages to be sent to the far reaches of Eastern Arizona and back to Army Headquarters at Camp Verde. The heliographs used only sunlight and mirrors mounted in special devices. Signals were flashed using Morse Code from mountain to mountain using trained heliograph operators. The name of the mountain (Picket Post) and a geology formation at the top (Heliograph) reflect the mountain's early military history.[9]

The Ill-fated Miner Expedition

About the time of the search for the "Sullivan's Silver" as it was referred to, Arizona Territorial Governor Safford of Tucson led the ill-fated Miner exploring party of about 280 men. They were called the Mogollon Mining and Exploration Company and searched for placer gold in the summer of 1871. They went through the Pinal Mountains, the Sierra Anchas and the Salt River area of the Superstition Mountains in search of "Miner's Gold." Tom Miner, a prospector, came into Prescott with a tale of that he had found a very rich area of placer gold somewhere in the mountains south of the Mogollon Rim.

The party gathered its men and went from Prescott to Tucson by way of Ft. McDowell. Tucson was the territorial capital at the time. The expedition was led by Governor A. P. K. Safford and organized into seven companies, each with a captain. These captains and well known explorers were Ed Peck, James Rogers, T. B. Kerr, Francisco A. Gandara, Jose Duran, Francisco Martinez and a man named Maxwell. Each man was well armed and the party had a pack train supplied with flour from the mill at Adamsville (now a vanished ghost town) near Florence. From Tucson the expedition crossed the San Pedro River and reached the Gila River. They prospected down the Gila to the San Carlos River, worked their way up to the river's head and crossed over to the upper Salt River. After several months of searching it was evident that Miner did not know where he was, and had no idea where he had claimed to have been, knew nothing of prospecting in this part of the country, and was at best a fool, and at worst a fraud. The prospectors were very angry with Miner, and he was lucky not to have been lynched. Miners Needle in the Superstition Mountains is supposedly named for him.

Some of the party turned back to Prescott, the Globe area, Florence and Tucson. Others including "Hunkydory" Holmes, frontiersmen Ed Peck and Al Sieber and the Dutchman Jacob Waltz, among other notable explorers of the time, went down the Salt River to Cherry Creek and up to the Tonto Basin. They worked their way up Cherry Creek and went into the Sierra Anchas without much success. They returned down Cherry Creek to the Salt River and followed the Salt to the area known as Wheatfields, about 10 miles northwest of Globe, on Pinal Creek. At this place the Miner expedition was dissolved. Safford went back to Tucson and Sieber and Peck and most of the others went back to Prescott by way of the eastern slopes of the Mazatzal Mountains, west of the Tonto Creek.

Others including "Hunkadory Holmes" prospected in the area near Globe, Florence, and the Stoneman Grade. They found some rich outcroppings of silver ore near a creek but several miles from the Stoneman Grade. They worked their claims the fall of 1871, but these claims appear to have been forgotten by history. One of these searchers was, "Hunkydory" Holmes, whose story is told later in this book. The Dutchman Jacob Waltz, along with others, went to Phoenix.

A Rifle in One Hand and a Pick in the Other Hand

It was too early for effective mining. Indian depredations disrupted the work of the prospectors, and ultimately the Apache threat resulted in abandonment of their sites. In those days, the early prospectors used to work their claims and prospect holes with a rifle in one hand and a pick in the other. Their trusty horse or mule stood saddled nearby in the event they had to make for the hills. There was a realization of danger at all times and many a lone prospector's bones lie in shallow graves while his hair and belongings was in some Indian's wickiup.

One must understand what it was like to live in Arizona Territory in the 1870s. The settlers' situation between 1861 and 1874 has been called "the most sanguinary and violent in the annals of the Southwest."[10] The "Apache Menace" was the great obstacle all over central and southern Arizona during the mining and prospecting era of the 1860s, and by 1870 it seemed that something had to be done. A Prescott paper listed 300 people killed in Indian raids since the establishment of Territorial Government. (President Abraham Lincoln signed the statute

Apache Scouts cir. 1870s
—Gila County Historical Society

establishing Arizona Territory on February 24, 1863.) Horse and cattle thievery had reached intolerable conditions.[11]

One cannot tell the story of the Silver King Mine without telling of the trials and tribulations that the early settlers had with the Apaches, especially the prospectors. The history of this area is like a two edged sword. The army was called in at the request of the settlers to enforce the peace between them and the Apaches, both of whom wanted exclusive rights to the land. The Apaches wanted to live as they had before the Anglos, able to roam free without any bounds. The Anglo settlers wanted to ranch and mine, wherever they wanted, without being attacked by hostile Apaches.

"Hunkydory" Holmes

In October of 1871, a 24 year old prospector named William A. "Hunkydory" Holmes, along with the Anderson brothers, T. B. Kerr and several others, were prospecting on their own where Queen Creek comes out of the Pinal Mountains. They had been searching for months with Tom Miner's large group of (about 280) prospectors. They had searched in vain for Miner's lost placer gold mine, somewhere in the mountains of Central Arizona, and had decided it was a lost causc and struck out on their own. Near where the Town of Superior stands today, they came across some deeply stained outcroppings. They took samples and built monuments. The claim they filed was called the Silver Queen Ledge and was recorded at the Maricopa County Courthouse on October 25,1871. Holmes got his unusual name from a poem that he had written and set to the melody of an Old Irish tune called the "Limerick Races." "Hunkydory" was to become an intricate part of Globe and the Pinal Region's history, and is also discussed in Chapter 11. The poem went, in part, like this:

Oh, I am a jolly mine lad,
Resolved to have some fun sir.
To satisfy my mind,
To Phoenix town I came, sir.
oh, what a pretty place,
And what a charming city.
Where the boys they are so gay,
And the squaws they are so pretty.
(Chorus)

Ma sha ring a ding a da,
Sha ring a ding a da di oh,
Sha ring a ding a da
And Hooray for Hunkydory. [15]

At the approach of summer 1872, the miners, who were farmers first, went home to attend to their farms and work at the Queen stopped. They said they would return after the harvest and would again work the ledges.

The Apaches however, had different ideas. When the miners returned to work the mine, the Apaches attacked and forced them to leave the mountains. The *fate of the miners* for the next few years was in the hands of General Crook.

General Crook was somewhat successful in subduing the Apaches, and the territory once again was inundated with prospectors, especially those in search of Sullivan's Silver. Word of the Silver Queen spread rapidly and soon a minor "silver rush" was underway. Scores of prospectors headed to the Pinal Mountains to see the strike themselves.

The Prospectors and Miners take to the Hills Again

As the Apache attacks subsided, prospectors and miners once again went back into the mountains. On September 9, 1873, the Anderson Brothers and some other miners were prospecting along Pinal Creek. They had among their group Benjamin Regan who also had the diverse dual occupations of *saloonkeeper and part-time preacher*. Their party also consisted of Isaac Copeland, William Long, J. E. Clark, T. Irvine, William Folsom, P. King, M. Welch, W. H. Sampson, B. Edwards and J. Riley. About 4 o'clock in the afternoon of September 11th, the Apaches attacked without warning, killing William Sampson and wounding Isaac Copeland in the leg. The prospectors beat back the Apache attack and then retreated back to Florence with their dead companion. In 1873, the twin brothers named Anderson, Bob and Dave, located two silver claims known as the "Globe" and the "Globe Ledge" in the newly formed Globe District.

On September 17,1873, the miners filed their claims on the Globe Ledges. The first claims were filed in the names of Folsom, King, Welch, Sampson, Edwards and Riley. The second claim was filed in behalf of Copeland, Long, Regan, Clark, the Anderson brothers and Irvine.[16] The claims of the above men were not worked for over a year after the death of Sampson.

They apparently feared more attacks by the Apaches.

True Story of the Discovery of the Silver King Mine

Charles G. Mason never gave up hope of finding Sullivan's bonanza and early in 1875 joined a party of five prospectors. They consisted of Mason, Benjamin W. Regan, William Long, Issac Copeland and a fifth person, whose name is lost to history, but may be the Sampson mentioned above. At the request of the Anderson brothers, they took a mule train to

March 22nd 1875. Pinal County. Arizona.
"Pioneer District."
The undersigned claim this Lead of Mineral
as follows- fifteen hundred feet in length and Six
hundred feet in width. the claim extends North West
from this Monument seven hundred and fifty feet;
South East seven hundred and fifty feet.
from each terminal point three hundred feet trans-
versely to a monument. the transverse line Easterly
is the Easterly boundary. the transverse line Westerly
the Westerly boundary. the Northeast and Southwest
parallel lines between these Monuments are the
lateral boundaries along the length of the Claim.
The claim is called "Silver King" It
lies South of Stoneman Road about seventy yards
Monuments have been placed at the four corners
and in the centre of the claim
Filed and Recorded at request of Isaac Copeland at 10 o'clock A.M. March 29th/1875
Jno. J. [illegible] Recorder

Isaac Copeland
B. W. Reagan
W. H. Long
C. G. Mason

Silver King Claim March 22, 1875 —Pinal County Recorders Office

the Globe Mine and Globe Ledge to take out some ore. Ben Regan had thought it safe enough to do some assessment work on the Globe claims at that time since the Apache attacks had somewhat subsided.

While returning on March 21,1875, they were attacked by Apaches above the Stoneman Grade near old Camp Supply, and the unknown prospector was killed. The unknown prospector may well have been William Sampson, as there are conflicting stories as to when he was killed. His name however, is not on record on the filing of the 1875 claim, and is on record on the 1873 claim. His body was then taken to a place called "Camp Supply," at the summit of the Stoneman Grade, and buried in an old stone baking oven used by General Stoneman's men for baking bread. Stoneman's camps all contained brick ovens built of bricks made of baked clay. Human remains were later found in one of the old earthen ovens at Camp Supply by caretaker William T. Brown of Pump Station Spring (the spring that provided water for the Silver King Mine) near the old camp. This was also reported by Dr. William H. Corbusier during March of 1874. Dr. Corbusier was the military doctor attached to Lt. Walter S. Schuyler's detachment of General Crook's punitive exedition against the Tonto Apaches, with Al Sieber as his Chief of Scouts.

As they were going down the "Grade" Mason asked if they could try once more to locate Sullivan's bonanza using a clue Sullivan once mentioned to him at his farm. He had been told by Sullivan to go to a large boulder and throw a stone as far as he could, and there he would find the silver nuggets. This time, however, Mason decided to throw the stone downhill instead of uphill toward the Stoneman Grade. He told Copeland to watch very carefully where the stone fell. Copeland went searching for where the stone fell and found some black silver nuggets, just like Sullivan's. He shouted that he had "struck it" and they need look no further. He said that he had located the silver nugget area that Sullivan had first found by accident years earlier. The long sought after bonanza had at last been found. They thought they had found the "King" of mines, in fact they called it the Silver King! They laid out their monuments fifteen hundred feet long in an easterly and westerly direction and six hundred feet wide in a northerly and southerly direction. This was on March 22, 1875. [17]

Fearing the return of the Apaches, the saddened yet happy prospectors gathered as much ore as they could carry to add to the already loaded pack train. (The ore was about 1500 pounds, which assayed at $4300.) They then left the area and went immediately to Florence where they went to the assay office and filed their claim. They stated that "The claim was to be called the *Silver King*, and that it lies south of the Stoneman Grade, about 70 yards, and that monuments have been placed at the four corners and in the center of the claim." The claim was filed and recorded at the request of Isaac Copeland at 10 A.M. on March 28th, 1875, listing Copeland, and

Benjamin W. Regan, William J. Long and Charles G. Mason as partners. Sullivan's name was not on the list. Since they had not heard from him for years, they thought he was killed by Apaches or died in the desert.

The next day they obtained supplies and hastened back to the site of their discovery. There they found many more of the nuggets on the surface, among the various other mineral stains of Azurite and Chrysacolla.

Word of their discovery soon spread throughout the region and the settlements were almost depopulated as all available prospectors went to file claims next to the Silver King strike. Clara T. Woody states in her notes that "Florence and all the Gila ranches were depopulated of all able bodied males." Fortunately, there were no Indian attacks. John Clum, the Indian Agent at San Carlos was still keeping the San Carlos people very busy and also very contented."[18]

There were several romantic tales regarding locating the Silver King Mine, the following tale is just one of many. This story comes from the *Arizona Gazette,* September 17, 1985 and reads as follows:

"Romantic Account of the Discovery of Silver King-
Dodging Indians and Catching Silver"

"Another sensational mining discovery was made in 1875 in Arizona, but this was a pure silver production. It was one of the most interesting finds ever made on the coast. Three prospectors, Copeland, Mason and another, were one day dodging Indians in the neighborhood of Queen Creek, Pinal county, about fifty miles from Florence. The Indians had been very bad that year in that region, and no prospector felt safe a moment. One evening Copeland and party were looking for a place to camp, when all at once one of their pack mules gave a short and with ears pointed forward stood stock still. A mule scent for an Indian is keener than that of any dog for game, and the party knew there was danger ahead. They dismounted and reconnoitered.

While Copeland prepared to scout on the side of a little raise, he tied his mule to a clump of sagebrush, which grew on a ridge of float rock. While he was gone something scared the mule, and the latter jerked the sagebrush up and started down the canyon. The party located a rancheria of Indians a mile off and wisely beat a retreat undiscovered. A half-mile away they picked up the mule, the sagebrush still hanging to his bridle. Copeland detached it and was about throwing it away when something clinging to the roots attracted his attention. It was a piece of shattered white quartz, as big as a walnut, the disintegrated (sic) mass being held together by a perfect network of pure white silver threads the size of a number eight wire. Some of the roots of

> the bush ran through the quartz and firmly attached it. The party was greatly excited, but they did not then dare go back on account of the Indians. They staked the locality, and Copeland fairly cudgeled his brain to impress upon it the exact spot he had tied his mule.
>
> Two weeks after they ventured back and to their joy found the coast clear. For several hours they searched among the rocks and the scoria of the vicinity, and at last Copeland found the place where the sagebrush had been torn up. A few minutes digging revealed the crown of the most beautiful silver quartz ledge any of the prospectors had ever seen. They dug for several days on the spot and laid bare a section twenty feet long and ten feet wide. The vein was without foot or hanging-wall, was of pure white quartz, with streaks of native silver (pure white silver) and blotches of black sulphurets running all through it."

This was the discovery of the famous Silver King Mine of today. This mine has a peculiar interest, from the fact that so many public men of note invested in it."[19] It was later said that the mule was retired and put out to pasture, never to carry a burden again! Mules or burros found many a mine if you believed in the old stories.

Marshall Trimble writing in *Arizona-A Calvacade of History,* describes the Silver King as being the richest single silver mine during the years he called Silverado, about the boom years of silver in Arizona Territory. He also writes about a wandering mule that was found on top of Sullivan's silver find.

Another early tale of the Silver King comes from the *Arizona Daily Citizen,* Tucson, Arizona Territory, April 3, 1875.

> "Pinal County stands head now in the matter of a silver deposit, according to all accounts. B. W. Reagan informs us that himself and Isaac Copeland made the discovery and C.G. Mason and W. H. Long are joint claimants in it. The discovery is about 32 miles by trail and 35 by wagon road from Florence, at the foot of the Stoneman grade. It consists of a deposit of ore worth thousands of dollars per ton, in fact, some pieces of pure silver weighing pounds are said to have been found, and that Dr. W. W. Jones has one of two pounds weight which he intends to exhibit at Philadelphia in 1876. Mr. Reagan showed us a piece with pure metal in it and like samples could be seen in Congress Hall in Tucson. He brought a ton of the ore to town, a sample of which was assayed here showing it worth $4340 per ton."[20]

"Dump" Wilson and the Silver King Mine

Right after the four original finders of the Silver King Mine located the mine, they dug out the first load of ore to be shipped. Apparently not knowing too much about silver ore, they separated the ore and shipped the bright shiny lead galena, and threw the dull looking sulfide of silver on the dump. They shipped this load of ore and lost $1200 on the deal. The great pieces of silver they left on the dump. About this time, before the news of the loss, four Nevada miners came along and found them throwing the silver sulfide on the dump while carefully picking out the galena. They asked the owners if they could work the dump and agreed to split the profit 50-50. The miners hauled the "waste" silver ore down to Queen Creek where they rigged up a crude concentrator. By using this method and hand picking the ore, they obtained enough for a good shipment. One of the miners was named Wilson, later always known as "Dump" Wilson, took the concentrates to the Shelby Smelter in San Francisco, California. When he returned he had a certificate of deposit for $50,000, even after paying expenses.

It was said that "Dump" Wilson was a queer duck. He married a girl and took her to San Francisco with him. Later he became somewhat of a doctor in Arizona, and was labeled as a "quack doctor." He was the manufacturer of the famous *Apache Salve*, a cure-all salve, which was very popular in Arizona at the time.[21]

No Space Left for New Claims near the Silver King Mine

Within a few days of the discovery, there was no space near the Silver King Mine available for filing, for the land was filed on for 10 miles in all directions from the site of the Silver King. In an unbelievable short time the whole slope of the Pinal Mountains were covered with mining locations. Late arrivals were obliged to follow the Stoneman Grade up and over King's Crown Mountain, thus began the second stampede. Prospectors located some claims at the top of the Pinals, but more went into the Pinal Basin, where they located several silver claims in the gulches behind the Globe Ledge claims including the Big Johnny and the Copper Gulch.[22]

Some prospectors recalled the old Silver Queen locations of 1871 and a new and larger Silver Queen claim was recorded soon after the Silver King was recorded in 1875. In fact, on March 29, 1875, right after Mason had recorded the Silver King, he recorded the Hub and Irene Claims, as Shaft # 1 of the Silver Queen location, and organized the Silver Queen Mine, in the Pioneer Mining District of Arizona Territory. Very little recorded work was done on this claim at this time, however, records indicate that only about $27,000 was shipped from this location. [23] Meanwhile, mining men rushed to their sites and settled down to do the required assessment work. They erected tents and rough shacks and were very glad when a couple of tent stores opened up. They now were assured of food and other necessary supplies and could devote their time to mining.

Mines and Mining Report of 1877

While traveling in Nevada during the summer of 2004 Jack Carlson found some old reports on the condition of mining in Arizona, Territory. From the files of the DeLamarre Library in Reno, Nevada comes an excerpt from pages 345-346 of *Statistics of Mines and Mining in the States and Territories West of the Rocky Mountains;* Eighth Annual Report of Rossiter W. Raymond, United States Commissioner of Mine Statistics, published in Washington, DC, 1877. It discusses the finding of the Silver King Mine and more importantly discusses the early works of the mine and the method of extracting silver from the ore.

> "The Silver King shaft is now down 42 feet, with a drift from the bottom 12 feet. The shaft is 6 by 9 feet and the drift is 5 feet wide by 61/2 feet high. The shaft started on mineral, and as it goes downcuts numerous small seams of rich ore, all pitching toward the main mountain at an angle of about 55°. These seams vary in width from 3 inches to 18 inches. The hill is a formation of brownish stone, which the miners think is a kind of granite. The vein-matter is quartz. The mineral consists chiefly of chlorides and black sulphurets. Great quantities of nearly-pure silver is found in little black nuggets in the quartz; the nuggets are soft, have coherence like bar-lead, and can be chewed between the teeth without feeling any grit; they assay about $20,000 to the ton. The first lot of ore worked was about 500 pounds, taken from the first 15 feet of the shaft. It was worked in a little furnace built at Florence by Messers. Airy & Hughes to work ores from this district. This lot of ore yielded over $8 a pound. To work this, they brought in Tucson pig-lead produced from the Patagonia Mine, which is about 80 miles south of Tucson. It is estimated that the ore taken out of the shaft of 42 feet depth, 6 by 9 in size, will taken as a whole, yield about $50,000; or in other words, that the original prospecting shaft on the mine has in the first 42 feet given a yield of over $1,000 a foot.

"The mine produces mainly a milling-ore, but the richest portions may be best reduced by smelting. Many silver nuggets have been found, which only requires the application of heat to reduce them to merchantable silver.

"The Silver King has been examined by experts from San Francisco, and it is said that a company will soon be formed in that city (where the ore has attracted much attention) to work it upon a scale commensurate with its size and richness. The ore in sight warrants the immediate erection of extensive works." [24]

"Too Late for the Ball"

A story is told about a prospector who was there at the Silver King site on the second day, but had to hurry on to Florence on other urgent business. He says that there were only a few prospectors there. When the prospector went back a few days later, he says that there were over 50 prospectors there, and that all the good claims were taken. He moved on with several others up and over the Stoneman Grade into the Pinals in search of a site to stake his claim. He had gotten there "too late for the ball" or as the saying goes, "A day late and a dollar short."[25] Meanwhile, the prospectors who discovered the Silver King Mine laid out the mining district now known as the Pioneer Mining District. This district extended 20 miles square, extending 10 miles each way from the Silver King Mine location. The district was properly made in accordance with the mining laws of that day, and recorded at Florence, which was, and still is, the county seat. Pinal County had been newly created on February 1,1875 by legislative act, and the Silver King claim was the first claim recorded at Florence. The ownership was equally divided among the four prospectors.

The Original Finders of the Silver King Mine Sell Out

On the June 30th, 1876, Isaac Copeland sold his one-quarter share to Charles G. Mason, and William H. Long sold his one-quarter share to Benjamin W. Reagan. On the 9th of January 1877, Charles G. Mason sold his one-half share to Col. James M. Barney. On the 5th day of May 1877, the Silver King Mining Company was incorporated under the laws of the State of California. On the 9th of May, 1877, James M. Barney and Benjamin W. Reagan deeded their entire interests in the mine to the corporation."[26] Col. James M. Barney was a Yuma merchant at the time, but had been Aide-de-Camp to Governor Anson P. K. Safford, with the military rank of Lieutenant Colonel.

Copeland and Long had sold out to their partners for $80,000, under the impression that the mine was too good to last, but this amount was made from the net profits in less than six months. Then Charlie Mason weakened as the glory hole began pinching out, and sold his interest to Col. James M. Barney of Yuma, Arizona for $250,000. At this time the first-class ores assayed $8,000 to $20,000 per ton, and were shipped to San Francisco by way of Yuma. Soon after, Reagan began to suspect the mine had a bottom to it, and sold out to Barney for $300,000. *What they did not know was that the Silver King Mine would make millions!*

Col. Robert Williams reached the Silver King Mine just as this deal was consummated and opened a hotel. His was the first substantial building erected outside of those belonging to the Silver King Mining Company, but as the permanency of the mine was demonstrated other buildings were erected. The company's payroll seldom fell below $40,000 per month, and the camp was prosperous until the decline in silver and the scarcity of readily abundant, high grade ore in the mine caused a cessation of work and the practical abandonment of the camp.[27]

False Boom Town Mining Stock

Robert Williams, one of the first settlers in the new mining camp of Silver King, and who later built a magnificent hotel there, recalled this story about those heady times when everyone thought they would strike it rich at any time. Among Colonel Williams' stories is one that illustrates the methods employed to *boom* mining stock in Silver King and Comstock days, where there was a "live superintendent at one end of the line, and skillful financiers at the other." Colonel Williams had an old militia title, and many of the men in those days used the title Colonel.

> "The Seventy-Six Mine was located near enough to the Silver King Mine in Pinal County to have some savor of goodness in the eyes of a gullible public, but industrious and expensive excavations produced nothing but 'holiness.' In fact, as the colonel expresses it, 'there wasn't a smell of ore.' "If the company couldn't get ore, there was nothing but common sense to prevent them from procuring a mill and a five-stamp plant was erected to grind out hope for the stockholders. But even this hope failed, and the superintendent was called upon to cooperate with the directors of the Seventy-Six Company to get them out of the hole."

Meanwhile, Copeland had sold out his interest in the Silver King Mine to Mason and Reagan, but retained possession of the dump, from which $20,000 in ore had been selected and shipped. When he had sorted this over again to his own satisfaction he sold the rest of the dump to four Frenchmen for $1,000 and they also found fine pickings. Two of them bought out the others for $10,000, and still had a small fortune left.

The superintendent of Seventy-Six saw the possibilities of this dump and proposed to the owners to run the balance of the ore through his mill at so much per ton, the bullion to be marketed through his company. The ore was now too low grade for shipment. They readily agreed and the stock market was soon impressed with the fact that the Seventy-Six was turning out bullion at the rate of several bars a day. Without stopping to investigate the question of ownership, the public made such a rush for stock that it went up to $7.50 a share. Of course, it wasn't worth the smallest fraction of a cent, but this was not discovered until the inside stockholders and the superintendent had unloaded. The mill was afterwards sold to the Silver King Company, but "the hole was too deep for a grave and not long enough for a well."[28]

Sullivan Returns to the Silver King

Professor Blake describes the return of John Sullivan (the original finder of the rich silver ore) to the place where he located the silver nuggets. "One day two years ago (1881), an aged man came slowly into the thriving settlement at Picket Post, and with great interest wandered about the Silver King Mill, where twenty stamps were day and night merrily pounding out silver from the rock. The man was evidently in need of help, and soon went to the office of the company and announced himself as Sullivan, the old soldier, the original discoverer of the vein, and humbly asked for work. Although long before he had been given up as dead and very few of his old acquaintances survived, he was identified beyond a doubt, and was immediately taken into the company's service by the day. His story was briefly told as follows. After leaving Mason's ranch he crossed the wide deserts to the westward as far as the great Colorado River, and beyond it into California. Being penniless, he sustained himself by working as a farmhand in California. He was always hoping to obtain sufficient means to return to Arizona and secure the benefits of his discovery. He had labored on year after year, looking vaguely forward, and keeping the secret of the locality to himself, until one day he heard of the discovery of the rich deposit of silver by Mason and others. He was convinced that the place had been found, and that he had lost his chance of making the location for himself. Although without any ownership or right in the location, as made by Mason and his friends, and disappointed in his long-cherished hope, he could not resist a desire to return and see the result of the opening of the mine."[29]

Well, the above story is partially true; however, Sullivan did not immediately leave for California. After the Silver King was located and filed upon legally, Sullivan realized that he missed a magnificent opportunity. He then filed mining claims on locations not far from the Silver King in hopes of also striking it rich. On May 8, 1875, (just 3 weeks after the Silver King was filed on) Sullivan filed on the 1st Extension on the Silver Brick Claim, along with Jon J. Devine, Pat Holland and Lewis Starar on a location in Pinal County. They stated in the affidavit filed in the Pinal County Recorder's Office that the claim was a lode claim or vein of silver bearing rock. M. Rogers, Andrew Starar, Jacob Starar, and Patrick Morgan filed on the Silver Brick Mining Claim, also located on May 8, 1875. Rogers filed several claims in the) Rogers Canyon area on Iron Mountain in Pinal County, but the Silver Brick does not give any specifics as to the location. The Starars were known to have filed claims near Rogers Canyon and one can only speculate. Rogers Canyon was named after James Rogers. The Starars were also known as Starrars and Starrs and at one time were associated with one Jacob Waltz, of the Lost Dutchman Mine fame. They were neighbors in Phoenix. This leads to some interesting speculation that Jacob Waltz could have found his famous mine in the eastern Superstition Mountains as he was known to have visited Silver King Town to buy supplies.

On June 1, 1875, Sullivan filed a claim on the 2nd extension on the Sacaton Mine. On July 22, 1875 Sullivan filed a claim on the 1st Extension East of the Black Republic Lode, along with Richard Powers and F. M. Griffith, which was stated to be 2 miles northwest of the Silver King Mine. The Black Republic was also filed on July 22, 1875 by R. Morgan, P. Holland, Jon J. Devine and M. Rogers and lists its location as 2 miles northwest of the Silver King Mine and 5 miles north of Picket Post. Apparently John Sullivan, the old soldier who found the silver nuggets at the site of what was to become the Silver King Mine, left for California sometime after 1875, when he became disillusioned that he let the Silver King get away and could not strike it rich on the claims that he did file. Sullivan's claims did not get much notoriety in the papers or mining reports. Although I did not exhaustively research all the old mining claims of Pinal County, I did not find any further claims filed by Sullivan, nor any mention of him in the newspapers of the day.

Fate of the Four Prospectors Who Struck It Rich
Charles G. Mason

Charles G. Mason, one of the four founders of the Silver King Mine, was one of the earliest settlers of the Town of Florence, Arizona. In 1867, he along with Thomas A. Ewing purchased a property at the corner of Quartz Street and Ruggles Avenue in Florence. A house was later erected there, supposedly the first house in Florence. Their partnership was one of ranching and farming. However, Mason also prospected for minerals in central Arizona along with other rancher-prospectors. Prior to coming to Florence he was a hunter and prospector in the Weaver Mining District near Prescott, Arizona. He also was said to have been a pony mail rider on the route from Tucson, Arizona to San Diego, California. This was a distance of 110 miles with only one water hole on the way. Mason established a hotel in this building. Legend has it that a woman named "Frenchie" ran a house for "soiled doves" in the hotel.

Clara T. Woody and Milton Schwartz, writing in *Globe, Arizona,* describe Charles Mason. "Charles Mason was a Colorado prospector who came to Arizona after enduring an exceptionally hard winter which had left him snowbound for months. His brothers Aaron and Solon had given him up for lost, but when the spring thaw reopened the trails, the indefatigable Charles emerged alive and well." This winter's isolation may have determined him to seek a milder climate, for soon after he moved to Arizona.[30] Mason had taken residence in Arizona Territory in 1864 and was one of the early pioneers. He was a man of exceptional versatility. He had been a stone cutter, hunter, mail rider, rancher and miner. He also operated a pack train and freighting route between the farming region around Florence and the mining camps in the Pinal Mountains.[31] Charles Mason did not live long after he obtained the great wealth of the Silver King. He died shortly after selling out his share of the mine.
The following news article from the *Arizona Sentinel,* September 8, 1878, reported his demise as a probable heart attack, but others theorized that he met with foul play.

> "Charles G. Mason died at the Grand Hotel, San Francisco, about one o'clock in the morning, July 29th. Mr. H. C. Austin, of Los Angeles, had

accompanied (sic) him to San Francisco. Mason had complained, the day before his death, of pain and oppression in his breast. He had called Mr. Austin from his room and was talking with him, when he suddenly dropped dead. A doctor had been sent for, but arrived too late. Heart disease is hereditary in Mason's family, and was, without doubt, the cause of his death. Mason was the first farmer on the Gila near where Florence now stands. He went there nearly fourteen years ago, and fought off Apaches as gallantly as any of Arizona's pioneers. He was one of the discoverers of the famous Silver King Mine; and he acquired a full half interest in it, which he sold to Col. James M. Barney for $300,000. At the time of his death he was improving a fine residence tract of 17 acres in the City of Los Angeles. He left a wife, but no children. He was a native of Maine and possessed the industry and enterprise characterizing the people of the State, together with the big-hearted generosity developed by frontier life."[32]

Sometime later, Freeborn Mason, Aaron's brother dropped dead suddenly in a hotel in Florence, Arizona Territory, reportedly of a heart attack.

Ben Regan—-A Man of Many Hats

Benjamin W. Regan, one of the four prospectors who located the Silver King Mine, was a very versatile man. He wore many hats at various times. He was a saloonkeeper, a carpenter, prospector, miner and an ordained minister of the Disciples of Christ Church (a Campellite Minister). He pursued each of these vocations separately usually without conflict, even closing down the saloon on Sunday mornings to offer a strong sermon for all sinners present. On one Sunday morning a soldier walked into the saloon/church, well under the influence of some spirits he had imbibed elsewhere, and demanded a drink during the sermon. He was really creating a disturbance with that bellyful of fighting whiskey, and Ben tried to quiet him down but to no avail. The disturbance finally led to fisticuffs and many verbal threats from the

Ben Regan & wife cir. 1870s

—Greg Davis Collection

soldier. As Ben tried to throw the fellow out of the church there was a tremendous melee between Ben and the soldier, ending up with Ben fatally stabbing the soldier. Ben then had a body on his hand, and no one would claim the soldier as a friend or even acknowledge knowing him. Ben then put on his carpenter's hat and swiftly made the dead man a coffin. Then he put his Man of God hat on and presided at the burial and funeral. He then spoke the few appropriate words of the time, as the body was being lowered into the hastily dug grave, for it was summer and the body would not last long in hot weather. Ben Regan was truly a man of many talents and today would be called a man for all seasons.[33]

Ben Regan and his wife moved to Oakland, California soon after selling his share in early 1879. Regan built a Campbellite Church at Oakland with his share of the Silver King money. Regan and Barney incorporated and sold some shares in the Silver King Mine. Regan also soon established an estate in Oakland. Driving out of the grounds one day with his wife the young animal hitched to the carriage became frightened and ran away throwing both occupants to the ground. Regan died on July 26, 1879, but his wife recovered. An interesting aspect of Regan's mining career notes that before his death he bought out his partners' Globe Ledge mining interests, and filed patents for two additional claims. Those patents were not approved until 1881, (two years after his death).

William Long cir. 1877
— Greg Davis Collection

William Long

William Long was also a farmer-prospector from Florence, Arizona. He married Mary Elizabeth (Molly) Whitlow of the Whitlow ranching family of the Superstition Mountains and Florence area. William Long took his money, his wife, and left to live in California. At Yuma, Arizona Territory, he was stricken with smallpox and died in California in 1877. He left Molly a widow. Her story is told in the chapter on pioneer women.

Isaac Copeland

Isaac Copeland was also a farmer-prospector from Florence. He was married and had two children. After he sold his interest in the Silver King there is little trace of Isaac Copeland. He did surface in 1905, however, to give a deposition during the Dr. Jones/El Medico Claim vs. Silver King Mining Company. *Thus within four years of their finding their bonanza, all four of the discoverers were either gone from the Silver King or dead.*

When the four men recorded their claim, they did not record either the name of the prospector who was with them and was killed by Apaches, nor Sullivan, the original finder of the rich silver lode. Superstitious people claim that these four men met their untimely deaths as a result of this neglect. Perhaps there is a ring of truth in the saying: "As you sow, so shall you reap." In any event, these four men did not live long enough to savor the rewards that the Silver King, one of Arizona's richest mines, produced.[34]

Isaac Copeland cir. 1880s — Greg Davis Collection

James Mitchell Barney—-Silver King Owner During 1870s-80s

A native of New York State and former resident of Nantucket Island, Col. James M. Barney had come to California in 1854, just after the Gadsen Purchase. He had prosperous mercantile stores at Santa Cruz and Watsonville, California. He also had a rope manufacturing business. He went to Ehrenberg, Arizona Territory, where he was appointed a Wells Fargo agent. Barney became associated with the development of Arizona Territory, and was a partner of George F. Hooper & Company, who was the post sutler at Fort Yuma. In fact, Barney was the Quartermaster's agent at Yuma, expediting military shipments for the Army's Department of Arizona. He performed so well at this task that he was subsequently appointed to the military staff of Governor A. P .K. Safford, which provided him the title of Colonel. After the Civil War ended, he participated in mining and agriculture ventures. He was named postmaster of the trading centers of Maricopa Wells, Arizona City (later called Yuma) and Ehrenberg. He then acquired the Silver King Mine. During his tenure of running the Silver King, he created a sensation on the San Francisco

Mining exchange by paying big dividends on the Silver King Mining stock. After he sold the Silver King Mine, Barney moved to London, England, where he lived until his death in 1914. His namesake and nephew, James Mitchell Barney, became a leading historian and writer of thc pionccr history of Arizona.[35]

"All you'll find is your Tombstone"

During the early days of the Silver King, one of the most famous prospectors of the old west came to the mine. Ed Schieffelin (the founder of Tombstone) rode alone from his diggings near present day Tombstone in the summer of 1877, heading toward the Silver King Mine through 200 miles of hostile Indian country in search of his brother Al who was supposed to be working at the King. As it turned out his brother was not at the King but 200 miles further north at the McCracken Mine. Schiefflin was broke and threadbare and picked up work locally. When he had earned $13, he bought new clothes, re-shod his mule, packed up some supplies and headed back to the yet to be named Tombstone area with Dick Gird, a geologist. When he encountered soldiers on patrol in the past, they always told him "you'd find your *tombstone* if you don't stop running through this country while the Indians are so bad." Schiefflin later said the warning always stood out in his mind, and when he struck the Lucky Cuss Mine he called the small mining camp that sprung up, *Tombstone,* after the soldiers' warnings.[36]

Ed Schieffelin cir. 1880s
— Superstition Mountain Historical Society Collection

Chapter Two

When Silver Was King—1880s Silver King Mine

The following paragraphs take the reader back in time, to the 1880s, when the Silver King Mine was at the pinnacle of its success. To rewrite these paragraphs, taken from the *Pinal Drill*, September 15, 1883, would not give the reader the true flavor of the period. I am including this very poignant description, just exactly as I found it, that takes us back to those halcyon days of the Silver King Mine, *When Silver Was King!*

Mule Trains leaving ore house cir. 1880s — Bowen Family Collection

"Our late visit and journey through this immense subterranean Silver Palace, gave us further assurance of its great future. The impression first received in now entering on the premises is imposing. The approach to the mine is over a hard road, running around the ore-house. The three trains of wagons, each hauled by 18 mule teams, pass up on the right side, then swing under the large ore chutes below the rock-breaker, and are loaded in a few minutes. Each of these trains consists of 3 wagons, each carries about 25 tons, and thus about 75 tons pass daily down the easy grade from the mine to the mill, and are there worked. Each train has two men, a driver and brakeman. The wagonmaster, like a super-cargo, accompanies the train. People who have not seen such an enormous caravan can form no idea. Think of 3 great wagons, coupled together like a railroad train, hauled by 16 strong mules and two horses, all hitched to a long chain from the tongue. Each wheel weighs 1000 pounds. Each wagon loaded weighing over 10 tons, moving like a crushing avalanche over the road.

You enter the upper part of the ore-house over a bridge, leading from the surface platform of the main shaft. As the ore comes up from the mine in iron cars, they are wheeled over the railways upon the bridge to the ore-house, and there dumped over a grizzly, through which all the fine ore passes into the bins below, the large pieces passing to the ore-breaker, from which it drops into the same bins, and thence through the chutes into the wagons when they are to be loaded. (A grizzly is a grate through which the broken ore must pass before entering the chutes.)

Mule Trains Leaving Silver King cir. 1880s — Bowen Family Collection

This ore-house is 60 feet high, 30 by 60 feet, 2 story, capacity 800 tons, strongly built, a neat imposing structure. The engine room contains the now historic little engine, which has followed the King in its progress from the 5 stamps and 2 Frue tables, to the hoisting works, and now does the rock breaking. The building is protected against fire by the big tanks and by water pipes, which also form the railings along the bridge. The enormous jaws of the Blake ore-breaker crush the large pieces of rock, as if they were mere bubbles. And here stands the strong box, in which are gathered the exquisite specimens of rare beauty and great value. (High-grade native and wire silver.)

The neat walls formed by waste rock, which line the roads and walks, are signal marks of order. The engine room is near the main shafts, the hoisting works, sawmill, machine shop and blacksmith shop. There is a separate engine for the sawmill, close to these are the change rooms, where the miners change their clothing, the washroom and other prerequisites. At a short distance and between the engine rooms and the stables, there are the large water tanks, filled through pipes from the grand reservoir on the hill, 500 feet above, ready at any instant to spout water over any part of the buildings and works. The miners are trained and exercised as firemen, to act at any moment. They are perfectly secure against fire.

We have yet to describe the beautiful new office in front of the works, overlooking it all. This is 30x30 feet, 2 stories high with porticos. Here are the Superintendent's office and headquarters, also the Doctor's office. It commands a charming view over the country even down to Casa Grande. The old works, the first big excavation of ore, and which yet discloses thousands upon thousands of tons of ore, is left undisturbed, above to the east, 200 feet.

These are the main objects attracting the attention of the visitor upon the surface, and we cast a retrospect upon the busy scene of sunshine, and take our stand upon the platform in the shaft, sinking, sinking, in the darkness, down over 700 feet, at the dictate of the signal bell. We were in the same cage with the polite Superintendent, and when we got our sight again, we discovered the solid form of the foreman of the mine, under the beaming rays of stearine (candlelight). We are used to 'follow our conductor and fear no danger,' but before starting on our journey, we lit our candle and looked around. We were on the 714 foot level. There stood at once before us the pure white quartz, full of streaks and large lumps of solid metal, on both

Silver King Office cir. 1880s —Bowen Family Collection

sides of the drift, bearing in direction between west and north. This is a large excavation, from which over 20,000 tons of ore have been taken."[1]

"We are on the 714 foot level. We should judge that the depth of this opening is at least 200 feet, and the width 100 feet. There is ore all round, above and below. The brilliant dark streaks of metal, like figures on a frescoed wall, stand out in bold relief, and glisten to the eye. The rock is porphyry, quartz and spar, and the metal is in all. Every bit of it assays, and the free silver is distributed throughout, making the whole mass rich. We observed that between the timbers, they are building walls of masonry, which make the support solid. They are progressing with this work, and as these walls are built of waste rock there is also much saving of labor obtained. It is difficult in passing through so large a space, and through so many sets of timbers to know your exact directions, but the turnings to the right and left were many, and can not be very interesting to the general reader, for whose comprehension we write, avoiding all scientific terms and technicalities.

But, following the right hand side of this grand chamber, our general course was to the left around, until we again reached near the shafts, and the same easily blasted rock was all around, apparently without a limit. It reminds one of some Oriental grand market house supported by massive pillars, forming a labyrinth, through which one may lose his way. The air is fresh and the ventilation perfect. The air is forced through large galvanized iron pipes by a Sturtevant blower on top, and the winzes and air-shafts between the levels cause a gentle, pleasant circulation throughout the mine. We were lifted to the 6th and 5th levels and upwards, but we run the risk of tiring our readers by repetition, for the same story is to be told over again. We passed up ladders, and from timbers to timbers, with the deep open gaps below us, climbing like a sailor on the yard arms, and we felt like singing with Jack Tar: 'With the blue above and the blue below' for the peculiar, blue colored silver bearing porphyry was there, and the long sides of solid white quartz and crystals shone in the light like the silvery crests on the breaking waves in phosphoric sparling jets, and 'as an ocean plentiful.'

From the 7th level and up to the 1st level, which is 114 feet below the surface, there is the remarkable firm body or quartz and ore to be seen all the way; the solid side of which is the main support of the mine. Much ore has been taken down in the various levels, leaving this standing, and it is full of wonderfully rich streaks and lumps of mineral all along. Now, let us calculate the quantity of ore in the breast of the mine alone. Supposing it to be only 100 feet thick, it being estimated that 13 cubic feet form a ton; there is 600 feet in highth by 200 feet in length, in round numbers, a million tons, and how many more tons of ore are in sight? There were about 14 men at work on this shift actually mining and some 16 laborers and car-men; they were scarcely observable in the vast cavern. There is work for 1000 men, and 200 stamps cannot crush the ore thus to be taken out. It seems to us that the Company ' make haste slowly' in realising the millions in sight. Of that we have no right to speak; but we do say that there is in every nook and corner of this establishment. Visible effects of knowledge, in the system, order, economy and quiet progress made. There is as little needless noise as in a drawing room, as little coarseness and vulgarity as if the workmen were officiating priests.

The mine is perfectly dry throughout, a little water seeping from the surface only, and which is conducted mainly into the shaft, the timbers of which are preserved thereby. We doubt whether amongst the mines there is a parallel of order, repectability and skill, from the highest to the lowest, in every department of the work, in office mine and mill. It is well named 'the Silver King' amongst the mines."[2]

This article puts in perspective the opinion of people in the 1880s, of the exalted position and esteem which the Silver King Mine was held.

Mine Site and Accessibility

The Silver King Mine is located 4 miles north of Superior on an unpaved Forest Service Road. The road winds upwards from Route 60 through the rugged canyons, past an earth watertank with a windmill, through desert mountain landscape, on to the old mine site. As you pass some stone foundations of the old buildings, you can imagine how life was in this isolated mountain wilderness for the miners and their families of the 1870s and 1880s who made the Silver King mining camp their home. Only a few old ruins remain where there was once a thriving town. The old mine buildings are no longer there, (some old mining equipment and mine dumps, and shafts still exist), but the terrain is basically the same. The Silver King Mine is situated in a small valley surrounded by mountains. It

King's Crown Mountain with crown 2000 — Author's photo

appears to be the bottom of volcanic activity that resembles a cauldron. It is dangerous to walk unescorted around this area due to the numerous old shafts and dig holes still open. King's Crown Mountain is above and just to the east of the King. On King's Crown Mountain, just above the main shaft, is a large rock formation jutting out from the rest of the rocks that looks like a king's crown. During the late afternoon this "crown" actually takes on a gold color and a regal look.

The Old Stoneman Grade is visible from the old mine site, and with a little imagination you can almost see the soldiers pursuing the Apaches up the mountains toward Globe. The glory hole (open pit area) is clearly visible where several tons of high grade silver ore was taken in 1876. This was before a tunnel (drift) was started to the main shaft area (called the engine shaft area on various diagrams). The glory hole measures about 100 feet long and 90 feet wide and about 70 feet deep, although early reports had it between 78 and 120 feet deep. It is partially filled in now from exploratory work over the last century.

King's Crown Close Up — Author's photo

The *Pinal Drill* May 14, 1881 describes the original small brown hill that was found by the discoverers of the Silver King Mine.

> "The hill or cone on the Silver King location where the discovery was made, angles like the two sides of a regular square with the point erect, or a right angle triangle. It is a right cone from the apex or vertex regularly to the surface base, or in other words, the sides are somewhat pyramidal in form." This rather confusing description goes on to describe the underground

> workings. "Thus it will be seen that the ore—dips into the earth conical or latitudinal, and is supposed to extend underground below the surface of the adjoining claims. Hence, these adjoining claims expect to get into the hoped for submerged mass of metal by depth, and to obtain virtually the same ore supply which the King has now—here are chances for incalculable fortunes, worth the risk of millions in the effort to obtain."[3]

It seems that this is why all the claims were placed so close to the Silver King claim, and the nearby Bilk Shaft, did in fact, tie into the Silver King mineshaft at a depth several hundred feet below the surface to the west. According to J. B. Tenny, the original shaft drifted downward along a network of stringers ranging 3 to 18 inches wide in quartz or porphyry rock, similar to granite. The material in the stringers was quartz and the ore minerals were chlorargyrite (chloride of silver), argentite and native silver. This ore when sorted ran about $2000 to the ton. In order to treat the raw ore, Curry and Hughes erected a small furnace, of the cupel type, at Florence. Pig lead for collecting the silver was obtained from the Mowry Mine in the Patagonia Mountains which was 150 miles south of the Silver King Mine.

In 1879, Aaron Mason was the mine superintendent. The ore was then crushed under his direction in a Blake Crusher (named after the inventor Eli Whitney Blake) and reduced to corn-kernel size. The Blake machine was made out of cast iron and was considered cheap, light and very effective for it's day, and in fact, lasted nearly to the end of the frontier period. The Blake Crusher was run by two flywheels and delivered the power by belt from a line shaft. The crankshaft, which ran between the wheels, was hooked by a connecting rod to the center of a toggle bolt which greatly multiplied the leverage applied to the inner jaw. A small spring-loaded rod opened the jaws at the end of each stroke. Moving parts were few and rugged; tolerances were easy. Any lump of ore tough enough to jam the jaws had to be small enough to be cleared by hand from the crusher.[4]

Silver King Mine Location

The following description of the Silver King Mine is taken from official reports of the early 1880s period, when the Silver King Mine was in its heyday. Much of this information is credited to Professor William P. Blake, former Professor of Mineralogy and Mining, College of California, and the Territorial Geologist at the time. He inspected the Silver King

Mine on several occasions and wrote reports at the request of the ownership of the mine.

The specific location of the Silver King Mine is at the base of the southwestern slope of the Pinal Mountain range in Pinal County in what was then Arizona Territory. The mine is located at an elevation about 3700 feet above sea level. The road from the mine led for about four miles down the foothills and then over the plains to the County Seat at Florence, on the Gila River, about 35 miles away. In the 1880s the road then continued to the Casa Grande Station of the Southern Pacific Railroad. The railroad then went to San Francisco, a distance of 913 miles by rail. The road from the mine to the railway station usually took eight or nine hours by horseback or buggy. Blake said that the trip was 53 hours to San Francisco and 5 1/2 days from New York where the investors were located.

The mill for the mine company was located at the Pinal City, located on Queen Creek, that had year round water, five miles from the mine (The Town of Pinal, often referred to as Pinal City, is described later in great detail.) The trip was downhill from the mine as Pinal was at an elevation of 2400 feet above sea level. In those days there was a good road that connected the mine with the mill town and the mule loads of crushed ore steadily were hauled to the mill. There was also a telephone that connected the mine to the mill. At Pinal, the offices of the Silver King Mine were also linked by the Silver King and Florence Telegraph Company to the Western Union telegraph lines which allowed cross-country communication, even in those days. A plan was devised to bring the Southern Pacific Railroad northward to the Silver King Mine, but later reports indicate that it was never completed. However, during the 20th Century a railway was later completed, after the close of the Silver King Mine, to the nearby famous Magma Copper Mine, below King's Crown Mountain in Superior.

Geology of the Silver King Mine Area

Blake describes the geological structure of the Pinal Region as "some what complex but very interesting." He further states: "The whole country appears to have been covered, in comparatively recent geological time, with volcanic rocks, partly in sedimentary form as volcanic sandstone's and conglomerates, and partly, and lastly, in a molten or lava form. A considerable area of these rocks has been swept away by denudation and erosion, leaving the older and foundation rocks exposed to view, particularly in the valleys and canyons. Remnants of the lava flow cap the summits and in places remain as flat-topped mountains; as, for example, the mountain known as Tordilla, (Picket Post Mountain) at Pinal, back of the mill.

The Pinal Range, above the Silver King Mine, is formed chiefly of Paleozoic strata with heavy beds of quartzite at the base, overlaid by massive limestone dipping eastward. The strata is seen along the Stoneman Grade, as it ascends the mountain, and can be seen in great thickness, probably not less than 3000 feet, by ascending Queen Creek further to the southwest. Above the limestone

Silver King ghost town 1997 — Author's photo

at the King Mountain (later called King's Crown Mountain), which is the highest point above the mine), the volcanic outflows form the capping, consist largely of pitchstone porphyry carrying obsidian, geodes of quartz crystals, masses of chalcedony and semi-opal.

Below the limestone and quartzite series, gneiss and hornblende rocks of the Archaean Age crop out and appear to be the foundation rocks on which the sediments rest unconformable. Still lower down the slope the rocks become sienitic, and are then replaced by a distinctly formed feldspar-porphyry, with small white feldspar crystals, and an abundance of iron pyrites is finely disseminated. The formation appears to have the position of a dike, cutting the other rocks, but it is extensive, and continues to and beyond the Silver King Mine. It may be regarded as the enclosing rock of the Silver King, though in one place some 200 feet above the mine, there is a strongly defined outcrop of a dense, hard porphyry, with hornblende crystals and brilliant glassy feldspar forming a sienitic porphyry. The porphyritic structure in the mine, generally becomes obliterated, and the rock has more the appearance of a granular quartzite with a large amount of earthly admixture, and obscure

> fragmentary crystals of silvery mica. The larger portion of the rock beyond the open cut (glory hole) has a dark greenish color, and is known by the miners as "dark porphyry". In some places it appears to be chloritic. This, with the light porphyry, is the ore-bearing rock of the mine. It is penetrated by veinlets of quartz and ore."[5]

Blake called the porphyry of the mine; altered porphyry. Porphyry is an igneous rock containing conspicuous large crystals in a fine-grained matrix. Hornblende is a common dark green, black or brown mineral occurring as rod-like crystals in igneous and metamorphic rocks. Gneissic rocks are course grained metamorphic rocks with alternating light and dark minerals. Feldspar is the most abundant group of light colored rock-forming minerals. Metamorphic rocks are formed from older rocks that have been subjected to great pressure and heat, or to chemical changes. Mica is a group of minerals that separate into thin, shiny plates or flakes. Silica is silicon dioxide, which occurs as quartz and a major part of many other minerals. Quartz is a hard glassy mineral composed of crystalline silica, one of the most common rock-forming minerals. Quartzite is metamorphic rock formed of sandstone, cemented by silica.

The small hills (sometimes referred to as hillocks) surrounding the Silver King Mine were light brown in color. They were sometimes called blowouts and were shaped like (upside-down) truncated cones. The country rock of those hills consisted of much-weathered porphyry or andesite, whose original color had been leached away and replaced by the brown stain of iron said to have originated from the oxidized silver ore within the hills.

Metallic ore bodies take two shapes. One is thin and vein-like and the other is massive and globular. Typically, silver is manifested in both forms, and the silver at the Silver King Mine is both vein-like and massive, in ore bodies of different values or percentages of silver within the ore body. At the 114-foot level in 1999, veins of high-grade silver were found within the massive bodies of rock that also contained other silver values.

The Vein Formation

The vein formation of the Silver King is in a mass of ore bound quartz, surrounded by porphyry, seamed with veinlets of quartz. The veinlets vary from the thickness of a sheet of paper to one-quarter inch to one inch thick, accompanied by the ore having quartz on each side of it next to the rock. In addition to these veinlets there are masses of ore, and, starting in the upper levels, a large and compact body of mostly white quartz which extends to the lower levels. The heavier bodies of ore have been cut from the footwall side of the quartz body. This mass of quartz presents itself as a true fissure, although it does not have the usual sheet-like or tabular form. It is, instead, a columnar chimney-like mass, some 80 feet in diameter in places, but irregular and without longitudinal extension. In other words, this quartz-vein, instead of forming a sheet-like mass, is *cylindrical* or columnar in its form, filling a spirally formed cavity, as if it had been smoke rising from a chimney. [6]

> According to Blake's report of 1883: "Although the massive quartz does hold bunches of rich ore, it is not, as a rule, so rich and profitable to work as the rock adjoining. The ore is more abundant in connection with the small branching veins in the outside rock, than in the mass of the quartz itself." Blake went on to state that the quartz bodies had not been fully explored in the upper levels, and that further it was his opinion that the quartz veinstone did not carry the best part of the ore. "It (quartz chimney) appears rather to have been the main channel of the mineralization, the main artery or feeder to the thousands of veinlets branching from it into the wall rock, following the clefts and penetrating the substance of the rock, depositing and diffusing native silver and sulfides, throughout the whole mass of rock, for an indeterminate distance on each side."[7]

Blake said that the silver was found impregnated into the rock, which he classified as *stockwork,* which required the rock to be mined en masse. What Blake described essentially, was that the veinlets and porphyry had to be crushed together as it was impossible to separate the rock from the ore. He also said that the white quartz in the upper levels contained gray copper ore which looked like black powder, but that it was rich in silver.

Mineralogy

The mine contains minerals and beautiful silver specimens that even today are of great interest to mineralogists and collectors. The surface ores are partially oxidized and decomposed; however those taken from inside the mine are more brilliant by contrast. The surface ores in the mine dumps contain light green chrysocolla, dark malachite green and azurite blue, due of course to the carbonate of copper and small amounts of silver. Also found here are the black nuggets of silver. Blake composed the following list of principal minerals that he found at the mine. Native Silver, Stromeyerite, Argentite, Sphalerite, Galenite, Tetrahedrite, Bornite, Chalcopyrite, Pyrite, Quartz, Calcite, Siderite and Barite. Also located were Horn Silver (Cerargyrite), Malachite, Azurite, Native Copper and Lead.

Native Silver- It looks like wire or can be fine as thread, very white in color. It can also look like coarse wire as in a wire bundle. Its size ranged from hair like to knitting needle to a mass one-half inch thick. It is found on quartz crystals.

Stromeyerite- It is highly metallic, with a bluish-black color, and has a brilliant luster. It is generally contained in quartz crystals. It contains about 51 percent silver and about 30 percent copper ore. It was generally found at the third and fourth levels.

Argentite- It is metallic, with a shiny luster, and lead-gray to black. It contains a high content of silver, sometimes as much as 70 percent. It was found down at the third and fourth levels.

Sphalerite- It is zinc-sulfide, which presented itself as transparent, brittle oil green crystals or as black masses. It was abundant at the seventh level of the mine and is closely associated with silver.

Galenite- It is lead gray and crystalline or granular in nature. It did not appear to be rich in silver content.

Tetrahedrite- It is described as gray copper and was abundant in the upper levels. It was found in the porphyry as well as in quartz above and in the third level. In 1880 it was regarded to be one of the most important ores of the mine.

Bornite- It is copper-red to bronze brown in color or metallic grayish-black and is associated with Chalcopyrite.

Chalcopyrite- It is Copper-iron sulfide and is often associated with some silver or gold. It is usually brass-yellow to golden-yellow. One of the distinguishing features from gold is that it is brittle whereas gold is not. It is found in small quantities in quartz at the mine.

Quartz- This was the chief veinstone of the mine and is found in massive amounts at all levels. It is crystalline in nature with a milky-white color, although there is some clear quartz. It also has a light purple or amethyst color, which was considered by the miners as a good sign of rich silver ore.

Calcite- Although it was not abundant, it did occur on occasion as bead like crystals with a whitish color. It can sometimes be confused with quartz.

Siderite- It is iron carbonite and appears light to dark brown or reddish brown. It occurs with chalcopyrite or barite, and can occur with some of the other minerals above. It is crystalline in nature and a minor ore of iron.

Barite- It occurs in several colors, such as white, gray and yellow. Next to quartz, it was the most abundant veinstone, but not associated with the rich ore as quartz. It comes as crystals or granular in form.

Horn Silver or Chlorargyrite- It derives its name from the mineral's chlorine content and silver content. It is a silver ore. Its color is pear gray to brown. It has a waxy appearance and high gravity and occurs often with argentite, (Acanthite) barite and calcite.

Malachite- It is described as dark green to emerald green in color. It is copper carbonate in crystal form and is associated with Azurite and Chalcopyrite. It is sometimes prized as a gemstone.

Azurite- It is basic copper carbonate in crystal form. It is azure blue to dark blue in color. It commonly occurs with Malachite and chalcopyrite and is sometimes used as an ornamental stone.

Chrysocolla- It is copper aluminum hydrogen-slicate hydroxide hydrate. It is usually bluish green in color, although it can also be bright green or mixed color due to impurities. It is a secondary mineral often associated with copper. It is also described as harder than turquoise but softer than chalcedony. It occurs in areas that include malachite and azurite. Miners that find outcroppings of chrysocolla also then search for other more valuable minerals.

Native Copper- It is copper red in color, metallic, shiny and highly malleable and sometimes can occur in masses. Native Copper is used as an ore of refined copper.

Gold- It must be noted that gold was never mentioned in any of the 1870s-80s reports, but numerous assay reports during the 1990s and 2000s revealed various amounts of gold at a payable amount.

Silver King Mine Ore at the 1882 Denver Exposition

This following description was given of the Silver King Mine ore, as listed in the Denver news of 1882:

> "There are among the hundreds of exhibits of fabulous richness or striking beauty, none which more comment, than the array of Arizona, and a prominent feature of that is the Silver King Mine case. It is filled with specimens of rare attractiveness, which not alone draws the eye with its glittering silver threads, but fascinates, with the thought of the marvelous wealth it represents in the veins of this great western territory. Wire silver is the metal in its native state, and in a form, which is handsomer than any article the most skilled workman with his tools and his genius can produce. The Silver King Mine makes the display of all displays in this line."[8]

Jewels made for Odd Fellows Assoc. — Bowen Family Collection

Pinal City cir. 1880s. — Bowen Family Collection

Chapter Three
Development of the Silver King Mine

The development of the ore body at the Silver King took place by first digging into the surface ore. Then the company drove shafts sometimes referred to as *adits* or *inclines* deeper into the ground to begin underground development. Once the Silver King struck the high-grade ore body, which was a pipe, the miners followed the pipe downward. The internal workings of following the high-grade ore bode consisted of horizontal tunnels, also called *drifts,* or *crosscuts* extending off the drifts. Internally inside the tunnels there were shafts known as *winzes* which dropped down from the tunnel floor. Internal shafts, which were driven up were called *raises*. Once the engineers had determined the ore body boundaries, actual mining began in earnest. Digging out the ore in huge amounts from one area created rooms. These rooms were called *stopes*. Stopes, tunnels and shafts were worked in the flat faces of the rock areas and these were called *headings* or *working faces* by the hard-rock miners.[1] The Silver King was initially developed to a depth of 830 feet. This included the open cut or glory hole and a water sump reservoir of 36 feet. The main shaft by 1883 was dug to a vertical depth of 714 feet. There were two compartments, each about 4 feet square, side by side, that were fitted with cages. These cages were hooked to a head frame, attached to cables and powered by steam. They brought out the ore and transported men and supplies into the mine to the various levels.

Silver King Two Compartment Shaft
— Bowen Family Collection

At about the 114 foot level there is water from a spring that runs into the mine and accumulates at the sump area at the bottom of the mine. Even though this spring produced about 2000 gallons a day, there was not sufficient water from this source to run the mine's equipment. The water was hoisted by means of a tank fitted into the cage. The purpose of the sump was to act as a holding tank while the water was accumulating. A description of the water problem in the Silver King was discussed in the *Mining Library, Volume V. - Details of Practical Mining* compiled from the *Engineering and Mining Journal, 1916.* Some 12 years after the mine opened, more water for operation the mine was obtained from a site known as the Pump Station Spring. The Pump Station Spring was located at the former site of General Stoneman's Camp Supply, at the summit of King's Crown Mountain, where bread was baked in the stone ovens. While developing this site it became necessary to remove one of the ovens. During the opening of the old bake oven, the skeletal remains of a man were found there. Robert Bowen of the Silver King Mine directed Jack Fraser to remove the remains. The remains were then reburied near some of the large Emery oak trees nearby.[2]

Report of the Silver King Mine 1879
Secretary's Report San Francisco, December 31st, 1879

To the President and Board of Directors of the Silver King Mining Co.
Sirs: In compliance with the by-laws of the Company, I have the honor to submit the following report as Secretary of the Company.

Yours, truly,
EDWIN B. BOOTHE, Secretary.

FROM MAY 5TH, 1877 TO DECEMBER 31ST, 1879

	Product
From Sales of Ore	$819,141.58
	Disbursements
Supplies:	$47,179.27
Wages at Mine:	$80,559.22
Wages at Mill:	57,357.92
Freight:	
Per wagon on supplies:	$14,931.46
Per rail on supplies:	8,267.96
Per wagon on concentrations:	30,863.77
Per rail on concentrations:	13,453.81
Subtotal	$67,517.00
Portage of Ore from Mine to Mill:	44,101.87
Wood:	19,440.26
Expenses of "76" Mill:	27,944.35
General Expenses:	10,346.51
Sampling and Assaying:	2,182.74
Lot and Offices at Pinal:	2,000.00

Taxes:	2,902.44
Insurance:	610.00
Rent of San Francisco Office, 32 months @ $50 per month	1,600.00
Salary of Secretary, 32 months @ $50 per month	1,600.00
Expenses of San Francisco Office, 32 months	355.00
Dividends Nos. 1 to 9, each $50.000	450,000.00
Cash in Treasury	3,444.50
	$819,141.58

Six inch Pipe for Airlift

While sinking the shaft at the Silver King Consolidated, trouble was had from a flow of water, about 15 gallons per minute from the 110 foot station, and about one set (set of timbers) below. Instead of installing a pump, it was decided to rig an airlift. The water had to be pumped to a winze about 180 feet in from the shaft, the winze collar being about 20.5 feet above the station level. In the shaft, 34 feet of 6-inch pipe was hung from a point one set below the station and blanked at the lower end except for a one-inch drain with a gate valve. Within 4 inches of the end of some 2 inch pipe, several ¾ inch holes were drilled, and just above this a ¼ inch nipple with an ell on each end was tapped in. On the outside ell a ¼ inch pipe was screwed. The ¼ inch and the 2 inch pipes, bound together, were lowered into the 6inch pipe. The ¼ inch was connected to the compressed air line, with compressor pressure being 100 lb. The 2 inch pipe was connected to an old 2inch airline which led to a winze. At about 100 feet in a tee was put in on the 2 inch line and a valve to relieve the air pockets. The flow of water was directed into the 6 inch pipe. With the air inlet valve open about one-sixth of a turn, the water kept about 14 inches below the top of the 6 inch pipe, except when the miners turned on the air to blow smoke from the shaft bottom.[3]

As of 2005, there was water still accumulating from this spring into the bottom of the mine, and the mine had to be pumped out before anyone could go down into it. The author went down into the mine to the 114 and the 256-foot levels several times during 1997 to photograph existing mine conditions and improvements to the tunnels or drifts.

Description of the Main/Engine Shaft

The mineshafts were usually timbered vertically, however, while inspecting the main or engine shaft I found it to be lined with reinforced concrete, about 75 feet downward from the top. Below the 75 foot level the rest of the main shaft is a two-compartment shaft, but timbered by 4x6 inch beams. One shaft was apparently initially used to transport ore to the surface, while the other shaft was used to transport men and supplies to the various working levels of the mine. However, this changed with the acquisition of the nearby Bilk Shaft.

Silver King Engine Shaft 1880s— Bowen Family Collection

The seven main levels of the mine were:

Number	**Depth from Surface**
Level I (One)	114.6 Feet
Level II (Two)	256 Feet
Level III (Three)	353.9 Feet
Level IV (Four)	408 Feet
Level V (Five)	570 Feet
Level VI (Six)	612 Feet
Level VII (Seven)	714 Feet

There was also an intermediate level put in between Levels II and III at a depth of 302.9 feet below the surface of the main head-frame, sometimes called the Engine Shaft on various charts. At each level various amounts of work was done, either in drifts (tunnels) or stopes (ore rooms) opened up where the ore was taken out. Tracks were laid at each level to facilitate the removal of ore in carts.

A brief description of work completed at each level, by 1883, is as follows:

LEVEL I. Level One goes downward 114.6 feet through 80 feet of quartz and extends by way of a main tunnel or drift back about 80 feet through the ore bearing vein. It extends into a large stope room about 60x80 feet wide and drifts downward to a depth of about 150 feet. Tracks extend out from the main shaft to this stope room.

LEVEL II. Level Two goes downward 256 feet through the ore body. It extends through the ore body by way of a drift. Tracks extend back through the ore body where it drifts back and becomes several drifts or tunnels. The white quartz is abundant at this level. Very rich silver ore was taken from this level.

INTERMEDIATE LEVEL. This level is at 302.9 feet and followed the lower side of the quartz for about 30 feet and was cross-cut extending at right angles for 50 feet through the quartz.

LEVEL III. Level Three went downward 354.9 feet through the ore body and two drifts went around the ore chimney and connected. At this level considerable stoping was done and a large body of high grade silver ore was taken out. The size of the stope was 63x43 feet or about 2700 square feet. *This ore was so rich it was hand sorted.*

LEVEL IV. Level Four went downward 408 feet through the ore body where a stope room was established. The area of this stope room was 2557 square feet. The miners cut through a body of quartz eight feet thick at this level.

LEVEL V. Level Five went downward to a depth of 570 feet. The 1883 report states that this level was only partially opened and showed dark colored porphyry and chlorite and contained amethystine quartz and galena ore in abundance. The area of the stope room was 3348 square feet.

LEVEL VI. Level Six went downward to a depth of 612 feet. The 1883 report stated that this level was only a drift, and was connected by a winze from Level Seven, and was also connected by winze to Level Five.

LEVEL VII. Level Seven went downward to a depth of 714 feet. As of 1883 this was the lowest level and contained the largest stope room, 5904 square feet. The miners said that this was a very rich level. This level is said to have yielded beautiful masses of Native Silver, associated with Stromeyererite and Hornblende.

SUMP LEVEL. This level went downwards another 36 feet below Level Seven, and was also reported to contain ore.

Later reports have the Silver King Mine main shaft down to 875 or 900 feet, and even as far as 1100 feet, but this may well have been the Bilk Shaft. It was said that the richest ore was found at the lower levels, and that the lower grade ore was laid aside and sorted and worked over later.

The mine during its heyday employed from 300 to 500 miners and mill workers and kept the mule trains running daily to the mill town of Pinal five miles away. A large crushing mill was set up at the mine, where the ore was passed through a large sized Blake Rock Crusher, and was stored in bins with chutes and gates so that it could be directly loaded into the wagons and then taken to the mills at Pinal. There were three mills, along with a mercury amalgamation plant, operating at Pinal City during the profitable days.

Silver King Mine timbers 1882— Bowen Family Collection

Report of the Silver King Mining Co. Jan. 1, 1881

SILVER KING MINING COMPANY

Principal Office:
Safe Deposit Building, 328 Montgomery St.,
San Francisco, California
Capital Stock, ----------$10,000,000
Divided into
100,000 Shares of $100 each.

Directors

James M. Barney	B. A. Barney
William H. Stanley	Geo. L. Woods,
	J. L. Jones

Officers

Geo. L. Woods------------------President
B. A. Barney --------------------Vice-President
James M. Barney----------------Treasurer
Aaron Mason---------------------Superintendent
Joseph Nash----------------------Secretary
James M. Barney, General Manager

Silver King Mine Values Reported

The *San Diego Union,* on January 19, 1877, reported this about the ore values. "Two trains, one drawn by 18 and the other by 20 mules, bringing 36,050 pounds of rich silver ore from the famous Silver King Mine of Arizona, arrived in town yesterday. This ore is consigned to W. W. Stewart & Co. by Col. Barney, and will be forwarded to San Francisco for reduction. We predict a first class sensation among mining men when it arrives there. The freight trains presented quite an imposing appearance as they came to town. The immense wagons are the biggest we have ever seen here. The hind wheels are seven feet high, and look as if they must weigh a ton each."[4] The *San Francisco Stock Report* of 1877 listed a sale of 33 tons of Silver King ore, the best of which brought $4650 per ton, and the lowest grade $1230.[5] On June 30, 1877, the *Arizona Citizen* stated that the Silver King Mine had been shipping 30 tons per month for about a year, no ore being shipped that yielded less than $1000 per ton. "Actual receipts from sales of ore in San Francisco have averaged $1400 per ton, a few selected lots going as high as $15,000 to $20,000 per ton."[6]

The Silver King Mining Company

This description of the Silver King Mining Company comes from the *Pinal Drill Newspaper,* December 17, 1881:

> "The Company was organized in San Francisco in May 1877. The capital is $10,000,000, with shares $100 each. During its entire existence, James M. Barney has been its manager, and Aaron Mason it's superintendent. The mine is five miles from Pinal, at Silver King Village. The amount of ore they crushed a day was 50 to 57 tons. In November 1882, they crushed 1532 tons, or an average of about 51 tons per day. The stamped ore was wet with water and went to the Frue Concentrators (Frue Vanners), which were designed to handle the product. The concentrates were then dried and sacked for shipment to the smelters in strong canvas bags. During 1883, the sacks then went to the Castle Dome Mining and Smelting Company

by rail to Melrose, California, the Selby Works at San Francisco, or to the Omaha Smelting Works. We have heretofore described the Silver King Mine and its apparently boundless wealth. Last week we had the pleasure to accompany Mr. Doran, on his visit to the mine, and in company with him and the foreman, Mr. Robert Bowen, passed through the stopes, amongst the timbers, from floor to floor, in the windings and descents of this monstrous excavation. The main shaft has reached a depth of 730 feet. There are levels at 110, 250, 300, 350, 408, 510, 612, and 714 feet. All the ore that has been worked for one year past has been taken out between the 300 and 408 levels, and masses are standing in view on all sides."

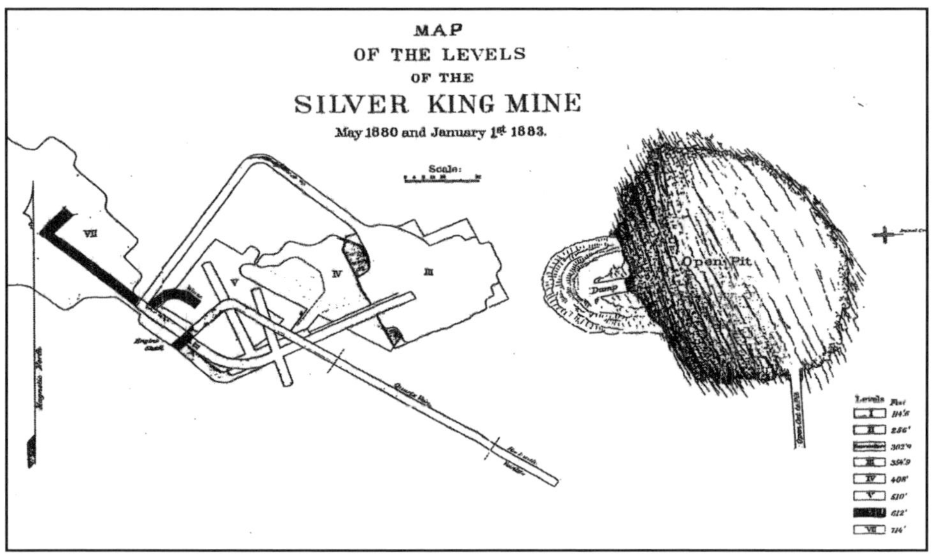

Map of Silver King Levels— Author's collection

"This large chamber, or stope, is timbered, as also the mine elsewhere, with pine from Oregon and Truckee, California, which costs here $45 per 1000 feet. The posts are 12X12 inches and 14X14 inches and the sills 10 by 14. In this large chamber, which is 60 by 75 feet, we climbed 100 feet upwards, amongst these timbers that appear solid and adamant. Nothing but ore has been taken out of this large now empty space. There are also several drifts from this chamber to the west and south, all in ore. On the 714 (foot) level there is one drift 54 feet to the north, all in ore bristling at you wherever you look. You find the ore in the quartz, in spar and in porphyry. From the 612 level, there is a winze, connecting with the 510 level, all through white quartz and porphyry carrying rich ore, and masses of pure white quartz lower grade ore. The 510, we think the richest level. There is native silver, antimonial silver and zinc-blend prominent, and forming itself, a grand bonanza. The ore in the large 408 chamber is chiefly native silver, with polybisite and galena, carrying antimony, of great value in silver. On the 300 level there is a crosscut to the west, of 35 feet in porphyry, and a drift to the east 60 feet along the sides and then porphyry beyond. There is a body of spar running from the 408 level to the 250 level all the way up on the south east side of the porphyry, which will probably run into the old works on the surface, from which so much rich ore was taken and shipped. Quartz, spar and porphyry appear distinct and also intermixed.

"*The ore above the 250 level to the 110 level has been left intact for future use.* In fact, the levels down to the 300, although much work has been done, are yet but for prospecting purposes. We have tried to see and find walls, but really we must say that none exist, for what appears to be a wall, is but a seam with porphyry on the other side, carrying rich ore. Down in the depths, at 730 feet, it appears that the real ledge is found, the backbone, thus surmounted with an enormous mass of ore, of which some of the richest has been taken, yet so abundant that it seems inexhaustible. But there are also immense bodies of low-grade ore, which will in time encumber the work, and must even now be inconvenient. This must be worked in a different manner and on a more economical plan, as also the tailings about the mine. There is money enough in these lower grade ores to make the King forever famous, apart from the rich ores therein, and with suitable and more extensive machinery, the amount of yield is only a matter of industry."

"The near future will most probably confirm what now appears to us demonstrated. That the seams and courses of the mine, with its rich veins, run west of north, and east of south, and that the present mine is but a surface development in the heavy covering, that rests upon a vast quartz body below. Through which this volcanic maelstrom of molten mineral has been thrown, containing all varieties of silver ore, combined with baser metals, as if it were a witch's cauldron."[7]

I was able to locate copies of the 1880s maps of the different levels of the mine, as well as an 1883 underground map, detailing the work areas of the seven different levels which also showed the extent of development by that date. I color-coded each level for the Deens which clearly defined the large stope areas dug out by 1883. I also photographed all the open tunnels, drifts, raises and winzes on the 114 and 256 foot levels shortly after meeting the Deens.

Author at 256 foot level of Silver King Mine— Joe Deen photo

Silver King Mine Report of 1883

The *Pinal Drill* on September 22, 1883 published a report of the activities and a tour of the mine. The report was very complimentary to the mine and I have excerpted some of the following paragraphs to show the richness of the mine.

> "We should judge the opening is at least 200 feet and the width 100 feet. It is ore all around, above and below. The brilliant dark streaks of metal, like figures on a frescoed wall, stand out in bold relief, and glisten to the eye. The rock is porphry, quartz and spar and the metal is in all. Every bit of it assays, and the free silver is distributed throughout, making the whole mass rich. We observed between the timbers, they are building walls of masonry which make the support solid. ....Following the right hand of this grand chamber our general course was to the left around, until we again reached near the shafts, and the same easily blasted rock was all around, apparently without limit. It reminds one of some Oriental grand market house supported by massive pillars, forming a labyrinth, through which one may lose his way....the particular blue colored silver bearing porphry was there, and the long sides of solid white quartz and chrystals shone in the light like the silvery crests on the breaking waves in phosphoric sparkling jets, and an ocean plentiful."

"From the 7th level and up to the 1st which is 114 feet below the surface, there is the remarkable firm body of quartz and ore to be seen all the way; the solid side of which is the main support of the mine. Much ore has been taken down to the various levels leaving this standing, and it is full of wonderfully rich streaks and lumps of mineral all along. Now let us calculate the quantity of ore in the breast of the mine alone. Supposing it to be only 100 feet thick, it being estimated that 13 cubic feet form a ton; there is 600 feet in heighth by 200 feet in length, in round numbers, a million tons, and how many more tons of ore are in sight?

"There were about 14 men at work on this shift actually mining and 16 laborers and carmen; they were scarcely observable in the vast cavern. There is work for 1000 men, and 200 stamps cannot crush the ore thus to be taken out. It seem that the Company 'make haste slowly' in realising the millions in sight. We doubt whether amongst the mines there is a parallel of order, respectability and skill,from the highest to the lowest, in every department of the work, in office, mine and mill. It is well named the Silver King amongst the mines."[8]

Future Prospects of the Silver King Mine

The *Pinal County Record* of September 11, 1885 wrote an article about future

prospects of the mine. I have included the salient part of the article below.

"This mine has been working steadily for nearly ten years and there is enough ore in sight in the lower workings to run for another ten years at least. At present the company is taking out more and better ore and treating it at the 30 stamp mill at Pinal than at any other time in the history of the mine, and the probabilities are that additional machinery will be added to their present works within a year or two. They are now working nearly one hundred tons every twenty-four hours and employ in the mine and mill about 175 men, who draw salaries ranging from $3.50 to $20 per day. Indirectly there is about 200 more men who derive their support from this company, cutting, packing and hauling wood from the wood camp nine miles distant, and employed by parties who have contracts for furnishing the company with supplies, and all are paid good salaries and most of them have families, which most naturally place a large amount of money in circulation in the towns. This company never fails to pay off its employees the first of every month, and has paid between $1,500,000 and $2,000,000 in dividends to its stockholders, and pays the best wages of any company in the Territory."[9]

Silver King Annual Meeting Report of 1886

The *Pinal County Record* January 22, 1986, printed the report titled "Annual Meeting of Arizona's Great Silver King Mine-Review of the Mining Works," as presented by the General Manager James M. Barney and Mine Superintendent Arthur Macy. I have extracted the key elements out of this lengthy report for the reader.

> "The work of the mine has been the extraction of ore from the 700, 600, and 500-foot levels, and in keeping all the underground workings, including the shaft, in good condition. No new work has been done below the 700-foot level during the past year and previous to again doing so it would be well to have anew shaft, fitted with a pump, to take care of any water we may have to contend with. This new shaft properly connected with the old one will also assist greatly in the ventilation of the entire underground workings of the property. I know of no outlay beyond this necessary the present year, other than that required in the routine work of the extraction of the ore and its proper reduction. The present condition, combined with the favorable outlook for the future, is a matter of congratulation to all interested, the only thing to be regretted being the low and declining price of silver.
>
> "The pumping station on the east side of the mountain divide has been maintained in good and working order throughout the year for the supply of better water for the boilers and various purposes of the mine. In the early months of the year the rains furnished a large portion of the water used, being caught in tanks specially provided, thus reducing the pumping requirements.
>
> "The old concentration mill of twenty stamps stand without change, and is now designated as Mill No. 1, while the new mill is known as Mill No. 2. The total amount of ore treated at Mill No. 1 during the year was 21,853 tons, which yielded 68 shipments of concentrates averaging about 11 tons each. The average assay value of the ore, as per daily assays was $56.79. Of the silver contents of the concentrations, 48 percent was in the form of native silver, 51 percent was in the form being combined of the various argentiferious sulphides."[10]

Chapter Four
Processing Ore at the Silver Mills at Pinal City

During the 1880s, the ore, once it was brought up out of the mine, was moved into a huge three-story ore house. It was then loaded onto three to four huge ore wagon, drawn by 20 mules, and taken down the mountain to the concentrator mills at Pinal City. There it was further crushed by a battery of 20 stamps (similar to a pile driver) and concentrated over 12 Frue Vanners. The ore was then roasted in a Bruckner Furnace before it was fine milled into concentrates. The concentrates were then sent to the Dome Mining and Smelting Company at Melrose, California, the Selby Works at San Francisco and the Omaha Smelting Works. In 1883, at the peak of the mine years, the mill at Pinal City treated 50 to 57 tons a day. In addition to silver, the concentrates assayed lead, zinc, copper and other base metals.

Ore wagons arrive at pinal cir. 1880s — Bowen Family Collection

Stamp Mills

The stamp mills used by the Silver King Mine located at Pinal City consisted of pillarlike gravity stamps whose ends were replaceable iron cylinders, each weighing up to 1000 pounds. Overhead was a drive shaft and cams which were designed to raise the stamps. The stamps were then released to pulverize the ore into a powderlike substance. The stamp mills were awkward and ear shattering, but survived for a long time

Pinal Stamp Mill — Bowen Family Collection

because they were easily transportable and relatively easy to erect upon wood framing and cheap to operate. They required few moving parts and could be repaired by any competent blacksmith. A properly run mill could pulverize the ore and classify it by running the crushed ore called *fines* through a screen. The oversized particles of ore were then bounced back into the stamps for another crushing. The size of the mills in the early western mining days was described by the number of stamps that they contained.

Frue Vanners

The Frue Vanners used by the Silver King mills consisted of a continuous belt that shook and washed the crushed ore. To *van* the ore is essentially to wash the ore with water. The machine was about 14 feet long, nine

Frue Vanners at Pinal Mill — Bowen Family Collection

feet wide and five feet high. The belt was about four feet wide. The top surface was about twelve feet long made of heavy-duty rubber. Twelve iron rollers supported the top of the belt. The crushed ore or pulp was then sprayed with water. The wash water helped flush the waste rock or *gangue* down a slope into a waste trough. The pulp (also called slimes) or good concentrates were collected in a copper box. It took very skilled mill men to operate the Frue Vanners.

Bruckner Furnace—-Roasting the Ore

The Bruckner Furnace consisted of a cast iron cylinder with a brick lined door on the side. Within the furnace were pipes that were designed to keep the ore from lumping up. The furnace also had four rollers that supported the cylinder and a firebox at one end. The purpose of the roaster furnace was to burn off the sulfides that contaminated the concentrated ore. The Bruckner Furnace roasted nine tons of concentrated ore in 24 hours and brought the sulfur content down from 30% to 5%. The roasting cylinder was 8.5 feet in diameter and about 18 feet in length. It weighed about 45,000 pounds. It consumed large quantities of wood and was not very economical but nevertheless practical and served a useful purpose.

Bruckner Furnace at Pinal Mill — Bowen Family Collection

Amalgamation Process

The mills at Pinal also used the amalgamation process. Amalgamation consisted of using mercury to adhere to the silver. The ore was finely divided and suspended in water (called pulp) and was passed over the surface of, or agitated, with mercury to form an amalgam. This material was then subjected to a fire-refining process to recover the silver and gold if present. The percentage of silver that was amalgamated was variable and subject to differing conditions. If the silver was exposed on the surface of the material (small particles from course-sand to flour-like), the mercury would wet it, but sometimes the mercury would not amalgamate the entire mass of particles. This usually accounted from the small losses in mercury from the stamp milling process. If the particle of silver was coated with another substance then the mercury would not amalgamate to it. Thus this process was not entirely efficient and relied upon the stamp mill men and the amalgamate men. The Silver King Mill that used the mercury process had large vats for this process.

Retorting Process

After the silver pulp had been subjected to the amalgamation process, the product called amalgam was usually strained through a chamois like material to free the excess mercury. The product was then ready for a process called retorting or heating it in an iron pot until the mercury was vaporized and became a gas that passed through a pipe. This mercury was then collected or condensed under water called a water jacket. It took quite a bit of time to vaporize the mercury. Small retorts took about 7 hours and a lot of fuel to maintain the required heat. The presence of selenium in the ore of the Silver King caused the mercury to become contaminated and black or frothy. The mercury then had to be cleaned with nitric acid or potassium cyanide and the surface wiped clean of all contaminants. When the retort process was complete the retort container was cooled and opened. The mercury was then left in the form called a retort sponge or collector of the shape of the retort container. These sponges or collectors would then be stored until there were enough gathered to melt into silver bullion.

Concentrator Process at Pinal Mill
— Bowen Family Collection

As you can see, this process, with all the dangerous chemicals used, could cause serious danger to the retort workers. Exposure to mercury could and did cause some of the miners to have serious medical and mental problems. If breathed in the mercury fumes could cause serious damage to lungs or even death. If handled without proper care the mercury could cause the miners to hallucinate. In a mine near Phoenix, Arizona, located on Squaw Peak about 100 years ago, the miners that used mercury did in fact hallucinate, causing the area of the mine to be called Dreamy Draw, and this name still exists today on the Phoenix maps.

Smelting the Silver

Smelting the silver of the Silver King became popular during the 1880s and a smelting furnace was erected nearby. The silver material was placed in a container called a crucible and placed in the smelter and subjected to about 2200 degrees Fahrenheit. Fluxes (soda ash, silica and borax) were added to the silver material to form a material called slag. This slag collected the impurities in the material. The melted silver material was then placed into conical molds or rectangular molds made of cast iron. Once the silver mass was cooled off, the black or reddish black slag was chipped off the top and the buttons from the conical molds, or the bars of bullion from the rectangular molds, were collected and stored for shipping to a refinery in California. The Silver King usually smelted their silver into rectangular bars of bullion. These were protected by a Wells Fargo guard and then shipped via Wells Fargo on a stagecoach.

Pinal City Mill cir. `1880s — Bowen Family Collection

Active prospecting of adjoining ground closely followed the success of the Silver King. In 1883, fourteen groups were being actively worked and three mills had been created. The largest of these was the Windsor Consolidated Company. In 1884, this mill was leased by the Silver King

Company to treat part of the ore. The mine then went to a depth of about 800 feet with most of the ore being taken out at the 700-foot level.[1]

In order to better understand the mining and milling process of the 1870s and 1880s, I have included some of the articles written during that era.. I hope to give the reader a better picture of mining and processing ore during that time period . Some of the articles *may seem redundant but each article adds to the explanation of the process* and the last article seems to pull together the extinct process of *vanning* and *roasting* the ore. Also, refer to the explanatory terms in the glossary.

According to the *Mining and Scientific Press*, July 20, 1878, the following Silver King ores were adapted to concentration at the Pinal Mill.

> "At Picket Post, Pinal country, are situated the concentration works of the Silver King Company, this being the nearest point to their mine, at which water in sufficient quantities for the use of the works could be obtained. Owing to the success that has attended the working of the ores of the Silver King Mine by the Frue concentration process, after failure to treat them by amalgamation, concentration has come to be looked upon as the only method by which the ores of this district generally can be successfully worked. This method of handling ores has the merit of being cheap, simple and expeditious, the objections to its employment being that are incomplete and applicable only to certain varieties of ore. It is used with great advantage to separate, by means of specific gravity, all ores having a crystalline structure, their gangue being especially well adapted to the concentration of sulphurets. The Silver King ores contain large quantities of sulphurets of silver, lead, antimony and copper with a sprinkling of carbonate of latter, about 20% of the silver existing as a chloride or 'dry chloride,' as distinguished from horn silver. These sulphurets are crystalline and readily break from the gangue under the stamps. Thus, they are practically all saved, but the dry chloride being light and exceedingly friable is at the same time, carried over with the tailings and lost until rescued by another process. Were the metal all contained in the sulphurets, not more than 3% or 4% would be lost. With the mines occurring in the limestone formation, however, entirely different conditions are presented.
>
> "There is a large group of this class of lodes bordering Queen Creek on the north, the ore of which are characterized by the presence of gray carbonate of lead in considerable quantities. This mineral contains the silver which blackens it, but does not otherwise alter its appearance or fracture. It is amorphous or entirely homogenous in structure, and therefore cannot be

concentrated, because no matter how fine it be pulverized, each grain still possesses the same specific gravity, and the only result of concentration will be a separation of the coarse from the fine particles. It is proposed, by parties interested in these mines, to employ this process in the treatment of the ore, but experience will, I fear, demonstrate to them, as it has to the writer, the folly of the undertaking. How this ore really should be handled remains to be determined."[2]

The Silver King ores had a peculiar problem. They were coated with sulphur compounds, which sealed them off so that they would not agalmate with mercury. As a little song of the 1800's miners put it:

Our German Fathers, working mines
First exorcised the devil
While we affirm that sulphurets
Are the sole cause of evil.[3]

To exorcise this particular evil a number of chemical processes were devised over the years. The mill engineers and metallurgists would treat the ore by roasting it, and by then steeping the ore in tanks of chlorine gas, which drew off the sulphur and converted the ore into chloride form. This was then soluble in water and could be extracted by a percolation process called leaching. Another process later developed in Scotland used sodium cyanide. Cyanidation, it was called, made possible the extraction of almost all the metal from the ore. The large tailing dumps were then run through this process. The process of using cyanide today is generally frowned upon because of its effect on ground water and the environment. In fact, many communities disallow this process today.

The Lixiviation Works at Pinal.

This process was described in the *Pinal Drill,* May 28, 1881.

"The works consist of twenty stamps, six concentrating tables, three roasters, one large roasting furnace, and five tanks for lixiviation. Webster defines lixiviation as 'washing or percolating the soluble matter.' The process involves separating the silver in the form of a wet precipitate, which a few hours is subjected to a steady heat in a furnace, and from there to the final melting of this precipitate in a cupeling furnace, and the molding of bars of refined silver, averaging 99 % fine. About 50 tons of ore are worked daily, hauled by teams from the mine, and on an average $20,000 in silver bars is shipped every week, as the product of the establishment. Wells Fargo & Co.'s agent then transported the bars of silver by stage to the railhead at Casa Grande. From there, the bars were shipped to San Francisco. An enormous quantity of tailings covers acres of ground at the mill, containing large quantities of silver and copper. The base metals have, until within a few days, not been extracted from the ore, but works are now in operation extracting the copper.

"Larger, more extensive works are most imperatively called for. The mill has an abundant supply of water for 100 stamps or more. We have fuel plenty for considerable time from the desert, but the coalfields call loudly upon us, and the timber close by these will supply all our anticipated wants of that kind. A narrow track railroad could be built, for a three months' yield of this mine, connecting Pinal with the coalfields and the Southern Pacific Railroad. Millions in silver lay glaringly before the feet on the surface and in the mine. By enlarging and remodeling the present works, and adapting a process for the profitable working of the lower grade ores, that is, ores below $100 per ton, the yield of this mine will be at once quadrupled."[4]

Metallurgist's Report on the Silver King 1881

ASSAY OFFICE, SILVER KING MINING CO.,
PINAL, ARIZONA TERRITORY, JANUARY 1, 1881.

Aaron Mason, Esq., Superintendent
of the Silver King Mining Co.

Dear Sir: I take pleasure in submitting to you the following report, regarding the reduction of the Silver King ore by the lixiviation process at the Company's mill.

THE ORE

Which is subjected to the lixiviation process is very base, and consists of the following silver-bearing minerals:
1. *Native Silver,* in close contact with Fahlore, Zincblende, and in some instances with Galena. It is brittle enough to be pulverized in the battery; is a bright white color and 975 thousandths fine, the impurity being copper; contains no gold.
2. *Silver Copper Glance,* with 70.3 per cent, or 20457.3 ounces per ton, silver, 9.8 per cent copper; 17.4 per cent, sulphur.
3. *Antimonious Fahlore,* containing over 3000 ounces of silver per ton. This mineral is the most important constituent part of the ore.
4. *Zincblende.*
(a) Zincblende, found in large and perfect, transparent crystals of a lustrous green color. This is the poorest of the silver-bearing minerals of the Silver King, but is highly interesting from its beauty as a specimen. It contains only 10.2 ounces of silver per ton.
(b) *Brown Zincbende,* occurs more in solid masses and in large quantities, frequently intersected with wire silver, and contains 97.7 ounces of silver per ton.
(c) *Black Zincblende* is scarcer, and contains 40.8 ounces of silver per ton.
5. *Peacock Copper Ore,* with 430.62 ounces of silver per ton.
6.*Galena,* containing antimony, and with from 29 to 185 ounces of silver per ton.

7.*Copper Pyrites.*
8.*Iron Pyrites*
The gangue consists of Quartz, Heavyspar, and some Porphyry.
The average value of the ore, as it was delivered from the mine to the mill during November and December, proved to be two hundred and eight dollars and twenty-six cents ($208.26) per ton. This I ascertained by daily samples and assays.

Extracting the Ores at the Pinal Mills

The *Mining and Scientific Press* of April 28, 1883 gives further clarification of exactly how the silver was extracted from the ore. The Silver King reduction works at Pinal in 1883 consisted of 20 stamps and 12 Frue Concentrators. The ore was crushed and then wet and then casually roasted in Bruckner cylinder furnaces of the class then known as the Pacific Chloridizing Furnace which was arranged with a fire-box at each end so as to more equally heat all parts of the ore. The roasting process consisted of placing 5 tons of ore in the furnace along with about 10 percent salt and then roasting the ore for about 10 to 14 hours.

Pinal City Mill cir. 1880s — Bowen Family Collection

The character of the ore gradually changed until 3 tons were roasted for about 24-40 hours. This roasting was a very costly part of the extraction process and during 1883 about 19 men were engaged in drying, repulverizing, roasting and elevating 9 to 10 tons of wet crushed ore per day of 24 hours. At that time a cord of wood cost from 7 to 8 dollars and it took a cord per ton in the roasting process. A major improvement during 1883 was the introduction of a Blake Rock Breaker, which immediately caused a 20 percent increase in the crushing capacity of the stamps. The ore treated during this period consisted of native silver dispersed through quartz, heavy spar, altered porphyry, zinc blende, galena and small portions of pyrites and some other minor minerals not noted.

Work of the Concentrators

The *Mining & Scientific Press* in April 1883 reported this about the mill concentrators at Pinal.

> "The concentrators saved from 80-84 percent of the value of the ore. According to the report, it was no fault of the machines or management that they did not get better results, but was due to the onerous conditions under which they worked. The ore as stated above, was charged with native silver. It was crushed somewhat coarsely through a number two needle-punched screen a little courser than thirty-mesh wire sieves. It also contained a notable quantity of heavy *gangue* matter of at least as great specific gravity as the zinc blende and copper pyrites. The headings were required to be worth $1000 per ton by assay. The report continued by stating that under such conditions it would be absurd to expect any machine to yield extremely poor tailings by a single operation. Repeated experiments, by the most careful vanning, demonstrated the impossibility of obtaining a rich product from the tailings, unless they were previously reground. They also proved that 65 percent at least of the value of the tailings was contained in from 10-12 percent of their weight, in the form, chiefly of native silver enclosed by particles of zincblende, copper pyrites and heavy gangue. This material assayed from $30-$50 per ton. The tailings also assayed about $140 per ton in black sulphuret when cleaned."[5]

The *Mining and Scientific Press* of January 19, 1884 further explained the silver milling process in a manner that I believe summarizes the *vanning* and *roasting* process used by the Silver King Mill.

> "The intention when the mill was erected was to concentrate all the ore, hence wet batteries only were put in. But it was found that the heavy spar interfered with the concentration so materially that leaching was adopted for that class of ore. The mill now consists of a *Concentration Department*, with ten stamps, whence the pulp is conveyed to six Frue Vanners; and a *Leaching Department*, with ten wet stamps, from which the pulp is caught, in slime-pits, then dried, roasted, and leached. The wet stamps were being replaced by dry batteries.

> "The furnaces are the notable feature of the mill. They are Bruckner cylinders, fired from both ends. The spiral diaphram, proposed by Mr. Bruckner, was not found to move the ore so effectually to and fro from end to end of the cylinder, as to secure a perfectly uniform roast. The

cylinder is therefore, lined with a smooth layer of fire-brick, but at each end is built a fire-place, and between each fire-place and the cylinder a flue provided with dampers leads into a common stack. The fires are quickened alternately, and simultaneously the damper near the brisk fire is closed, and that in the opposite end of the cylinder opened. The result is pronounced altogether satisfactory, though the consumption of fuel is heavy. There are three cylinders, sixteen feet long and 6 feet internal diameter. They roast on an average 22 tons daily, and consume 8 cords of wood. Five tons are introduced to a charge, with 10 percent of salt. Antimonial compounds of silver roast slowly; sulphur compounds quickly. The duration of the roast is very variable, as the character of the ore is subject to marked changes. The Bruckner is, therefore a more suitable chloridizer than the White-Howell or any automatic furnace, inasmuch as a charge need not be withdrawn until chlorination tests show it to be done. They aim at extracting 90 to 95 percent of the silver in the roast. The roasted ore is damped and charged in a layer about 1-½ feet deep into tanks 8 feet in diameter, and which hold about 2 ½ tons to a charge. The base metal chlorides are, as is customary, washed out with water, the livixium containing a little copper, which will henceforth be precipitated. The silver is then extracted with hyposulphite of lime, and is recovered as sulphide in the usual way. The bullion is .995 fine."[6]

Report of the Silver King Mine 1886

OFFICE OF SILVER KING MINING CO.,
328 Montgomery Street,
San Francisco, January 12, 1886

To the President and Directors of the Silver King Mining Co. San Franciso
GENTLEMEN: Herewith I beg to submit my annual report of the receipts and disbursements of your company for the year 1885. I am, gentlemen

Yours respectfully,
Joseph Nash, Secretary.

RECEIPTS

Balance on hand January 1st.	$33,008.87
Sales of bullion during year	92,550.53
Sales of concentrations during year	650,649.91
	$776,209.31

DISBURSEMENTS

Assaying	$ 564.00
Expenses	9,056.43
Insurance	1,065.75
Freight	19,439.14
Merchandise, machinery & supplies	37,374.28

Dividends No. 46 to No.53 inclusive	200,000.00
Superintendent's drafts account of mill and mill expense	402,568.95
Balance of hand Dec. 31st.	106,140.74
	$776,209.31

Joseph Nash, Secretary

Silver King mine cir. 1880s —*Bowen Family Collection*

Chapter Five
Mining Men of the Silver King Mine

Prospectors in the west followed different paths than other settlers. Although the placer miners followed the creeks, the hard rock miners clambered over remote hillsides looking for telltale signs of quartz outcroppings. When a strike was made, hundreds of miners usually hurried to the new location in search of work, and the prospect of new found riches if they could find a claim nearby to call their own. The Mining Act of 1872 was adopted as the law governing each mining district. The district by-laws provided that each claim must be recorded within 30 days of discovery, and that $25 worth of location work must be performed within three months of the date of the location, and failure to do this work was deemed abandonment of the claim. All claims that had location work performed then had to record this additional information in the mining district's records within 60 days.

Silver King Miners cir. 1883 — Bowen Family Collection

The two stages of development of a mine are as follows. First the prospector or miner makes the "big strike." Rarely did the prospector have the equipment or finances to develop a strike. Usually the prospector seldom stayed around long enough to see the mine developed. He usually celebrated by getting drunk and then sold his rights to his claim to outside interests with money. The second stage is the actual development of a working mine, with working capital and equipment and seasoned mining men.

Munson's Chunk of Silver

The following tale typifies what happened to the prospector who struck it rich. In February of 1876, a miner named Munson, who was prospecting in the area of Pinal Creek, stumbled upon a massive chunk of nearly pure silver, weighing almost a ton. With only a mule for transportation, Munson was in a quandary. He quickly scrawled the notice "This is Munson's Chunk," on the back of an envelope, and left it at the silver boulder. He then hurried to his camp to secure enough help and pack animals to break up the boulder and haul it away. Munson succeeded in getting his treasure broken up and took it to Florence, where it was weighed, assayed and sold for $20,000. Munson then went on a drinking spree that ended when he was hit over the head and tossed out of a saloon in the cold winter's night. The next morning he was found dying of the head injury and exposure. He did not live long enough to really enjoy his "strike."[1]

As soon as news of the big strike was spread around in the local papers, mining engineers followed virtually in the footsteps of the prospector to assess the claim. If it proved to be worthwhile, outside capital from the mining syndicates usually arrived to finance the development of the mine and a new mining district was developed. Such was the case in many strikes and this included the Silver King.[2] For reasons of security in the 1870s and the 1880s, miners established small settlements or camps as they were called. The miners clustered their makeshift dwellings near the claim and in an area near water. In the hardship and isolation of the early mining camp, the miner sought solace in whiskey and the saloons were the first commercial establishments to be set up. At the Silver King Mine a camp was soon established, along with saloons, "palaces of pleasure" and merchants to equip both the miner's social and material needs. The army soon followed a large strike, as conflicts with hostile Indians were usually eminent.

Otis E. Young, in *Western Mining,* states that "the western miner was something of a tramp." It was said that most mines needed eight men to handle the normal two-man task underground, since there were "two a coming, two a going, two a rustling, and two a working." However, he did state that the tramp miner was a good hand when he took it his head to be one. He also said that the miner was very good at resolving problems. But when payday came along he might decide to suddenly depart. It was just that type of individualist who made up the work crews at a typical mine as the shift began.[3]

Employment and Pay at the Silver King

During 1885, the Silver King Mine was working heavily and was taking out nearly 100 tons of ore every 24 hours. They were employing 175 men at the mine and mill who were drawing wages ranging from $3.50 to $20 per day, depending on the job. Indirectly, there were about another 200 men cutting, packing, and hauling wood from the wood camp nine miles distant to the mine. Among this group were the packers and freighters who packed in supplies to the mine and mill towns. The mine also always paid their employees on time which generated a lot of money circulation in the two towns of Silver King and Pinal. The Silver King Mining Company apparently was a very good employer.[4]

Cornish miners underground cir. 1880s— John Swearengin Collection

High-Grading at the Silver King

The Silver King days during the 1880s were heady and extravagant days! Casual practices led to more high-grading than usual, and the company then employed guards to deter theft. From the owners point of view it was outright theft but to the miner, who put his life and limb on the line every day, it was part of the job. The miner's metal lunch pail was often the receptacle for the removal of high-grade ore. Once the lunch pails were subject to a search the miner's improved the concealment of the ore. They developed a number of devices including the false crowned hat, capable of holding up to 5 pounds of high-grade; a leg pocket which usually was a sock tied inside the pants leg; small pockets sewed in the waistband; and a corset made of canvas to be worn under the shirt. Once these devices were found the mine bosses then instituted change rooms. The miners had to leave their work clothes at the mine's change room, could take a shower and then leave the mine with their non-mining clothes on. It was only after the miners made concessions, like agreeing to be searched, that Mason reopened the mine. There was no way to determine how much high-grade ore the miners took. When change rooms at the Silver King did not seem to be enough Mason took more drastic action. On two occasions Mason closed down the mine due to high-grading.

Andy Dudley's Wonder Horse

Andy Dudley was a young miner, who was the proud owner of a large Norman horse. He was always talking about the wonderful strength of that horse. Once when there was a large smokestack to be raised, Andy and his horse was called upon to help put it up. The stack was very heavy and where the horse was pulling was solid rock which gave poor footing for the horse. After slipping and pulling, the horse got a good pull and raised the stack. In telling about it afterwards, Andy would say, "By gosh, he pulled so hard that he sunk six inches into the solid granite."[5]

Silver King miners cir. 1883 — Bowen Family Collection

Cornish Miners and "Cousin Jacks"

Cornish miners migrated to work in the United States during the 1800s and stayed to work the mines in Arizona. Three Cornish families among others came to work at the Silver King Mine in 1875. The Richard Trevathan, George Lobb, (Lobb married Trevathan's daughter) and the John Knight families were from Penzance, Cornwall. Trevethan and his wife are buried in the Silver King Cemetery along with Emma

Knight. They were often called "Cousin Jacks," because it seemed that every Cornishman had a Cousin Jack willing to work in the mines, and they were described as very clannish in nature. In the evening, after their customary drinks, they would often gather and sing for the benefit of themselves and others. These Cornish miners were often recruited from their home in Cornwall for their legendary knowledge and skills in underground mining. Cousin Jacks carried metal lunch pails, which invariably held a Cornish meat pie which they fondly referred to as "letters from home." Their wives were called Cousin Jennies. Much of the terminology of underground mining was derived from Cornish terms. *Whims, cobs, spalls and kibbles* are Cornish terms.[6]

Gassy Thompson and the Tommyknockers

Gassy Thompson was one of the old time prospectors who drifted all over Arizona in the 1880s with his burro in search of the elusive "mother lode." He was a teller of tall tales and told this story many times around the campfires of how the Tommyknockers got to America. "Tommyknockers" as everyone who has been around the mines knows, are unseen spirits who flit about the underground passages, the stopes, winzes and crosscuts, never seen by man, but making their presence known by mysterious knockings on the rock walls. These "Tommyknockers" were said to have been introduced into the mines of America by imported Cousin Jacks, Cornish miners, in whose baggage these spirits from the tin mines of Cornwall concealed themselves and then escaped into the American mines. Some of the cousins were superstitious about the Tommyknockers and said that when the mysterious knockings were heard in the depths of a mine, they were warnings of a cave-in or some other disaster to come.

But somehow, Gassy Thompson got on more intimate terms with the Tommyknockers than the Cousins ever did. He learned the meaning of their knockings, which was something like the Morse Code. Gassy specialized in pocket hunting, in out-of-the-way places, where he occasionally found small but rich gold deposits. He was known to have come in from one of his desert trips with a canvas bag full of free gold, valued at $2,700.

> "Them Cousin Jacks what was raised with "Tommyknockers in them tin mines over in England never did get to know that the knockers was trying to tell 'em where the rich ore was. But you never could tell no Cousin Jack nothin', nohow," said Gassy.

"It was in the hot summer of 1885", said Gassy, that he holed up in an old mine in the mountains and learned how to understand "Tommyknockers" language. In a crosscut near the face of the main drift he was picking into a metallic vein in the roof when a "Tommyknocker" commenced sounding off. Tap-tap-tap-three knocks-came plainly to his ears. Tap-tap-tap came the signal again. "It sounded just like no-no-no", said Gassy. He dug into a metallic shoeing again and it all fell to the floor. There was nothing left but worthless vein matter. "Right then I figgered the "Tommyknocker" was tellin' me no-no-no, that was not the place to find good ore", went Gassy's story. "I tried another place," Gassy went on, "and the 'Tommyknocker' told me no-no-no. I went all over that old mine, sampling everywhere for three days and got no pay; just no-no-no from the Tommyknockers. Finally on the fourth morning, I went right into the face of the drift, where the miners had fired their last shots, and looked at it close. It was loose ground and looked like coarse sand bound together with a little red clay, not a sign, even of sulphides. I was going' to turn away and leave it, but at the last minute hit it a lick with my pick. Right then the 'Tommyknocker' got busy and through the rocks came, knock-knock, knock-knock, knock-knock, two knocks repeated eight or ten times. I knowed right away I had it. Two knocks meant *thats's it*."

"And do you know," said Gassey, "that damned 'Tommyknocker' kept pounding on the walls with *that's it,* for five minutes? He celebrated along with me and the dog, though I couldn't see the gold would do him no good. Just a friendly cuss, I guess."[7]

Mining Men and Mining Methods of the 1880s

A long beard, floppy hat and muddy boots is a popular picture of an old time miner. But this was not really accurate, as not all miners looked alike, talked alike or had the same customs and habits. They came from different ethnic groups and different cultures. The miners of Silver King were Americans, Mexicans, Germans and other nationalities as well as Cornish. Miners of the 1880s generally wore a battered felt hat of some sort, and carried either an oil lamp or candles for light. They carried only enough oil, or usually three candles, to get them through a shift. Their candleholder had a spike on the end to be driven into a timber or crevice in the hardrock. The candleholders were made of iron rods, ¼ to 3/8 inchs thick with a looped handle at one end. The lengths were between 6 and 12 inches.[8] They got a half hour lunch break and if they had an extra candle they would use it in the manner of a chafing dish to heat their coffee, or if Cornish, their tea.

The Silver King Mine's internal workings consisted of digging *vertical shafts* and *horizonal tunnels.* There were extended by following the ore body by using *drifts* and *crosscuts* extending off the drifts. The internal shafts were called *winzes* which went downward from the tunnel floors. Internal shafts known as *raises* were dug upwards. The Silver King Mine also dug large rooms or chambers called *stopes,* or stope rooms, where the ore bodies were deposited in large quantities. These stope rooms resembled caves within the mine. On the 114 foot level there is a winze that drops down to one of the large stope rooms, which is about 150 feet long by 75 feet wide and 40 to 50 feet high.

Cornish miners warming their meat pasties — Buck O'Donnel Shaft Development Company

Mining consisted of drilling blast holes, loading the holes with explosives, setting off the explosives and then cleaning out the ore, called mucking out the shot rock. A cage or bucket would deposit each work crew at their work level, and work began by mucking out the waste left by the blasting of the previous shift. As the muckers were at work, the shift boss would point out the new drill holes for blasting, by either marking the places with his hammer or soot from his lamp. Drilling would begin for the next round of shots. Blasting methods are discussed in depth later in this chapter.

Single Jacking and Double Jacking

The Silver King Mine was a hard rock mine, where the silver ore was imbedded either in quartz or hard rock called porphyry, and had to be removed by hand drilling with hammers and drills and then blasting with dynamite or blasting powder. Drilling by hand, before pneumatic drills, consisted of the single jacking method or double jacking method. *Single jacking* consisted of a lone miner using a four-pound sledgehammer and a steel drill. *Double jacking* consisted of two miners, one turning or shaking the steel while the other pounded away in a disciplined, rhythmic succession with a nine pound hammer. Either method required muscle, coordination and skill in order to drill the holes in a short amount of time.

The steel drills were merely round or octagonal rods of steel with a chisel tip. They came in different lengths. The bull, or starter drill, was about a foot long and each subsequent drill was about six inches longer. It was designed so it would not become wedged but would "follow the hole." A three-foot length was about the maximum for drill steel. When the miners descended into the mine they carried several drill steels, enough to replace the ones that got dull during the shift. At the end of the shift, the bundles of steel were carried up so that the blacksmith could re-sharpen and re-temper them.

Silver King Miners cir. 1883— Bowen Family Collection

The Cousin Jacks were muscular and keen of eye and superb at single or double jacking. Drilling contests were held on the Fourth of July at various mines and the Cornishmen frequently won them. The world's record for the "straightaway" (two men, but no change in position) was won on July, 4th 1903, in Bisbee, Arizona Territory, by a Cornishman named Sell Tarr who drilled 28 and 5/8 inches in 15 minutes.

During the Silver King Mine days the drill steel was made of Black Diamond round or octagonal bar stock, cut into desired lengths. The bit end was heated and then hammered into a width greater than the prospective diameter of the hole. The Cornish miners had an archaic but very effective system of drilling, said to have been learned from the Phoenicians or Egyptians, which allowed them to produce a conical cavity when shot. "Where ventilation was good, the Cornish miner could blast his cut-holes just before lunchtime on the shift so that the smoke and dust would clear while he was consuming his tea and pasty and peradventure doing a bit of highgrading."[9] The *Cousin Jacks* had been working in the tin mines of Southern England for centuries, and their knowledge of hard rock mining and deep tunneling was immense. They were said to be skilled, tough and tireless, and many found work in the mines of Arizona Territory.

The Cornishmen were very popular, not only for their skills, but for their customs, amicability and flavorsome use of the English language. They always referred to everyone they knew or met as, "my son, my ansome, my beautay," or some other endearing term. They were robust drinkers and melodious singers, filling the saloons with tunes from home. They were also full of ironic wit, and loved telling tall or colorful tales, usually both.[10]

Miners double-jacking underground — *Buck O'Donnel Shaft Development Company*

In addition to the talented Cornishmen, the mines also attracted many other ethnic groups, such as Tyroleans from Austria, and workers from Ireland, Italy, Germany and Serbo- Croatia. The Chinese workers were always put at the bottom of the rung, by the other workmen, because the Chinese were willing to work for lower wages and would undercut the pay of the Europeans and Americans. This did little to make them welcome, and they suffered racial discrimination and lived in their own small neighborhoods or block. In fact, Pinal City had it's own Chinese block, and Silver King had its own Chinese section.

Timbering the Mine

As the miners sunk the shafts and created tunnels and drifts, raises or crosscuts and stope rooms, the mine required shoring or timbering to support unsafe ground or *bad ground.* Hard rock mines usually required timbering only to shore up loose rock. Stope rooms 30 feet by 60 feet sometimes required no timbering. At the 150 foot level of the Silver King there is a stope room 50 feet wide by 75 feet long that has no timbering except at the opening to the stope. Timbering called *stulls* were used to keep narrow passageways from collapsing. *Cribbing* was a method used to line shafts and hold the timbers against the walls of the shafts. *Square sets,* established by miners of the Comstock Lode during the 1860s, were used in the Silver King to support enormous amounts of weight. This method required three timbers joined together using a timber called a vertical *post* and two other horizonal timbers called a *girt* and a *cap.* These *square sets* had to be held together very tightly to be effective and were used throughout the mine to support drifts, tunnels, and stopes. This

Silver King miners and Robert Bowen — Bowen Family Collection

method of timbering is still in use today. Shafts were supported by using a method called *head sets.*

In Boom Times the Miners Led a Colorful Life

From *Territorial History of the Globe Mining District* comes this picturesque description of the miners' life after working hours. Life became colorful for the miners and residents of the mining camps, but life was also said to be cheap and garish, flimsy, tinsel and gaudy.

> "Rustling silks and satins hid the hearts of the ladies. Tinkling glasses, soft clicking of poker chips, the whirling whim of the roulette wheel, the shuffling of many dancing feet, the raucous voice of a singer, the hoarse

sound of drunken voices, the soft pleadings of a saloon entertainer, and the punctuating crack of a shooter, were all blended and mingled into a wild, riotous song of life. Bedlam sometimes reigned in the saloons where dance hall girls were entertainers. After paydays, the saloons were wide open, full of men and women and noise, fighting, cursing, drinking and gambling all night long. It was here that the miners found rooms and it was here that the 'soiled doves,' sometimes called 'variety girls,' also roomed."[11]

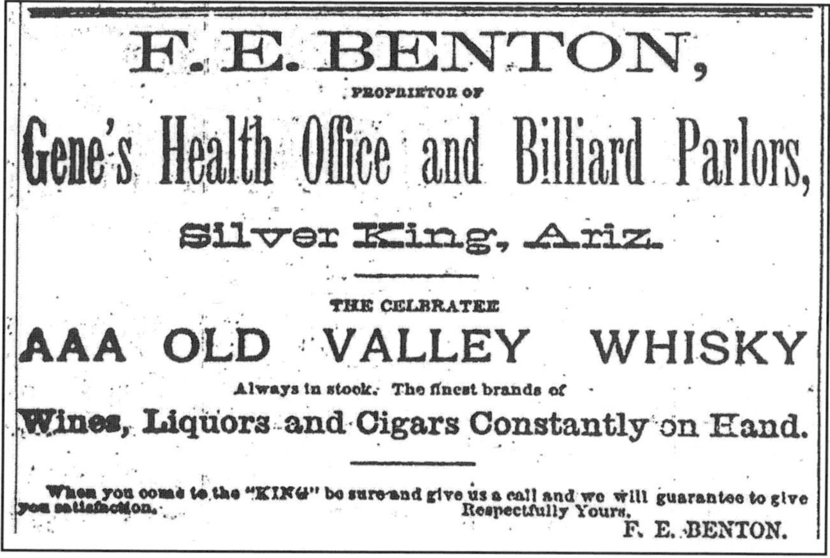

Gene's Health Office and Billiard Parlors advertisement
Silver King — Pinal Drill 1883

A Prize Fighter comes to Silver King

The 1880s brought prizefighters from town to town, in search of fights from among the town's toughest men. A prize was usually attached to whoever stayed so many rounds with the prizefighter. On one occasion a much-heralded bare-knuckle fighter came to Silver King looking for challengers. Miners being drinkers and known as tough men always took the challenge. This time no one could last more than one round with this fighter, and the drinks went round and round and bolstered the spirits of the would be challengers. A certain young miner named Tom Meehan took up the challenge, much to the applause of his fellow miners. But alas, Tom was a better miner than a fighter and was pummeled pretty soundly. Tom's father, John "Blackjack" Meehan, thought that the prizefighter took advantage of the young miner and jumped into the ring. He proceeded to give the prizefighter a good thumping and then knocked him out. Seems that "Blackjack" was a bare-knuckle fighter in his youth in Cornwall, England. Shortly after this, the prizefighter and his manager left Silver King somewhat short of funds.[12]

Miner Bill needs a Drink

At Silver King there was a miner named Bill Elliot who always was in debt, but sought innovative ways of getting a free drink. When the Apaches were at their worst, guards were put around the towns to warn people of an attempted attack. Old Bill was on guard duty one day when he came running into town, bare headed, out of breath, and showing every evidence of fright. Some one said, "what's the matter Bill?" "For God's sake someone give me a drink," was Bill's reply. He got the drink and the question was repeated, but all Bill would say was, "for God's sake, give me another drink." He got the second one, and the question was again repeated, at which Bill said "noth'in." Old Bill had taken the only way he knew to get a drink when his credit would not bring it in. He tried to create the appearance of something exciting happening, such as the sudden appearance of Indians.[13]

Lady Luck and the Miners

The element of chance always held the prospector and miner enthralled. Travelers throughout the mining camps usually considered it one of the most thrilling experiences of their lives, especially after passing through Apache country and risking death, sometimes by a hair's breadth. Miners and prospectors were always taking chances. *Lady Luck* would sometimes smile, sometimes frown. Every now and then, one of the prospectors would strike it rich. In those early days, hope and faith never died. In the land of chance, the miner, prospector, cowboy, "soiled dove", mine owner, muleskinner and mine superintendent, all became as apt with cards as they were with their picks, shovels, lassoes, whips and other skills. The gambler of the frontier days was said to have been usually a square shooter, and gambling in those days was thought of as an honorable profession. It must be noted, however, that this honor only applied to the acknowledged professional gamblers, and not the card sharps, who were as vicious and ruthless then as they are now. Gambling and liquor furnished about the only diversions available to, or understood by the pioneer and in whose pockets money was burning a hole.

The "Gold-Leaf" Ledge

Someone was always trying to "jump" someone else's claim or "flimflam" them out of their holdings in those days (and in fact that still goes today). However, this is a little story of a practical joke that backfired on the claim jumpers. Albert S. Neighbors, an old timer, related this story in 1926.

> " Between the Magma Mine and Pinal— I did not call it the Magma, but I be damned what I called it—I found a quartz ledge that was nice and porous, but absolutely barren. I took a piece of it down to Florence, stuck a little glue in the holes, sprinkled some gold leaf on it, and laid the thing on my desk. Several old prospectors questioned me about my find, and I told them where I got it. Several months later those old fellows drifted in one by one, and accused me of sending them on a wild goose chase. As a matter of fact, they had treated my specimen in a deprecatory manner, thinking I did not realize its value, and then they had quietly sneaked off to try and jump my location. Years afterward, one of them had a restaurant on Congress Street in Tucson, and when I entered, he recognized me and began to laugh. He told me all about trying to put one over on me, and the results, for of course he found nothing but barren quartz!"[14]

Jack Fraser and a Future President

One of the miners who worked at the King and became successful was Jack Fraser. Fraser came to the Silver King Mine in 1883 and worked there while the mine was booming, then bought the old Reavis Ranch (after Reavis was found dead) for $600 and began raising cattle. Actually Fraser paid for the Coroner's Inquest, thereby allowing him to file a claim for grazing rights to the Reavis. Since Fraser was not a US citizen he could not legally file for title on the old ranch. Billy Knight's wife (who was a US citizen, Billy was not) filed for ownership. He had a very prosperous ranch and raised thousands of heads of cattle. While at the Silver King Mine, Fraser met a young mining engineer, named Herbert Hoover, who had been called upon to make a survey and report on the property. From that time on, Fraser was a booster for Hoover, who was later to become President of the United States. Jack Fraser was also one of those who searched for the Lost Soldiers Mine.[15]

"Fire in the Hole"—The Dangerous Art of Blasting

"One round in and one round out" was the old expression used by the miners of the 1880s that summarized the early day mining process. Mining was basically a process of drilling blast holes, loading them with explosives, shooting off the explosives, mucking out the shot rocks and going through the same cycle on the next shift.[16] Black powder was used as the blasting agent, although dynamite was available and may have been used. Early dynamite and its use of nitroglycerine, as founded by Alfred Nobel was highly unstable and subject to freezing and leaking, thereby making it subject to premature detonation.

WAR DEPARTMENT,

Signal Service U. S. A.—United States Telegraph.

Dated Silverking Mine 12, 1877. M.

Received at TUCSON, May 12, 187 8 55 P.M.

To L M Jacobs & Co

Send by Stage immediately
two thousand fuse
Silver King Mining Co
7 pd

Order for blasting fuses 1878 — Sam Michael Collection

The blaster on the mining frontier used the Bickford slow-match fuse for the delicate and critical function of igniting the round of shots. Ignition of the blast was always a chancy and dangerous problem. There were many deaths and injuries during the blasting process. Dynamite, after being exploded, was very nauseous which caused problems in poorly ventilated mines. Occasionally, a shot did not fire or failed to explode, and this created a special hazard for the oncoming shift. The unexploded charge could be located easily enough, but it had to be removed with great care, usually by a miner with a pick who dug with extreme caution around the charge; however, sometimes the charge exploded in spite of the care, injuring the miner or miners nearby.[17]

Black powder caused fewer problems than dynamite, but was not as effective at splitting the rock away from the mass.

Several holes were drilled into the rock wall to be blasted in patterns called the *center-cut* or the *wedge cut* method. The patterns were designed to maximize the force of the blast in order to break the rock. The design of the holes in the *center-cut* method was devised to set off the blasts in a sequence with the center mass of rock being blasted out first. This allowed a cavity to be blasted out allowing the necessary space in the rocks, which enabled the other charges to be blased out in a sequence causing the maximum amount of rock to be blasted from the rock face. All patterns required a set of holes called the *trimmer* holes, which defined the wall or ceiling of the drift, and were blasted next in the sequence. The other holes drilled were called the *reliever* or *lifter* holes. When the charges were fired they heaved the broken rocks away from the rock face and was called *muck.*

When the blast was ready to be set off the blaster hollered *"fire in the hole,"* and lit the ends of the fuses called the rat-tails. After the blast there would be a choking cloud of rock dust hovering about the area just blasted. The oncoming shift cleared the muck and separated the ore and waste, loading each in individual cars or skips, to send it to the hoist station on their level, to be taken over by the cager, for the trip to the surface. Muckers sometimes shoveled the broken rock into wheelbarrows or *Irish Buggies,* then transported the muck to ore carts to be lifted to the top of the mine. At the surface, the top-lander pulled the cars off the cage, replacing them with the empty cars, then deposited the waste on the tailing dump and took the ore to the sorting shack and bins. It was then loaded on the ore wagons and taken to the mill. In the case of the Silver King Mine, the ore went to the Blake Rock Crusher at the mine where it was crushed to smaller pieces before it went down the hills to the mills at Pinal City.[18]

If the previous shift indicated that there were misfired charges it presented emminent danger to the miners. The unexploded charges could be located in the muck waiting for a miner's pick or drill to set them off. These unexploded charges caused many injuries and deaths in the early days of mining. However, the main cause of early death of the miners was silicosis, caused by prolonged breathing of rock dust.

The Tenderfoot Miner

This amusing story comes about a tenderfoot comes from the files of *Arizona Eighties*.

It seems that two experienced miners named Joe Taylor and Jim Bassett were sinking a new shaft and they needed more help, so a new man was put on to help with the windlass (a type of hoist) work. Since he was a tenderfoot, they decided to play a joke on him. One day at noon, when they had him filling buckets, they told him to climb in and they would pull him out. They pulled him up about halfway, fastened the windlass and went off and left him there. He would either have to climb up the rope or stay there, which he did until they came and pulled him out. The tenderfoot took it with good nature, smiling and said nothing. A little later when they had their holes drilled and asked him to send down some powder, he got even with them. He fixed up some imitation cartridges and put a fuse without the cap in them and hung them in the bucket, lighted, saying as he lowered the bucket, " I thought I would save you boys some time, so I lighted them for you." He sent them down to a point just too high for them to reach and said: "I am going to supper boys, I will pull you out after I get back from supper." They were the two most frightened men in the world at that moment. Neither coaxing nor cursing would bring a reply from the tenderfoot, but they never played any

Silver King underground 1880s — John Swearengin Collection

jokes on him anymore either.[19]

Death and Injuries at the Silver King Mine

Miners usually first lived in tents, shacks or shanties before more permanent housing could be built, which usually were unsanitary conditions, and disease caused problems. There was a smallpox scare at both Silver King and Pinal in the 1880s. In addition to disease and mine dangers, the early Silver King miners often dodged Apache arrows shot from the top of surrounding hills and mountains. The miners were somewhat protected by a sentry placed on the watchtower located on top of the Silver King Mine guesthouse. There were also numerous injuries and some deaths in the mine. As the town of Silver King developed, the Silver King Mining Corporation hired a doctor to be at the mine full time. They ministered to the miners as well as to the miners' families. Dr. Thomas Kenniard was the Silver King's doctor; however, during 1886, he had the assistance of Dr. William W. Wardwell who came west for his health.

The following exerpts are stories about injuries and deaths taken from the *Pinal Drill* and *Pinal Record* during the 1880s.

Injuries Due to Blasts

The *Pinal Drill,* September 16, 1882 reported the following injuries: "A son of John Binkley, about twelve years of age, played with a giant powder cap, this week and lost the ends of two thumbs and of two fingers. J. Enyart had his hand badly shattered, by the premature explosion of a blast, at Silver King. H. F. Mitchell, who was working with him, was damaged in the face."

Miner's foot crushed in rock fall at mine

The *Pinal Drill,* October 21, 1882 reported the following injury: "Samuel Oates, a miner at the Silver King Mine, met with a fearful accident this week. A stone fell from above, crushing one of his feet; it had to be amputated; then as gangrene set in, it was found necessary to amputate the leg. He is doing as well as expected under such circumstances."

The Death of Joe Deering at the Silver King Mine

The *Pinal Drill,* October 2, 1885 reported the death of miner Joe Deering:

> "On last Saturday night, at the King mine, Joseph Deering met with a terrible accident, which terminated in his death Sunday morning. It appears while he was working in the 500 foot level a large boulder rolled over him literally mashing a portion of his body almost to pulp. He was immediately conveyed

to his room, and Dr. Kenniard called in to attend him. Amputation was necessary, but he was so weak from the loss of blood that he died while the physician was in the act of taking off his leg. No blame is attached to the company. Coroner's inquest rendered a verdict of accidental death. The company suspended work and attended the funeral. Mr. Deering was a good man and well liked by his friends and fellow workman."

Joe Deering was a key figure in the search for the Lost Dutchman Mine. (Also see the chapter on the Lost Soldiers Mine.)

A Dog Survives 250-foot Fall in Mineshaft

The *Pinal Drill,* October 16, 1885 writes: "A dog fell down the Old Dominion mining shaft, a distance of 250 feet. The dog isn't hurt, but it is thought the bottom is knocked out of the mine."

The *Pinal Record,* March 26, 1886 detailed the following mine fatalities:

A Small Cave-in at the Silver King Mine Kills Two Men and Injures Two Others

"Between the hours of 2 and 3 p.m. Sunday, Silver King was thrown into a state of feverish excitement. A report that there had been a cave-in the mine and that several men had been killed and others wounded, and later developments proved that the report was only too true. The circumstances attending the accident, as near as we can learn, are about as follows. Mr. L. Bowman and Mr. Johnny Stanfield were working in what is called the 'Cock Pit' on the 8th floor of the 500 foot level, when a large body of loose ground fell and killed the former instantly, and fatally injured the latter. Mr. Stanfield's injuries were found to consist of a broken leg at the ankle, broken hip, and several severe bruises about the head and internal injuries. He lingered until the following morning, when he succumbed and passed away. He was barely 21 years of age, and was the mainstay of his parents and brother and sisters. He had only worked in the mine for about 7 or 8 months. The 'Cock Pit', is a sort of an incline shaft several feet deep, and when the rock began to fall it was impossible for the men to dodge the rock in any way. There was about 10 tons of rock fell altogether, and it took five or six men to remove some of the boulders from off the men. Nearly everybody in Pinal who could hire or borrow a conveyance attended the funeral of the killed miners at the King on Tuesday morning."

Powder Magazine Explodes

From the *Arizona Silver Belt, June14, 1884,*came this report.

> "We learn from Dr. Hendrix, who recently returned from the King mine, that a magazine at Pinal belonging to Mr. Martin, containing 1400 pounds of black and giant powder, exploded on the 2nd instant. The report was heard at a great distance, and the concussion was sensibly felt at Silver King, five miles distant, being so severe as to cause the houses there to shake."[20] Giant Power was a brand name of explosive black powder used in the 1880s.

The Mine Superintendents During the Silver King Boom Years

The Mine Superintendents during the 1870s and 1880s, were Aaron Mason, brother of Charles Mason (one of the mine's original discoverer's) and a practical mining man, Richard M. Phillips, who was not a mining man but was private secretary to President James Barney, was the second superintendent. Phillips did not run the mine very long but was well thought of by his peers. Andrew J. (Jim) Doran, who later became Sheriff of Pinal County was the third superintendent. Arthur Macy was the last superintendent during the 1880s boom years, and Robert Bowen was the last superintendent of the 1880s. (See the chapter on Bowen Family History.)

Aaron Mason first superintendent
— Greg Davis Collection

Aaron Mason came to Arizona at the request of his brother Charles Mason. He was at that time the manager of the Little Molly Mine, near Colorado Springs, Colorado. Prior to that, during the Civil War, Aaron Mason was a member of the US Secret Service, then went to Colorado where he was a Marshall, then the County Sheriff. He had two other brothers. A brother named Solon, and a brother named Freeborn, who also came to Arizona, and a sister Teresa, who ran a boarding house in Colorado Springs, who also came to Arizona. Aaron was said to have spoken with a lisp and used the term *by quist* frequently, to emphasize his feelings on a subject. Mason would sometimes drive down from the

Silver King Mine to the mills at Pinal with a string of natural wire silver, several feet long, twisted around his sombrero. He often sent native silver to the mint to be made into silver dollars which the company gave away as souvenirs. A Pinal merchant, Emerson O. Stratton, once asked a teamster to bring him a nice specimen of ore to send back east. On the following trip the teamster tossed off a fifty-pound chunk. Stratton chipped off a small piece and placed the chunk on his counter. About a month later, when he saw Mason, he said, "You'd better send a man here to pick up this chunk of ore, it's too valuable to be laying there." Mason shrugged and said, "By quist, I didn't put it there"! Mason would not send it back. Stratton had to send it back himself.[21]

After his days at the Silver King Mine, Aaron Mason tried raising thoroughbred horses and cotton farming near his 900-acre ranch, at old Adamsville, near Florence. This was during the late 1880s. He eventually settled in Los Angeles, where he went through his fortune that he made at the Silver King Mine. He died, supposedly of apoplexy; however, his kin thought he was hit on the head and murdered. (Another story has it that a runaway horse killed him.) He was alleged to have one hundred thousand dollars on him at the time he died. The money was never found. However, Mason's wife Guadalupe had some money of her own, and the family was able to survive. She is supposed to have started the Los Angeles Opera House with some of the Silver King Mine's money.

Richard M. Phillips, the second superintendent, had been the owner James M. Barney's personal secretary for many years prior to coming to the mine. Although Phillips was not there long, he was well liked by the mine establishment and community and was said to have made many improvements while there. The miners and the town of Silver King had a dance in his honor when he left.

Andrew J. Doran, the third superintendent of the King, was a soldier who came to Arizona after he mustered out in 1876. He supervised the construction of the Silver King Mill at Pinal and was it's superintendent. He was elected to the 11th Territorial Legislature in 1880. He became superintendent of the King Mine about 1882. He made a record shipment of 22 large bars of silver valued at $70,000, the result of 15 days run at the mill. After the mill and mine jobs, he became Sheriff of Pinal County for one term. (See the story regarding Sheriff Doran in Frontier Justice Chapter.) He later became associated with the Reymert Mine, about 12 miles south of Pinal. He helped build the Arizona Pioneers Home, and later in life, resided there until his death. It is ironic that his most serious injury was not received from bullets when serving as a soldier or a lawman, but from an auto accident injury while he was in Los Angeles. He died at the Pioneer Home on February 14, 1918, at the age of 78.

Arthur Macy, Silver King Superintendent 1886— Bowen FamilyCollection

The last mine superintendent during the boom years was Arthur Macy. He was described as an educated man and the most efficient mine superintendent during the mine's peak years. Macy was also a metallurgist and a graduate of Columbia University. There was a conflict between him and a certain Professor Blake regarding the operation of the mine and expenditures when the mining company changed over its equipment from wood burning to oil. It took literally mountains of wood to operate the boilers, which powered the mine's crusher and hoist. Blake blamed Macy for the mine's closure, but the facts were evident that the cost of wood was getting too expensive, and Macy's conversion to diesel fuel was a better alternative than wood. Many coincidental factors caused the closure of

Sam Michael at Silver King miner's cabin ruins 1997 — Author's photo

the mine, not the conversion from wood as the main fuel. (See Chapter Fourteen for text of the letters between Blake and Macy.)

At the beginning of the year 1888, those in authority at the mine were as follows: Arthur Macy, Superintendent; Randolph Adams, Assistant Superintendent; Robert Bowen, Mine Foreman; Eugene Hoefer, Mill Foreman; E. N. Van Courtlandt, Assistant Mill Foreman; Arthur L. Walker, Engineer; George F. Smith, Chief Engineer at the Mine; George F. Labram, Chief Engineer at the Mills; Kenneth MacKenzie, William H. Benson and William C. Trueman, apparently supervisors. Robert Bowen was the last superintendent of the Silver King Mine during the 1880s. The last President of the Silver King Mining Company, during its peak years in the 1880s, was Benjamin A. Barney, of San Francisco, California.[22]

Superintendent Arthur Macy 1880s —Bureau of Mines

Chapter Six
Mule Skinners, Freighters and Cowboys

You can almost see this scene out of an old western movie. The muleskinner yells *git* or g*o Jinnie* or *Haw-Jack*, and the freight wagons pulled by 20 mules hitched in teams begin their journey down the hills and canyons. Only this is not a movie, but a real life scene from the 1880s, when mule teams pulled ore wagons from the Silver King Mine to the mills at Pinal, five miles away. Amid the pounding noise of the stamps, the mule trains arrive at the mill where the silver ore from the Silver King is to be crushed and made into concentrates. The heavy ore is unloaded and the mule trains prepare to make the trip back, uphill this time, to the mine for another load, and so it goes.

The following paragraphs describe the ore wagons and how they were loaded and a vivid picture of the ore house, as it was in the 1880s. These descriptions come from actual news accounts and stories of the Silver King Mine during its peak times. Four wagons make up a train, and the train is drawn by a chain to which the mule teams are attached by wooden double trees. Each wagon weighs about 3000 lbs. and carries between ten to fifteen tons of ore. The rear wheels weigh between 600 and 700 lbs. each and are about six feet high with metal tires four inches wide. You enter the upper part of the ore house over a bridge leading from the surface platform of the main shaft. As the ore comes up from the mine in iron ore cars, they are wheeled over the rails upon a bridge to the orehouse and there dumped over a device called a grizzly through which all of the ore passes into the bins below. The large pieces pass to the ore crusher from which it drops into the same bins, and then through the chutes and into the ore wagons. The mule teams pass by the right side of the ore house at the Silver King, swing under the large ore chutes, just below the rock breaker, and are loaded in a few minutes. Each team has two men, a driver and a brakeman called a swamper. A wagon master accompanied the train. This scene took place many times each day when the Silver King Mine was prospering.[1]

The ore house was 60 feet high, 30 by 60 feet, a three-story structure, with a capacity of holding 800 tons and was an imposing structure at the mine. The engine room contained the first engine, which followed the progress of the Silver King, from a small operation to a large one. The enormous jaws of the Blake Rock Crusher crushed large rocks as if they were mere pebbles. The engine room contained the strongbox in which the most exquisite specimens of silver ore, known for their rare beauty and great value, were stored. The driver of the train handles the 20-22 mules by a single line attached to the leader mule. A jerk on the line, and a word of command from the driver, gets the mule train moving.[1]

The *Arizona Graphic* in 1899 described the expert work of the drivers. To drive a team of 22 mules over such a road as coming from the Silver King Mine "necessarily requires training and experience peculiar to the conditions." The journal further stated that the driver would say one word in a conversational tone such as "Jinnie" and "she jumped as if a bomb had been exploded under her, and started on the instant. This is not a mule nature, and her attention must have been the result of a thorough education and the application of unlimited gad."[2]

Freighting Supplies to Silver King

The freighter's life was a difficult soul-wearing struggle. Every meal was cooked on a campfire, and every night they slept in their wagons or on the ground, dragging along through dust to the hubs and through occasional storms. They ate nothing but beans and bacon and biscuts made in a Dutch oven. No wonder the freighter's dream was of the time when he could sell his outfit and live on canned goods the rest of his life and have, for once, all the canned peaches he could eat!

There is an old freighting song that goes in part like this:

Moll we left at Bailey's Wells
And Nell we left at the Sub;
And if their bones could only tell
They died for want of Grub [3]

Wagon Ruts on the Old Pinal Trail

Even today, you can see evidence of the heavy loads that those mule trains of yesterday carried. Just west of Superior, south of State Route 60, there is a dirt road that crosses a cattle guard immediately off the highway. You proceed about 200 yards south of the highway towards Picket Post Mountain, turn left by the old rusty sign that says "wagon ruts," and you can find the wagon ruts from the ore wagons in the old rock road from Silver King to Pinal City. Some of the ruts are almost 2 feet deep in places. In the center of the ruts you will see some round holes about every 4 feet or so. These were made for the mules, so they could get traction while pulling these heavy loads of ore over the rocky road. Those were the days of the mule-skinner, the prospector and hard-rock miner.

Greg Davis at Pinal Ore wagon ruts 1996 — Author's photo

Camels at the Silver King

A soldier by the name of Lieutenant Ned Beale brought camels into service in Arizona as part of an experiment of the federal government in 1859. Apparently someone in the federal government thought that since Arizona was part of the Sonoran Desert, perhaps that camels could be used as beasts of burden and pack animals to traverse the vast waste areas without water. They even brought a camel driver from the Saraha Desert named Hadji Ali, Americanized to Hi Jolly, for the purpose of training the soldiers to pack, ride, and lead the camels. When this experiment fell short of its expectations the camels were tried for other purposes. Records state that a Frenchman purchased some of the camels and brought some to the Silver King Mine about 1876, where they were used for packing ore from the Silver King to Yuma. Their feet became sore from the sharp stones on the trails, as the camel's hoofs could not take this kind of rough ground. Hi Jolly made boots of rawhide for the camels, but they did not hold up and the boots wore out rather rapidly.

The Frenchman became disgusted with the lack of a market for his 20 camels and turned them loose upon the desert near Maricopa Wells. For a while thereafter, these strange beasts were seen roaming the desert valleys and mountains throughout the Gila Valley, and a large camel called the "Big Red Ghost" was seen often near populated areas where cattle roamed in search of forage and water. The camels soon passed into history, victims of a rocky and hostile terrain. A bronze plaque was erected in their memory and Hi Jolly's at Quartzite, Arizona, honoring their part in Arizona's history.[4] In recent years, I have stopped at Quartzite, and the monument (shaped in the form of a pyramid) and plaque are still there.

Hi Jolly Monument 2000
— Author's photo

Marshall Trimble writing in *Arizona-A Calvacade of History,* states:

> "Although Beale championed his illustrious camels, his hired hands scorned them. The muleskinners couldn't speak Arabic, and the stubborn beasts wouldn't learn English. The animals were said to be bad tempered, foul smelling, and whose strange ways and appearance caused pack mules and wagon teams to panic."

No wonder the camels were short lived at the Silver King, where hundreds of mules were being used as pack animals to and from other supply stations, as well as horses, cattle, oxen and other animals being present at Silver King and Pinal City.[5]

Mule Skinners and Highgraders

There were several reports of high-grading silver ore in the days when very rich ore was taken from the Silver King Mine to the mill at Pinal City. The recollections of James M. Barney in *Forgotten Towns of Arizona, Pinal and Silver King,* states:

> "It is said that an enterprising liquor man built an adobe saloon on the mesa about a mile out of the town of Pinal. By special arrangements, the team's drivers brought him a barrel of water each day. It was told at the time, that upon leaving the mine with a load of ore for the mill, the drivers would pick over the tops of their wagons, and arrange all the high-grade (best ore) lumps

of ore at the front end of the wagon. When passing the saloon on the mesa they would throw these chunks of ore at their teams. After they passed, the saloon man went out and gathered up the chunks of high-grade silver that the drivers had thrown at their mules. He sacked and freighted out two shipments before he was caught with almost enough high-grade ore sacked for another shipment."[6]

Nell Murbarger, in *Ghosts of the Adobe Walls,* writes about similar occurrences.

"In hauling ore from the mine to the mill, the drivers frequently had occasion to speak forcibly to their animals. Rocks were the most forcible arguments at hand and usually it was the very finest silver specimens which were thrown at the teams. This was especially apt to be true if some friend of the driver happened to be passing at that time, and the friend was always curious enough to see what kind of rock had hit or missed the mule. It was thought that considerable high-grade ore intended for the mill, passed into other hands by the muleskinner route."[7]

The Packers

A man named Saxe operated a pack train over the Stoneman Trail between Silver King and Globe. Later, Eugene Middleton bought this business and operated the pack train over the Stoneman Grade. This pack train only carried passengers and their baggage and sometimes bullion. The size of the train depended on the number of passengers, usually drummers with their customary enormous trunks, and the amount of bullion carried. A packer, and sometimes an Indian boy as a helper, accompanied each train. A regular freight pack train was much longer, consisting of about twenty animals to each packer and helper. The animals were large fine Missouri mules or burros. At that time the heaviest load carried over the Stoneman Trail to Pinal Ranch were two barrels of whiskey, weighing 510 pounds. The bulkiest load was an organ. Later, when the Silver King Mine received the pump for its pumping station, they found it weighed 610 pounds. They asked for it to be packed on muleback and Saxe agreed. When the pump was loaded, the mule walked off with the greatest of ease, and climbed over the mountain without a "babble." It was said that during the prosperous days of the Silver King, as many as 200 pack animals crossed the trail up Stoneman Grade and through Pinal Ranch daily on their way to Globe.

The *Pinal Drill* of July 23,1881, relates this amusing tale of the burros.

> "Large numbers of pack trains of burros have been in town for supplies for the mines. It is rather amusing to see these 'patient animals' stealthily scatter in every direction while their proprietors are taking parting drinks. The boards and lumber with which some of the animals were loaded, threatened destruction to many windows."[8]

The most colorful packer on the trail was Frenchman named Claude Battalieu, who afterwards settled in Globe and accumulated quite a fortune in real estate and mines. His name was supposed to mean "noisemaker" and Claude was supposed to have lived up to his name. He could swear fluently in both English and French and was so loud that he could be heard from Devil's Canyon to the Silver King Mine. He had been captured once by the Apaches, but they turned him loose, after they took all his clothes, because they thought he was crazy. Apaches were not known to kill a person they thought insane.[9]

There was one old freighter known as "Black Jack" whose capacity for whiskey and tobacco was phenomenal. Every morning he would hitch up his team, load the freight, and stock the jockey box (a place for the driver's personal supplies) with one quart of whiskey and one pound of plug tobacco. The quart and the pound were rations for one day.[10]

An Opportunity Lost

Eugene Middleton had an opportunity to go into the mountains with Sullivan in search of the rich silver strike (which became the Silver King Mine) but declined due to the Apache danger. In later years, he always expressed great remorse that he did not go in with Sullivan in search of the "fabulous silver lode." Middleton, later with his brother Leroy, drove the stagecoach between Globe and Kelvin and over Chalk Hill through Cane Canyon.[11]

According to Dan Rose, writing in *Pre-Historic and Historic Gila County*, the following is a good description of transportation modes of the day.

> "With the great productions of rich ores flowing from the mines in all the districts adjacent to Globe, one could find the proverbial $20 gold pieces rolling in the streets, naturally everybody was 'loaded with dinero.' Business was booming, so were the saloons. Buggies, saddle horses, mules,

the ever-faithful burros, and freight outfits of from six to eighteen mules, horses, and oxen, line the one main street of every camp."[12]

A Tale of Silver Dollars and Nails

Perry Wildman, who operated a general store at Silver King Town for several years during the 1880s, writes this humorous tale in his reminiscences of Silver King and Pinal City days. He states:

> "The Silver King Company as a rule paid their men off monthly in drafts on the Anglo-California Bank, San Francisco. As a business proposition, and in order to keep the money from leaving town as much as possible, I made it my business to bring cash into the camp and had to use all sorts of ways to accomplish my purpose. I very seldom shipped by express because of so many hold-ups of stages between Florence and the mining camps. Most of the time I would have my friend Don Carlos Hayden at Tempe put a bag of silver inside a sack of grain, mixed in the load of several like sacks, brought to me by the Mormon freighters. Although I pursued this course for a long time, and brought into camp many thousands of silver dollars, I never lost one single dollar. Sometimes I would get the coins from Los Angeles by freight, and packed in boxes with merchandise. It was somewhat tricky, but I felt that it was safer than by express.
>
> "I remember one time I was rather late in providing for pay day, and I wrote the Union Hardware and Metal Company in Los Angeles, to ship me by express five bags of nails. Each keg of nails was to contain about half its weight in silver dollars, and the other half in nails. As nails were only worth 12 1/2 cents per pound, it was a foolish thing for me to do, because the express charges would be paid per pound and most anyone would think it queer to see nails going through by express. After I had mailed my order, I regretted doing so, but I could not cancel it and had to await the results. The nails came through all right as far as Florence. When the stage came into Silver King the day I expected them, I asked the stage driver if he had any nails aboard for me. He replied that there were five kegs of nails on the sidewalk in front of the express office, but he could not bring them up that day as he had such a heavy load of passengers and baggage, but would bring them next trip. So the nails lay out on the walk all day and night, sat on and rolled about, but they came up on stage next day, and it turned out all right. No one was the wiser but myself and I never tried that mode of bringing in money again. After the Pinal Bank was started, and Mr. Venton, the cashier, relieved us very much in the way of cashing checks and collection of accounts."[13]

Perry Wildman, in addition to owning stores at Silver King and Pinal, owned a pack train of 100 mules and burros. He forwarded freight and merchandise to other towns and camps, much of it going across the Stoneman Grade and Pinal mountain ranges, to the Globe Mining District.

Perry Wildman's Silver King Store cir. 1880s — Gladys Walker Collection

The *Pinal Drill* on September 30, 1882 stated:

> "The packing business between Silver King and Globe has assumed large proportions. Perry Wildman alone ships 150,000 lbs. monthly. There are about 50 mules and 80 burros constantly employed. The freight is one and half cents per pound. Distance 28 miles."[14]

Since wagons could not travel over the mountains, the mule-trains carried supplies for the town of Silver King. A long line of mules tied together, walking nose to tail, was called a mule train. There was a lead mule, sometimes with a bell around the neck. The other mules were trained to follow the lead mule. The mules were loaded with heavy packs filled with supplies of all sorts. Packing the mules and driving the mule train took great skill to get them over the mountain trails. The packs had to be balanced very carefully. Since Apaches often attacked lonely muleskinners, guards often had to be sent along. This could be a dangerous task because the Apaches often raided the supply trains, and in addition, had a taste for mule flesh.

The Freighters

There were many freighters who hauled silver ore and concentrates from the mine to Yuma and Casa Grande to the railheads. Based on available records, listed below are many of the freighters: James Quinlin, William Fenton, Matt Caveness, Apolonio Martinez, F.O. Donnelly, S.S. Jenks, John J. Gardiner, Nathan B. Appel, Dolores Baldonado, Aaron Barnett, George Reeb, George Frisk, Samuel Finny, Joe Espinosa, George R. Emerick, A. Daguerre, George L. Field. F. Meyer, John C. Hancock, Ben Block, Brown and Calvin, Juan Noriega, Stephen Berthoud, William Morgan, James Caldwell, Field and Morgan, O. Buckalew, Y. Padilla, John Brash, Francisco Noriega.

Chief among the early freighters was Jose' Gonzales. We have a reminder of the old Silver King bonanza days and when ore was hauled between the Silver King Mine to a smelter at Florence. Gonzales Pass which cuts through the mountains, high above the dry ravine, on State Route 60, west of Picket Post Mountain, is named for him.[15]

Matt Caveness—Pioneer Rancher, Freighter, and Saloon Owner

Matt Cavaness crossed the Great Plains during 1861 by covered wagon, and made his way into California after battling Indians, storms and lack of water. In 1864, Cavaness came to Arizona Territory. At the invitation of A. H. Peebles, he worked an arrastra at the famous Vulture Gold Mine at Wickenburg. Cavaness became a guide for the army and then ran a blacksmith shop in Phoenix in 1872. He became a freighter next, freighting ore for various mines. Matt Cavaness had one on the first cattle ranches, if not the first Anglo cattle ranch in the Superstitions. He settled on a wide valley about 1875-76 where the Quarter Circle U ranch is now located. According to Matt Cavaness' memoirs, he was told by an old chief of the Maricopa Indians of a place near Superstition Mountain where there was a good spring of water. He moved his wife and children on the ranch and built the first board house in that area. Most houses at that time were stone with canvass tops or adobe brick. Cavaness brought the boards back from Yuma, after he freighted ore from the Silver King Mine there, to be shipped to the Selby Smelter in San Francisco, California. Cavaness stated that he had about 1000 head of cattle put in this small valley in the Superstitions with the spring. Cavaness was quite a businessman and kept his freighting business and the ranch, as well as other enterprises such as prospecting and mining. Cavaness and his first wife were divorced, and she ran the ranch for a short time. Matt and his wife Alice Rowe Cavaness were separated June 23, 1880, and Matt gave the ranch and cattle to Alice. They eventually divorced on October 12, 1881.

As the Quarter Circle U is is probably the most famous ranch in the Superstitions, and has such a varied history, I am including a brief history as to the ownership over the years, based on the records located at the Pinal County seat in Florence. Alice quickly remarried Randolph Purdue at Pinal City, Arizona territory on November 22, 1881. The records are not clear at this point, but it appears they sold the ranch shortly thereafter to persons unknown, who in turn sold the ranch to Leo Goldman shortly after that. The records then reveal that the ranch was sold to Alfred Charlebois and his partner George Marlowe on July 22, 1882 for a paltry $200. Charlebois and Marlowe soon apparently had a falling out, and Marlowe purchased full interest in the ranch on March 30, 1883, according the County Book of Deeds, # 7, page 569. George Marlow obtained the ranch and ran it until 1889. The Marlowe estate then sold the ranch to the James E. Bark, Frank Crisswell and J. L. Powell on May 18, 1891. Bark and Criswell soon obtained ownership of the ranch as Powell sold out. Frank Crisswell then purchased the ranch from Jim Bark.[16]

Matt Caveness cir. 1878—
Author's Collection

Cavaness moved on and later ran cattle in the Tonto Basin where he ranched until 1903. Cavaness eventually settled in Mesa, Arizona. Earlier, Cavaness was also involved in the saloon business at Silver King. Matt Cavaness was one of the early pioneer ranchers that had the fortitude to set up a cattle ranch in the Superstitions during the Apache Wars, when many a man would think such a venture was too foolhardy and dangerous. He was a true pioneer during the settling of the west. He died in 1929 at the age of 84.[17]

Matt Caveness, notable Arizona pioneer, in 1876, was the first freighter to bring in machinery for the mine and the first to haul out a load of silver ore, worth $ 2000, which he delivered to the Colorado Steam Navigation Company at Yuma. *The Arizona Daily Star,* Tucson, Sept. 9, 1910, in recalling Matt Caveness at Pinal City, states that Caveness had a contract to haul ore from the Silver King Mine. It said that the first load of ore that he hauled, about 40 tons, was hauled to Yuma, and thence by boat to San Francisco to the smelter. Several weeks were necessary to make a trip. On one occasion, the boat at Yuma had sunk and Caveness had to haul the ore to San Diego. At San Diego, there was a crowd waiting them, and a large man in the crowd insisted on having a sample of the ore. Caveness would not open the rawhide sacks, but instead gave him a silver dollar, saying it was an exact sample of what was in the sacks.[18]

Matt Caveness said that he made a lot of money hauling Silver King ore. That allowed him to own a blacksmith shop, the Palace Saloon, the Alice Mine, 12 miles southwest of Pinal (named after his wife), a wagon making shop, numerous ore teams and wagons, race horses, part ownership in a saloon at Silver King and a cattle ranch in the Superstition Mountains.

The Absconder

The *Pinal County Record*, on February 5, 1886, reported:

> "Mr. Silas Caveness, who recently brought a band of cattle for his brother Matt, and delivered them to some parties in the Superstitions. Silas pocketed the proceeds of the sale about $ 3000, and skipped the country, leaving his brother without a quarter of a dollar. Matt has written to his friends here requesting them to give him a situation."

Original QCU Ranch cir. 1931— Barkley Family Collection

I was unable to find further articles about Silas Caveness and could find no further record of Silas in the old newspapers. However, Matt did go on to establish a ranch in the Superstitions where he raised cattle. His wife raised goats after their divorce and provided goat milk for the miners at Silver King and Pinal before selling out.[19]

Freighting the Silver King Ore

At first, the extremely rich ore of the Silver King was sorted, sacked and freighted from the mine to Yuma, at a cost of from 3 to 3 1/2 cents per pound. On September 5, 1877, the last Silver ore was sent from the mine to Yuma. Thereafter, the wagon trains were loaded with concentrated silver ore and turned out at the company's mill at Picket Post. There was a good road between the two places and the cost of hauling ore from the mill was about $ 2.50 per ton. During the year 1877, 3528 cords of wood were used for fuel at the Picket Post mills that were then working on Silver King ore (Mills No. 1 & No. 2). Wood at the mine cost ten dollars per cord, most of it being hauled 18 miles. Wood for the mill was brought about twelve miles, and the price paid was from six to seven dollars per cord.[20]

The Cowboys—-Beef for the Silver King Miners

Gus Barkley at QCU Ranch 1910
—Nancy Barkley McCollough Collection

Mining wasn't the only occupation in those days. Raising cattle was also very important. Several ranches near Silver King and Pinal each had 300 to 1000 head of cattle. Some of the earliest ranchers were Matt Caveness, Charles Whitlow, Jack Fraser, Jim Bark and the Robles family in the Superstition Mountains and L. deArnett (also spelled Arnett) on Queen Creek near Pinal. Others were George W. Cole, Hank Bray, Joe Kelsey, Tom Grooves, Dave Pottance, and Jose Gonzales. Also listed in the Brand Books of the 1880s for Pinal County: Roman B. Arballo, R. G. Brady, John N. Brown, Bayless &

Berkalew, Lizzie Chamberlain, W. C. Davis, V. R. Lopez, Aaron Mason, John McGrew, The Ripsey Family, Brown & Wells, Asa Walker, J. W. Whitlow, Henry Whitford and Chris Whitford.[21]

When the mining industry began around the mid 1870s, including the finding of the famous Silver King Mine, the cattlemen came from Texas, New Mexico and Mexico to furnish the miners with beef. According to Swanson and Kollenborn, writing in *Superstition Mountain—A Ride Through Time,* "Not until about 1872 did cattlemen first make their way into the Superstitions. Those early cattlemen and miners had to face severe hardships, which delayed the development of both the mining and cattle industries. Hostile Apaches, intense summer heat, lack of water and rugged terrain tested a man's mettle to its limits."[22] Other sources also stated that the Silver King Mine started the cattle industry in and nearby the Superstition and Pinal Mountain Ranges. Those first cowmen had tough times to endure in order to survive. But survive they did, and many cattle still roam these ranges, however, under the watchful eyes of the Bureau of Land Management who regulates the number of cattle per so many acres. This is done in order to prevent overgrazing, which was said to have happened during the late 1800s and early 1900s.

Whitlow Ranch and the Whitlow Cowboys

On the banks of Queen Creek in Pinal County, not far from Queen Valley, stood the historic Whitlow Ranch and Stage Station. The ranch house was where the Whitlow Dam is now located. On the Arizona Territorial map is the town of Marysville, which is on the Salt River a few miles north of Mesa, Arizona. On this map is also indicated the Whitlow Ferry. This ferry was owned and operated by Charles Whitlow and was the only crossing between Tempe or Hayden's Ferry and Blue Point, where Poncho Monroy had a cattle ranch. It was also called Rowes Crossing, Morman Crossing and then Whitlow Crossing. The little settlement at the ferry crossing was called Maryvale for his daughter, Mary Elizabeth "Molly" Whitlow. (Molly later married William Long, one of the founders of the Silver King Mine.) About 1874, Whitlow abandoned his ferry.

On the banks of Queen Creek was an abandoned Spanish ranch, with an enormous stone corral and a well. It was unowned and unclaimed, so Charles Whitlow just moved in and took over. Charles Whitlow also established a milk ranch four or five miles north up Whitlow Canyon. The newly discovered Silver King Mine was attracting miners and prospectors by the hundreds. Pinal was fast becoming a town of importance. The Whitlow Ranch was an ideal stop between Marysville, Tempe and Desert Well. Heavy freight wagons were often seen winding their way along the trail with supplies for the Silver King Mine. One of those freighters was Mike McGrew, who twice a month made a trip from Tempe with his lead wagon filled with flour and bacon, and his trail wagon loaded with grain and baled hay. The first stop on the freight route was Desert Well, where water was ten cents a head for the beasts of burden. The next stop was at Whitlow's, where the price for water was two bits a head. Whitlow's stage stop was not only a place of business, but also a hangout for all the notable characters of that region, some of them infamous. Known cattle rustlers and desperadoes mingled freely with the more honorable gents. There you could have met Old Man Arthur, "Mexico," an older brother of Charley Whitlow, Zellweger from Tucson, Fatty Perkins known as the "Mayor of Pinal City", Dad Bellamy, the Spret brothers and an occasional Ranger with his heavy guns on each hip. It was at this ranch that Waltz and Weiser supposedly stopped on their way into the Superstitions in 1881, when they discovered a fabulous Spanish gold mine.

Charles Whitlow 1880s— Whitlow Family/Greg Davis Collection

Charles Whitlow also filed several mining claims in the area including the Silver Belle and was connected with the Belmont Mine, in Martinez Canyon, east of Florence and north of the Gila River. He died in 1886 and was buried in the old cemetery at the now abandoned town of Adamsville, Arizona, near Florence. His wife Sarah Jane ran the ranch for a while with sons Charley and James William. (Young William found one of the murdered soldiers of the Lost Soldiers Mine story.)

A Lucky Wager or How Jack Fraser Founded the J. F. Ranch of the Superstitions

During 1886, pioneer cattleman Jack Faser got into the cow business more by chance than choice. While working at the Silver King Mine in 1886, he made a wager with the blacksmith about the political election that year. They often wagered on political matters. Fraser wagered $500 against the blacksmith's 50 head of cattle that the Republican candidate would win. The Democrats were defeated, so Fraser found himself in the cattle business. He built himself a ranch on the old Reavis property after old man Reavis died in 1896. Billy Knight, Fraser's foreman, found the old man dead on the trail with his three large dogs guarding the body. (Other people were said to have also located the body.) The coroner's jury that responded to the location of the death reportedly had to shoot the dogs to examine the body. Reavis was buried near where he died, as was common in those days. Fraser paid the coroner's fee of $600.00 and then revived the rights to the property. He then constructed a house near the old Reavis shack, which he burned down for sanitary reasons.

Jack Fraser on right cir. 1930's— Greg Davis Collection

At this time many people were running a few head on the open range near the town of Silver King. Mexican cattle were the only kind to be seen in the Superstition area, and Fraser bought out many of these small outfits. At the same time, he was building up a herd of quality Durham and Hereford cattle. It was estimated that Fraser had about 10,000 cattle scattered on the open range from the Salt to the Gila Rivers. During those early years, the streams ran freely so there was little need for water tanks. He hired Mexican workers to build many stone corrals that are still seen today in the Superstitions.

Fraser kept the ranch until 1909 when he sold out to the W. J. Clemans Company. He took the money from the sale of the ranch and bought a block of ground, in what is now downtown Mesa. He was part owner of Everybody's Drug Store in Mesa for years. Fraser was truly one the pioneer cattlemen in the Superstitions. Jack Fraser was born in Nova Scotia and was said to be a true Scot.[23]

When the Cowboy was a Knight

While on a trip to the old Reavis Ranch in the 1890s with some friends, Will (William George) Knight stopped at the old ranch and became acquainted with Jack Fraser, the owner. Jack took a fancy to the 17 year old boy, and Will did not return to Tempe to live with his parents but worked at the Reavis Ranch for the next 23 years, where he learned the cattle business and eventually became the ranch foreman. This was an amazing feat for the young cowboy who had fought illness and rheumatism, and had one time been on crutches, he was afflicted so bad. Wills's father had worked at the Silver King Mine and Will had lived there while growing up, right at the foot of the schoolhouse. Will's mother Emma Bray Knight is buried at the old Silver King Cemetery, near the mine. She died in February 1888, after giving birth to a daughter Erma and suffering from tuberculosis. Richard Trevethan built the iron fence that now guards her gravesite. Richard Trevathan and his wife are also buried at the Silver King Cemetery. John Knight, Will's father, opened a general merchandise store at Silver King Town. John Knight was listed in the Directory of Business at Silver King from 1889-1892. He later moved the store to Tempe. When Will was only nine years old, he used to take the pack train with passengers to Globe, which was 26 miles away. He used to bring the horses back using the old Globe and Stoneman Trails, mostly traveling alone on the return trip. That is where he learned to become such a proficient horseback rider. According to Will, on one of these return trips he was hurrying to get home on the 3rd of July one year when a horse wheeled and kicked him, breaking his leg. This happened on the west-side of Devil's Canyon, which is a very treacherous place to cross over the ridges, before you get to Silver King. He said he still made it home for the Fourth of July celebration that year.

Eunice Ann Riggs came west from her home in St. Marys, West Virginia to visit her aunt, May Clemans, in Mesa, Arizona. It was there that she met Will Knight and he courted her. He traveled to West Virginia to ask for her hand in marriage and was given permission on the condition that they be married in the family home in West Virginia. On returning to Mesa, Will took his bride on horseback with a mule pack train, high up in the rugged Superstition Mountains, to the Reavis Ranch. This homespun place would be their home for several years. This was such a unique experience for an eastern school marm who married a cowboy ranch foreman and lived in the wild Superstition Mountains of Arizona Territory. Eunice kept a journal, which became a book called *Emigrant Knight of Cornwall.* In this book, she details life on the range, living in early Arizona where the mountain lions screamed at night, and the cooking was done over open fires or wood stoves. Eunice said that the Reavis cabin was in terrible shape, fit for rough cowboys, not women and had to put her touch on the place to make it liveable.

Billy and Eunice Knight wedding 1910— Ann Knight Rose Collection

Will was also a good blacksmith and would shoe all the horses himself for the roundup. Once on a roundup there was a stampede. On one occaision, Will was driving a big herd, down from the high country, to the lower J. F. headquarters ranch. The cattle were sore footed, and when they hit the big sand wash they started running. There was nothing but a sea of white-faced cattle. Stampedes were a fearful time for cowboys, and you did not want to get caught between the cattle and the walls of a canyon. The cowboys took to higher ground and the herd soon ran out of steam. This time no one was injured, but everyone was covered with two inches of dust.

What was amazing was the fact that Will and the other cowboys were able to ride over such impossible terrain without getting hurt. Will was a tough hard rider, but no bronc buster, and the cowboys often played tricks on Will to make his horse buck. Will left the Reavis Ranch (the upper Jack Fraser ranch was called the Reavis many years after Elisha Reavis' death) after 23 years, and became a farmer and rancher near Florence, Arizona, in partnership with Jack Fraser. William George Knight was a real cowboy, and lived on the open range most of his life. He was born November 18, 1875 at Cornwall, England and died April 28, 1956, at the age of 81 years.[24]

The Life of the Early Cowboy

This boom in mining in Central Arizona created an increase in the need for beef, and thus the cattle industry was started. In those early days, the cowboy had to be self-sufficient and a very rugged individual. The cowboy had to have an invincible spirit to conquer the Arizona Mountain ranges. There were said to be about 3000 head of cattle in the Queen Creek Basin area, and some were wild, or called outlaw cattle. The round-up then was called a *rodéo* in Spanish, but the cactus cowboys called it *rodér.* The contest riders of today call it *ró-deo. Ro-dáy-o* is the term used today by most people, but no matter what it was called, there were rules and regulations that were strictly adhered to by the cowboys and their bosses.[25]

Each animal had a brand on it along with ear crops, bits and underhacks. The standing rule was that no cattle could be branded except at this roundup and at the *parada-ground.* Cowboys caught branding cattle elsewhere with a branding iron, or running iron, was fair game for the

Early cowboys at roundup — Ann Knight Rose Collection

noose. More than one range war started over the failure to abide by these rules. Some of the unbranded cattle roamed free as deer and were called outlaws or mavericks, and mother cow and calf had better have the same brand on them at the round up. There were no permanent line shacks or range houses for the cowboys and no doctors or medical buildings nearby. His pistol, rifle and trusty steed were usually his only constant companions against his enemies, both man and animal.

The Apaches were known to attack lone men in the wilderness, and the ruggedness of the terrain itself was a constant reminder to always stay on guard. There was much more wildlife in the 19th century roaming the hills and mountain ranges of the cowboy. This included bears, mountain lions, bobcats, rattlesnakes, javalina and other wild animals to contend with. His horse was often his sole companion, and it better be a good one, because the cowboy's life depended upon it. Cowboys on a round-up usually had a remuda, or additional horses to ride, as they were constantly in the saddle.

When you "rode for the brand," you were loyal to it, even if you had a grievance with the boss. When round-up was over, it was time to head to town and "live it up", and live it up they usually did. It was customary to ride through town shooting their six-guns in the air or shoot out the lights in the saloon in celebration of the end of a long hard season. It was against range ethics to ask about a man's background or question his name. The cowboys only revealed what they wanted others to know and that was as far as it went. Although they might fight at the drop of a hat, cowboys were there when a neighbor or even a stranger needed help, and they had great respect for women.

The cowboy subsisted usually on beef jerky, bacon and beans and some flour to make biscuts. He worked not only with cattle, but repaired fences, windmills, took care of his horses and cattle, mended his clothes, and fixed his gear, so it was ready at a moment's notice. The cowboy was, and still is, a rugged individual. However, his life was not the romantic way of life that was portrayed on the silver screen. My interviews with today's cowboys, many which are descendants of those early pioneers, still exhibit those rugged ideals and values of the early cowboys.

The cattle for the Silver King Mine and nearby mining camps in the 1870 and 1880s were usually taken to Frank Maier's meat market in Pinal, where it was butchered and made ready for sale to the miners and settlers. Much of the "beef on the hoof" likely was turned into Cornish pasties, called "letters from home." The Silver King Mining Company itself was a customer, serving beef in the mess hall for the numerous single men. The cattle were driven in on the hoof, and penned up and butchered as needed. Most of the cattle in the Arizona Territory at that time were Texas longhorns, or Mexican cattle, but some shorthorn cattle, called the Durham and Galloway breed, were also introduced.[26]

Chapter Seven
Life at Silver King Town and Pinal City

As soon as a "great strike" occurred, out of nowhere and almost overnight a mining camp sprang up. These camps usually lasted as long as the paying ore was readily available and soon became *ghost towns* when the ore disappeared. Silver King and Pinal City lasted several years longer than most of these instant camps and there are many interesting stories surrounding them. Included are stories of the camps, and tales of the people who inhabited these camps. I have also used many of the local news stories of the 1880s printed, just as they were written, in order to give the reader the flavor of the language used during that time period in the west.

Silver King Town cir. 1885 — Bureau of Mines

The Story of Silver King Town

Within a year of the Silver King Mine being established, 975 mining claims were recorded in the Pinal Mountains near the Silver King Mine, and a mining camp called Happy Hollow Camp grew around the mine. This settlement, however, soon took the name of the mine and was called Silver King. The Silver King Mine was "smack dab" in the middle of the new Pioneer Mining District. Tents and shacks soon gave way to more permanent structures. Soon a company office and other mine buildings, merchants' shops, boarding houses, saloons and residences filled the town. It even had a small Chinese section. Later the Silver King residents built a combination school, church and dance hall.

In it's day, the town of Silver King had about 70 structures and 500 people. In 1883, there were 4 general stores, a butcher shop, a hotel and 4 saloons. Silver King also had a barbershop and boasted of its pool/billiard table in the Bowen and Jones Saloon. Most of the other stores necessary to a community were located in nearby Pinal City. Silver King also had their "ladies of the night."

An article in the *Pinal Drill* on July 7, 1883 spoke of the "arrest of the notorious tramp Jim Allen for kicking in the doors, and swaggering with a revolver around the heads of some prostitutes at the King. He was tried before Judge Benson and *mulcted* $100 and costs, and he was ordered committed to jail for non-payment."[1]

Silver King Town with school house in front, 1883
— Bowen Family Collection

At Silver King Everything is Lively

Mining at the Silver King Mine was booming in 1883 and so were the various businesses. This story in the *Pinal Drill* reflects the rosy outlook.

> "Perry Wildman's large trade has demanded more room. His immense stock fills every part of his large building. He has had to increase by adding an elegant, airy and comfortable office in the rear, overlooking the store. Thompson's new restaurant is booming. Williams and Caveness have enlarged their conveniences by removing the bar and fixtures to their saloon which is separate from the hotel. All is alive and business. Loffler's Saloon is doing a good business. O'Boyles's Silver King Hotel, is as usual, full of

guests and boarders. Ludke's Store is doing a solid trade. John McCourt's Saloon in the stone basement of the King Hotel is as cool and comfortable as you can wish, and he has lots of business. Everywhere and all over the business is lively and money plenty."[2]

Life at Old Silver King

The following narrative comes from Mrs. Georgia Moore, a lifetime resident of Superior, Arizona and former correspondent for the *Arizona Republic*.

"Saturday mornings at the King meant blissful freedom from school for the boys. Every male in town under 20 had his own saddle horse, and all the trappings to make such transportation safe and comfortable. Saddle, bridle, open stirrups, a good rope or riata, a small canteen, a good wool saddle blanket, saddlebag and gun scabbard, and maybe a gun.

Silver King school children, teacher Oscar Atwood — Ann Knight Rose Collection

"Saturday morning—not to sleep late today. People didn't in 1876—too many things to do, and they took longer then. The smell of bacon frying probably is the most fetching of all foods. Suddenly, and without fuss, Mother had fresh hot biscuits, bacon, mush, and coffee set out, and the family was ready to begin the day. From outdoors came the rattle of harnesses, the screams of the muleskinner, the shuffle and scrape of mule hoofs clattering up the rocky slope to the hopper to load up with ore for the first haul of the day to Pinal, or Picket Post, as it was sometimes called. Since hundreds of wild burros lived in that territory, it was no trick for a small band of boys to whoop and holler up a stampede in the general direction of Silver King (Town). After the chase lasting for several minutes, some of the donkeys would find themselves in the big corral at the livery stable. The boys risked life and limb until dark trying to ride the burros. Rarely did anyone get hurt. "Probably one of the oddest playthings was the cowhide sledge. A harness of sorts was attached to the edge of the hide, hair side up. Children sat on it, and it was pulled as rapidly as possible across rocky terrain, as fast as six or seven boys could run. Everybody spilled off eventually, and then another group would pull the sledge. Boys had wagons made from boxes and girls had lively looking dolls made from bottles. There were plenty of bottles available from the dumps behind the houses, as the miners liked their beer. The saloons in those days were like those in today's horse operas. The bar was in the lobby of the hotel, complete with a tinny piano. Gals in fancy clothes were hostesses to miners, freighters, gamblers and other hangers on.

"Fresh food was obtained from back yard gardens, or peddlers who raised mostly root vegetables on a small scale. Much fresh meat was made into jerky and corned beef. Otherwise, food was canned and freighted in at prices like a cent and a half a pound for potatoes, the same for dried beans and 15 cents for bacon. Work shoes cost about $3.50, and a day's wages were $3. Deserts were plentiful as all the housewives baked. An old custom was giving a *pileon,* when a bill for merchandise was paid. The merchant gave something *extra* or a little gift to the customer and his family in appreciation of the payment. During those days each man was given a demijohn of whiskey and a fistful of cigars, and the children were presented with little bags of hard candy.

"Water for the mining company operations was obtained from the pump station, several miles east of Silver King, higher in the hills. The pumped water flowed by gravity, down the hill into the storage tanks at the King. After a day underground, miners could shower in either hot or cold water, which was provided by the company. But mama and the kids bathed in a washtub behind the kitchen stove. Globe was about seven hours away by horseback. A wagon road to Florence, then just a freighter station, also needed seven hours to be covered. Antelope abounded in the desert between Superior and Florence. As many as 200 were seen in one herd."

The remnants of the old pump station are still there, and the old cast iron pipe that carried the water can also be seen in many places just off the old Stoneman Trail.

Meat and Potatoes for Silver King Town

In 1884, J. J. Baker along with a man named Griffin, who were residents of the Salt River area, went up on Aztec Mountain, north of the Salt River. They found an abandoned cabin there, and a "white man's" grave, at what was later called Peterson's Ranch in 1926. At this location, Baker grew twenty-two tons of potatoes that year, and packed them across the Pinal Mountains to Silver King. Baker had to build his own trail to get the pack train across, but once there he sold them for one-half cent per pound.

John Knight, Silver King merchant
— Ann Knight Rose Collection

Frank Mayer, the butcher at Silver King, brought the cattle to the town himself and provided beef for the miners and the townsfolk. He was said to be protective of his cattle and would not let them be abused. After the Silver King Mine shut down, Mayer bought the NB Ranch on the East Verde at the mouth of Pine Creek.[4.]

Although meat and potatoes was a staple of the miners, there were other vendors who came from Mesa and the little Mormon settlement of Lehi on the banks of the Salt River. They delivered much needed flour and vegetables to the towns of Silver King and Pinal. W. Earl Merrill, writing in *One Hundred Yesterdays*, tells of the Mormon farmers providing food staples to Silver King and Pinal:

> "Frank Pomeroy, describing the hog raising business of his father, has written: 'He has furnished live hogs to the mining camps, especially to Pinal and Silver King, where they were transported alive in freight wagons, with high beds, and hauled by six horse teams. The first venture of this kind was at Christmastime in 1881. (The writer of these memoirs, Pomeroy's son went along as a swamper.) The hogs were loaded in the lead wagon, which had a long and high bed, and 30 fullgrown hogs were loaded in. A trail wagon hauled the bran and alfalfa and water for the hogs and supplies for the trip. After telling of encountering a heavy snowstorm in the mountains, and of the difficulties of getting the outfit over the narrow winding road, he was able to conclude his journey.

The next morning they weighed out the hogs, got the money, a nice sum, and returned home, making the trip in three and one-half days."

Other commodities that were sold in mining towns by our local pioneers was told in the following story by Wright P. Shill.

"George Steele, a member of our first pioneer group, was an astute trader. He carried on a trading business with Pinal and Silver King. Through him the Lehi people found a market for their poultry and dairy products. Brother Steele would collect the products and freight them to Pinal, about every week. One time he was not well and he got my father to take the load of provisions to market that week. Father did not have the skill at trading that Brother Steele had, and housewives at the mining camp took advantage of his unskilled ability. As a result, the returns of the trip were only a fraction of what Brother Steele would have had. Needless to say, that was the last trip my father made."

CHARLEBOIS & MARLOW,
—PROPRIETORS OF THE—
PINAL MEAT MARKET.
—Dealers in—
Beef, Mutton,
Veal, Pork,
and Sausage,
—AT THE—
LOWEST RATES
BRANCH SHOP
—AT—
SILVER BELL MINE.
Orders delivered in all parts of
Pinal City,
Silver King,
Queen Creek.

Charlebois & Marlow Meat Market ad 1882 — Pinal County Drill

Other accounts tell of flour and supplies being hauled to the Silver King from the mill of Charles T. Hayden, an early Salt River pioneer.[5]

News Accounts from the Isolated Communities of Pinal and Silver King

Silver King Town and Pinal City had an early history of isolation. During their existence before the stagecoach and railroad, people and supplies were hauled in by horseback, packmule or freight wagon. The first area newspaper, the *Pinal Drill*, was established about 1880, by J. D. Reymert, followed by the *Pinal County Record*, established in 1885 by Fred R. Kittle, after the *Pinal Drill* closed its doors. The *Arizona Enterprise*, with Thomas Weeden as editor, started about the same time in Florence. There was a bitter rivalry between Reymert of the *Pinal Drill* and Weeden of the *Arizona Enterprise,* culminating with the arrest of Weeden for libel, in August of 1883. Judge Hackney came to the area in 1876 and started his newspaper, the *Arizona Silver Belt,* in Globc. These papers brought much desired news to the mining camps, not only of local news, but state, regional, national and international news.

The *Pinal County Record,* on December 25, 1885, describes Silver King Town as follows:

> "The Williams House, kept by Robert Williams, is an excellent hotel. A good table is always set, and the people of the town that has sprung up about the mine, generally patronize the house. Mr. Williams also keeps a fine saloon, which is one of the institutions of the place. Perry Wildman and the Ludke Bro.'s have stores here, carrying large stocks of merchandise, and they are doing a big business. Among the institutions so essential to the happiness of the people of a live mining camp, may be mentioned the saloons of F. E. Benton, Jesse H. Brown and Foley and Barrington, the senior partner of the latter firm being quite known throughout the southern portion of the Territory. They are all doing a good and paying business and are general favorites with the miners. The town also has a flourishing Lodge of Odd Fellows."[6]

The town's activities included dances, sometimes referred to as "hops," and had their own brass band, horse races, drilling contests, church

R Williams Hotel cir. 1880— Bowen Family Collection

services, as well as baseball games. The town of Silver King had its own baseball team and frequently played the other local teams, including Pinal City's.

Baseball at the Silver King

The *Pinal Drill* of Dec. 25, 1880 tells this story of the Silver King and Pinal Baseball Clubs. "The Pinal Baseball Club—Organized permanently on the 13th of December 1880, the following officers were elected: H. L. Myers, President; L. M. Cox, Secretary; M. L. Gross, Treasurer; H. E. Garlick, Captain—A challenge has been received by the Pinal Club from the Silver King Club, to play a match game of baseball at Pinal, on Christmas Day, at half past 9 o'clock A.M., for the championship of Pinal County. After the game a grand dinner will be given by the Pinal Club, at the Grand Hotel Pinal, at 1 o'clock P.M., which all the members of the Silver King Club are invited as guests of the Pinal Club." The Pinal Club won the game and the Championship by the score of 22 to 15.[7]

A Grand Ball at the King

On August 18, 1881, the *Pinal Drill* had the following local quip about a dance at Silver King. "The Grand Ball at the Silver King was numerously attended. A large and elegant party assembled. Representations from Globe, Queen City, Pinal and Florence were there. Everybody was highly pleased and felt that Silver King had done itself honor." It was called the Great Event of the Season. "The Silver King Schoolhouse Ball took place last night in the newly built schoolhouse at Silver King. The Ball was really a grand affair, to judge by the toilettes of the ladies 'who all looked beautiful, of course.'—About seventy in all, ladies and gentlemen sat down to the elegant supper provided by Mr. Robert Williams—Thanks can be given to the Silver King Band, contributed by their good music—The ball came to an end at about half past 2 A.M."[8]

The Horse Race

A horserace at Pinal City is described in the *Pinal Drill* on Oct. 28, 1882. "There was a horse race on Sunday between Potts' mare and Borden's horse. The mare got the start, and won by a length and a half, but it was evident she was not in condition, and had the race been any longer the result would have been very different. We understand another race has been arranged—Quite a large concourse was present, including several ladies." Wagering on horse races was quite an activity in those days and many races were held. From the *Arizona Weekly Enterprise* on September 5, 1885, another horse racing story went like this. "The Whitlow boys at Pinal had a race with the Leppy horse this week for $75 a side, and several horses raced. The Leppy horse won."[9]

Raising Money for a Church, School and Dance Hall

Perry Wildman, a merchant, recalls how the money was raised to fund a church, school and dance hall through public generosity. "I remember one day I think in 1881, when the Rev. G. H. Adams of the M. E. Church made a visit, and suggested that I volunteer to get some place where he could hold religious services. On the spur of the moment I said let me see what we can do. We first went to the Thompson and Bowen Saloon, where four or five Faro games were running and several were patronizing the bar. I suggested to the boys in the saloon that we needed money to build a church, which we could use as a school, and incidentally would be fine for a dance hall. It was wonderful the way the men threw the coins in the hat as we passed it around. From there we went to Bill O'Boyles and Bob Williams places (saloons) with like results. This with other donations tallied $1200—and we soon had quite a building opposite the Williams Hotel, which we called the "Hill of Science." We thus provided a meeting place for any minister who happened to come along, a place for our children, and a place where we could dance or have any social entertainment. Miss Fannie Jordan was the first schoolteacher and later on Miss Angie Doran. The expenses of maintaining the school and building was done by holding dances, etc."[10]

Gomez Band ad 1882— Pinal Drill

Church Services

The *Pinal Drill,* September 9, 1882, stated: "Rev. Mr. McIntyre preached on morning and evening of Sunday last, to a well filled church. The text was Isaiah, 29th and 19th." There were numerous citations in the local papers of church services at Silver King in almost every week's paper. Services were held at the schoolhouse usually by visiting ministers.[11]

Silver King Cemetery

If cemeteries could talk, you wonder what tales they would tell. Not far from the Silver King Mine, on a seldom used, washed out dirt road, I found the long forgotten Silver King Cemetery. It was difficult to locate, but by exploring the old roads near the Old Silver King Town, I was finally able to find it in in 1996. There were about 20 graves that I could locate, but it was so overgrown and I'm sure there were others. Old cast iron fences enclose three gravesites, but only one has a headstone remaining. It belongs to Philippa and Richard Trevethan. I found out that Trevethan was a Cornish miner who came to the Silver King Mine right after it opened in 1876 and lived at the Silver King Town, until the mine closed. He then opened a saloon in Globe until he came back to the Silver King Mine in 1902, when it was briefly re-opened, and helped to try and re-establish the old Silver King back to its glory days. But alas, it was not to be. He and his wife now rest there eternally.

Trevethan Headstone. Silver King Cemetery — Author's photo

Among the three cast iron fences that I found around the graves, there is a smaller one that belongs to a small boy who died at Silver King. One of the most touching stories of pioneer life in Arizona Territory is about this grave. This is the grave of a nine-year old boy who was bitten by a rattlesnake and died. The child's family put the small iron fence around the grave, and as a token of love for their child planted wild irises around

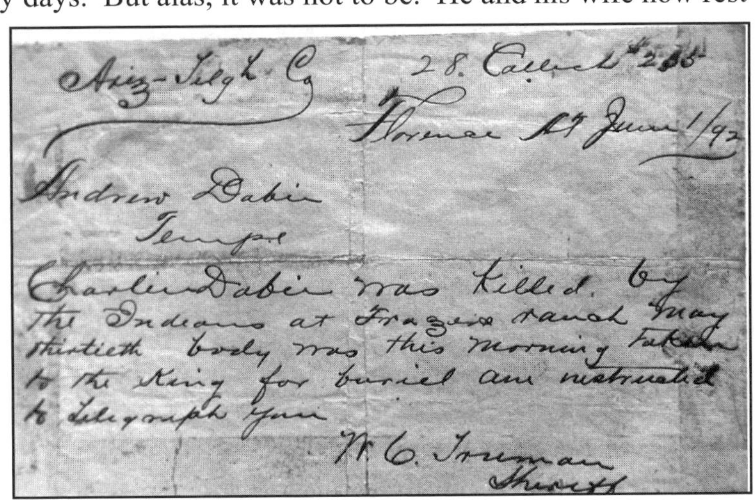

Ariz-Telgh Co 28 Collect $2.35

Florence AT June 1/92

Andrew Dabie

Tempe

Charlie Dabie was killed by the Indians at Frazier's ranch May thirtieth body was this morning taken to the King for burial am instructed to telegraph you

W. C. Truman
Sheriff

Charlie Dobie death telegram 1892 — Gladys Walker Collection the grave. The name of the boy is lost to history, but the wild irises are still there after 100 years. It made me sad to read this story in the old files, but happy to still see the iron fence, and know that the irises still bloom each year for that nameless little one.

Charlie Dobie, 13 years old, is buried at the old cemetery. A renegade Apache killed Dobie on May 30th, 1892, at the Jack Fraser Ranch in the Superstition Mountains. Charlie's friends often called him Adobe Boy. The infamous Apache Kid was suspected of the crime. Apparently the Kid had been seen a day earlier at Pinal Ranch. The cowboys at the ranch recognized the Kid. He had a scout's U.S. Army jacket tied behind his Army issued McClelland saddle. When last seen the Kid was heading northwest in a direction that would have taken him near the Jack Fraser Ranch.

In the evenings about twilight you can hear sounds like young boys playing at the old cemetery. Some people say it's just the singing of the high voltage power lines close by, but I prefer to think that those young boys are living out their youth that was cut short by those violent acts.

The *Pinal Drill* listed the following obituary on February 26, 1881:

> "DIED. At Silver King on the 20th instant, of pneumonia, John Valentine, aged 50 years. Mr. Valentine was familiarly known as 'Gayetta Jack.' He was an old timer, having resided in this territory about 12 years. Eighty-two gentlemen and eight ladies attended his funeral, and Rev. Mr. Wright performed religious services. His remains are deposited in the Silver King graveyard."[12]

Joe Deering, who was killed in a cave-in at the mine in 1882, is also buried at the Silver King Cemetery. Joe Deering is the prospector who allegedly located the Lost Soldiers Mine. (See the chapter on the Lost Soldiers Mine for the complete Joe Deering story.)

Emma Knight, age 32, was buried at the Silver King Cemetery in 1888. She apparently died giving birth to her fourth child, which lived. Emma and her three children came to Silver King in 1886 from Cornwall, England. Her husband John had come to the United States in 1878 and went to work at the Silver King Mine in 1880. Their story is told in the book *Emigrant Knight of Cornwall,* by Ann Knight Rose.[13]

Author with Mary Grass and Superstition Mountain Golf & Country Club guests at Silver King Cemetery 1999 — Author's photo

Charles P. Mason's *Tales of Silver King*

The next recollections come from the unpublished manuscript of Charles P. Mason, nephew of Charles G. Mason, one of the original Silver King Mine owners.

The Shotgun and the Play

"The young people used to have a pretty good time in those days in spite of the roughness, at least the boys did. I remember one time a traveling troupe got stranded in town. But they all got out but a man and his wife and a poodle dog. So we thought we would give a show to help them on their way. Of course, the man and his wife took part, and Hinton Thomas was to be the villain. I was to be the hero and rescue the girl. I asked Charlie Rapp to load the shotgun for me with a cap and a little powder. Everything went smoothly until the last half of the last scene when the rescue was to take place. 'Villain', I cried, as I rushed onto the stage and blazed away just over Thomas' head. There was a terrific report. That gun kicked me clean out into the audience; the lights went out, and then a spot about two feet wide in the curtain blazed up just back of where Thomas stood. If I had aimed at him I would have killed him sure, for that rascal, Rapp, must have filled the shotgun half full of powder. Well, that ended the show!"

Bootleg Liquor

"I remember one saloon keeper at Silver King, a little wiry man, as he sat on a block of wood in the back yard, directing the making of his special brand of whiskey. Every now and then this man would take a sip of the concoction which his helper was stirring in the old wash tub. Stick in another plug of tobacco, Juan, it needs more kick."[14]

Merchant Days—Reminicences by Perry Wildman

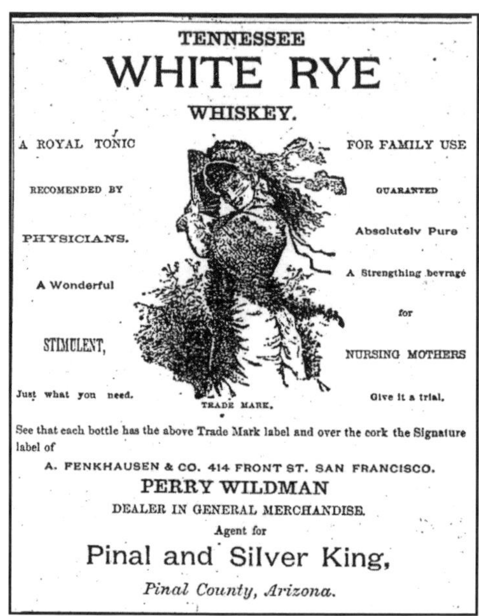

Perry Wildman Whiskey ad 1880s
— Pinal Drill

About 1883, Perry Wildman arrived at Silver King. He was to become one of the leading merchants of Silver King and Pinal. In recalling his days at Silver King, he relates one of his first days there:

"The inhabitants of the town were mostly Cornish, but there was a sprinkling of almost every other nationality. The whole town had been awaiting my arrival and the opening of the new store. When the freight teams had unloaded their wagons, and before I could open up the hundreds of boxes of merchandise and arrange them on shelves, many of the people came into the store. They wanted to know if I had this or that article, as they were in need of many things and did not want to purchase them elsewhere. My sales of that day were the largest of any day succeeding and there was at once established between myself and the community a feeling of good will that resulted in the success of the new enterprise."[15]

Mail Service to the Silver King

"There was at that time no telegraph or telephone service and being 55 miles or so from the railroad we were quite an isolated community. Ed Thompson of Thompson & Bowen, the largest saloon keeper was acting as postmaster, and when the mail arrived the sack was emptied on one end of the bar and we would pick out our own mail." The army first delivered the mail in those days, when patrols came that way, and then it was by pack mule or burro, often referred to as "jack-ass mail."

The mail came by way of the San Carlos Reservation, often delivered by Indians, who received 50 cents per letter for return mail. Later the mail came by stage when the road was improved up to the King."[16]

The Flood of 1883

"In August 1883, we had one rainstorm such as I desire never again to see. About four o'clock in the afternoon two little clouds appeared, and about a half-hour later they had assumed tremendous proportions. Water poured out of the clouds, not in drops, but bucketfuls, creating a perfect torrent that came down the side of the mountain with a roar like that of an airplane of today. Large boulders, five or more feet in diameter, rolled down the ravine like pebbles. An immense boulder lodged at the culvert of the road leading to the mine, turning water down Main Street. A miner in his cabin was caught by the torrent before he could escape to higher ground, and his body was afterward recovered about 10 miles down below. Houses were flooded and furniture of all descriptions were carried off. My store building was flooded with about 2 feet of water. About 20 rolls of barbed wire fencing—were carried out of the building—when we went down the ravine after the flood had subsided we found the wire a tangled mass stretching from bush to bush along the banks for miles."[17]

The *Pinal Drill* on August 18, 1883, reported:

"In Pinal Joe Lombard's house was swept away, walls and all, and that the Harrisses, the Brandenburg's and the Gossess, who lived near the creek, had their houses and goods swept away. The Chinaman below town lost 200 hens, his pigs, his houses, and was carried away for about half a mile." It also reported that on the creek where the Mexicans lived was complete devastation.[18] The *Pinal Drill* on August 25, 1883 reported this additional information on the missing men mentioned above, in the flood. "Mr. McCaffrey and a Mr. Donnelly were killed in the flood, Donnelly was found 15 miles down stream, and McCaffrey found between Silver King and Pinal."[19]

Wood for the Silver King

Wildman continues with this tale about how wood was provided for the Silver King Mine.

"About that time the business of freighting or forwarding goods had grown to considerable dimensions. I had about 300 burros and mules occupied in hauling wood to the mine, employing about 100 Mexicans, who with their families made quite a contingent of the population."[20] Wood was the main fuel for operating the mine and mills at Silver King and Pinal. There were two wood camps, one at Pinal City and another 11 miles north of the Silver King. Joe Champion operated the large lumber camp at Pinal, and lumber was hauled on large wagons to the mills." [21]

Tom Buchanan, also a contractor for wood, cut and packed it 15 miles south to the Silver King and Pinal. Joe Phy was also a wood contractor for the King and had a contract to provide 1000 cords of wood in six months. Joe Phy later became a deputy under Sheriff Pete Gabriel and was involved in one of Arizona's most famous shoot-outs in 1888.

Silver King wood cutters cir. 1880s — Bowen Family Collection

On Tuesday November 5, 2002, I took a 4-wheel trip up past Superior, and then on FS Road 342, which leads upwards past Fortuna Peak, and by the entrance to Haunted Canyon. Jack Carlson accompanied me on this trip. We proceeded east on State Route 60 and turned left at Oak Flats, on FS 342. We drove past the OMYA Marble Mine on FS342, where there are some steep, rough places on the road. Onward we drove to the old fence line, where Delbert Toney accidentally shot and killed himself, about 1915. Driving on to the Fortuna Peak overlook we stopped where the elevation was about 5293 feet. This is a great viewpoint that overlooks the southern part

Author at Silver King wood camp— Jack Carlson photo

of the Superstitions, and the main Superstition Mountain to the west. To the southeast there is a great view of the Pinto Valley Mine from here. We drove on to the entrance to the Haunted Canyon Trail. At this location, the trail leads downhill to a small corral and a spring about ½ mile away.

Next, we drove to FS 650, which leads to the entrance to the Wood Camp of the Silver King Mine days. The main old stone building is still there, with a roof and doors, although the smaller side buildings are now only stone ruins. The Herron Ranch used this old stone building in the past as a line shack during round-ups. We explored an old mine across the wash from the old stone house, and then drove south, out past the Cottonwood Corral."

Dr. William Wardwell's Letters from Silver King

Dr. William L. Wardwell, a young man about 29 years old, who came west for treatment of tuberculosis, spent 8 months at Silver King during 1885-1886. He wrote several letters to his parents back east. Dr.Wardwell's letters are an excellent first person account of life in the western frontier, and he describes several graphic incidents. He stood in for Dr. Kennaird, the mine company doctor, on occasion.

The *Pinal County Record* on February 19, 1886 states

> "A Mexican boy accidentally shot himself while fooling around with a revolver, at Mr. N. Nicholas, last Saturday morning. The ball went in on the cheek and pursued a downward course and came out on the side of his neck and back. Dr. Wardwell was called, during the absence of Dr. Kenniard, and dressed the wound, and the patient is now getting along nicely. The doctor says it was a mere scratch, and that the wound did not prove fatal."[22]

I wonder what the doctors of that time thought to be a serious wound?

The following exerpts from his letters are his impressions, written soon after they usually occurred, and published with the approval of his great-neice Clarissa Pell and the Arizona Historical Society, Tucson, Arizona.

Housing at Silver King

> "The town has about 400 population and a little mining camp, hung out up in the mountains. The houses are perched irregularly, around whenever enough space could be found to place them. Everything is uphill or downhill. The floor of the next house above ours is on a level with the ground, while from our front door 22 steps lead down to the street. The Silver King Mine is not

100 yards away. First at our side is a huge orehouse, into which the ore is dumped, and from there transferred to the wagons."[23]

Dr. Kenniard, the Mine Doctor

"Tommy, is the doctor for both places (Silver King & Pinal). He attends the mine and the mill hands by contract, and has besides, some outside practice. I think he is doing well. He drives to Pinal several times a week, and is often sent to go down at night. I came up intending to take his practice and let him go off on a vacation, but a night call to Pinal would entail so much of a journey, especially if it were raining, that I do not feel like undertaking it. I don't believe I can enter the practice for some time to come, on account of the night work."[24]

The Brave Stagecoach Drivers and the Highwaymen

"You know I wrote in my last letter that I was somewhat startled on the way up here by some men that looked like highwaymen. Well, on Saturday night three nights later, the stage was held up by a robber and the Wells Fargo box taken. There was nothing in it; however, in a box marked hardware was $700 in silver, bound for the Silver King Mine, but it was not disturbed. Two men were on the stage at the time, each armed with huge six-shooters, and according to the driver they had been boasting on the way up about what they would do if they were held up. When the time came, they didn't do it, but submitted tamely as possible."[25]

Regarding Indians

"Most of the statements made in the papers regarding the Indians I know to be false. The latest report is that the 'hostiles' wish to surrender. Every one of them is a murderer, many times over. They had no excuse to go out on the warpath. They had a large portion of the country in a state of terror for many months. If they are caught or surrender, they should be killed, every one of them. It will only be justice and it will soon save a repetition of these troubles. Eastern sympathy will prevent justice being done to these murderers, if their causes are in any way considered."[26]

Grubstaking Prospectors

"Generally you have nothing to show for money, time and labor. The way Tommy (Dr. Kenniard) does it is this: someone approaches him with accounts of a good claim and offers him a 1/4 interest. Tommy goes out to see it. He likes it's looks. He takes some ore samples and has them assayed. If it turns out well, he tells the other men to dig down 15 or 20 feet and he will put up the grub. They do so, get cash, twenty-five or thirty dollars. At a depth of 15 feet another assay is made. This one is not so good; at 20 feet, another assay; the ore is very poor. Tommy stops the grub. The men stop work. Tommy is out $215."[27]

The Toothache

Wardwell's letter dated February 15, 1886 describes the following humorous tale of tooth pulling. "The first day I went down there (Pinal), there were two or three men waiting for me in the drugstore, and as I went in I was immediately accosted by 'say Doc I wish you'd pull a tooth out for me.' I looked at the tooth and then told him I would put the forceps on the tooth, but they must get some strong man to pull it, as I didn't want to exert myself so much as would be necessary in extracting so large a tooth. He said, 'All right' and started out after some friends to help him through. A few minutes later my man with the bad tooth came in accompanied by his two friends, one a great stalwart muscular fellow, the other smaller. The men were presented to me as Tom and Bill. Tom, the bigger man was to do the yanking. The little man was to hold the patient's head. I placed the victim in a chair. Bill held his head. The crowd gathered around to see the fun and to cheer up the man with jocular remarks. I placed the forceps on the tooth, Tom, trembling slightly, took hold of them. I said, 'Now, when I give the word, you pull as hard as you can.' In a moment I said 'ready,' Tom yanked, and the tooth didn't budge. 'Pull harder,'said I. He pulled harder, not a move from the tooth and a sort of breathless silence had been observed in the crowd. Someone broke the quiet by saying something about a mule tooth. That set the crowd to laughing. Tom stopped to wipe the perspiration from his brow and then set to work again. 'Gracious' he said, and then he braced himself, set his teeth together, put faith in all of his enormous strength and in a moment he was lying on the floor and the patient and Bill were on their backs in the other direction. The tooth was out, the patient neither sighed or groaned. The tooth was passed around. It was commented upon. Reminiscences of tooth pulling were indulged in and I immediately set out to see some other patients."[28]

The Evil Wind

"The other day I went to see a Mexican baby who was 'mucho sick,' and no interpreter handy, so one was sent for. Every Mexican believes the source of all trouble is the wind. Wind in the head, wind in the feet, wind in the stomach, wind in the back. The root of all evil is the wind, and it is hard to convince them that it is impossible that any other cause of 'misery' exists. Now I am willing to give wind its due, and attribute to it a fair place in the creation of evil. But that a toothache should be caused by wind in the tooth, or a headache by wind in the head, or a leg ache by wind in the leg, or a heartache by wind in the heart, is giving the wind a prominence to which it is not entitled."[29]

This ends the Wardwell Letters.

O'Boyle-Willams Feud

Despite Perry Wildman's recollections of a peaceful Silver KingTown, as he describes those days, there were several incidents of violence, as in any mining town. According to Wildman:

"We were a peaceful, contented, happy people. Silver King was as quiet and orderly as one could wish. Each man was a law unto himself and anything unruly was not tolerated. We never locked our doors, and robbery was a thing unknown. Gambling, of course was wide open, but always on the square."

According to the *Arizona Weekly Enterprise,* April 16, 1883:

"Silver King had two hotels, one run by Bob Williams, and one run by Bill O'Boyle. They seldom patronized the opposition, because there was more or less jealousy between them, which finally resulted in a shooting match between the two men. Neither was seriously injured, though Bob Williams had to take his meals standing up for some time"

Silver King Hotel (white building) cir. 1880s — Bowen Family Collection

It was said that the fight originated over a waitress at Williams Hotel, who O'Boyle was paying too much attention to the chagrin of Bob Williams.[30]

Havestrom-McCaffrey Shooting

The *Arizona Weekley Enterprise* On September 5, 1885 reported a shootout at Silver King between John Haverstrom and James McCaffrey which resulted in the death of one of the participants. A quarrel broke out between the two men and they settled it with their guns blazing. The paper reported that Col. Robert Williams with characteristic generosity had a handsome coffin made for the deceased and rendered him a christian burial.

Silver King Town 1996

In 1996, only rubble, old rusted mining equipment and tailings existed where a king of a mine and a thriving community once stood. There are some remnants of buildings as you enter the old townsite of Silver King. The once majestic mining office with its captain's walk on top of the roof and elegant staircase is gone. No sign of the original huge ore building exists. There were, however, several huge ore dumps and tailings everywhere. When I first located the Silver King there was not only debris and broken glass at every step, but gloom and silence. It reminded me of a bombed out area after a war. It was difficult to imagine that amidst all this desolation and junk a bustling mining camp once thrived.

Silver King ghost town 1996 — *Author's photo*

The Story of Pinal City

After the establishment of the Silver King Mine, the mining company decided that the mills should be set up near available water and chose the area on Queen Creek near Picket Post Mountain for their operation. In 1877, the Silver King Company purchased the cattle ranch of L. D. Arnett and began setting up a mill.

The *Pinal Drill* on May 28, 1881 describes the founding of Pinal:

> "At the time there was but one house occupied, that by a Mr. Arnett, who was looking after his cows and goats. So as soon as the King commenced operations, other buildings were erected, and what had been an almost unknown valley in the foothills of the Pinal Range of mountains, frequented only by the occasional prospecting party or the wild Apache Indian, became a thriving village."[31]

At this time, the company mill at Pinal had 20 stamps that were kept constantly running. The mill town was located in Arnett Canyon on Queen Creek, which had been peviously located by William D. Arnett.

Pinal City 1880s — Bowen Family Collection

The location was ideal—only five miles from the mine and next to ample water. People began to flock to the area, and in 1878 the village of Picket Post was established near the mill on Queen Creek. The first post office was established on April 10, 1878, with William W. Benson as Postmaster. The name of Pinal was officially established on June 27, 1879. [32] In 1879, the first public school was built in Pinal. One teacher was hired and began with 50 pupils in attendance, and with the increasing population, a new schoolhouse was erected. The Methodists erected a church where services were regularly well attended. Soon, Lodges of Odd Fellows and Masonic Orders were established, as well as the Knights of Pythias.

The Arizona Business Directory of 1881 had this to say about Pinal.

> "Besides Pioneer, are the Summit and Mineral Districts, contributory to this place; also the neighboring villages of Silver King and Queen City. These with the many rich mines, the favorable location for residence and business, the salubrity of the climate, and other favorable conditions, justify the bright expectations of its citizens, that Pinal is destined in a short time to become one of the leading cities of Arizona. The elevation being about 3500 feet above the sea, tempers the summer weather to a delightful degree, and in winter brings it in the region of occasional snow."[33]

The Town of Picket Post Becomes Pinal City

In 1879, the name of Picket Post was officially changed to Pinal City.

Pinal City 1880s. Apache Leap seen in background.
— Bowen Family Collection

(The name Pinal means "*place of pines*" and is the name of the mountain range lying between Superior and Globe.) Pinal thrived, homes were built and streets were constructed. The *Pinal Drill* newspaper was begun under the direction of J.D. Reymert, Editor. Saloons and gambling places were erected to provide entertainment for grizzled miners. On Queen Creek a stamp mill was erected. The *Business Directory of Arizona for 1881* listed a blacksmith, butcher shop, a baker, a shoemaker, a drug store, a liquor store, two hotels, a drill company, four general stores and three saloons.

From Rags to Riches—George W. P. Hunt
Pinal's Saloon Owner and Future Governor

George W.P. Hunt was one of the early saloon owners at Pinal. They say he came to Pinal City with nothing but the "shirt on his back." The *Pinal Drill Newspaper* on October 30, 1880, advertised Hunt's Saloon as: " I have found it!—Hunt's Saloon—It is the place to find the finest whiskies and cigars." Hunt must have owned the saloon at Pinal during the boom years of the Silver King Mine, because I located another quip in the *Pinal County Record* on December 18, 1885. It stated: "If you want fine mixed or unmixed drinks, call at the saloon of that unsurpassed mixologist, Joe Hunt and quench your thirst." Hunt parleyed his savvy with the people into political

George W.P. Hunt cir 1880s
— Gila County Historical Society

strengths and would later become Arizona's only seven-time Governor. He was too busy to marry early as most men did in those times. But, he did marry beautiful Helen Duett Ellison aften a ten-year courtship when he was 44 years old. She was the daughter of Jesse and Susan Ellison, owners of the famous "Q" Ranch, about 20 miles east of Pleasant Valley (Young, Arizona), and aunt of famous Arizona author Slim Ellison, who wrote of his aunt in his book, *Cowboys Under the Mogollon Rim.* After Helen Duett Hunt's death in 1931, Govenor Hunt built a pyramid shaped tomb for her as a sign of his devotion. He died at age 75 in 1934 and was buried with his wife in the pyramid, near today's Phoenix Zoo.

Cold Beer for the Boys

If you wondered if they had cold beer in those days, they probably did at most places. The Globe Ice Company advertised continuously in the *Pinal Drill* that they had ice available, made from pure mountain water, through the process invented by S. D. Lount. The Ice Company stated that they would furnish ice to any of the camps and neighboring towns for a moderate price. This probably kept the miners happy![34] Pinal was described as follows in 1881 by one of the freighters.

Early ice plant cir. 1890s. — Gila County Historical Society

"Pinal had all the trimmings of the Arizona early-day town: saloons, "red-light" district, stores, hotels and residences; some of the *lowest* characters on earth, and some of the *finest*. Probably under no other conditions do the extremes meet, as in a mining camp. Alike attracted by the wealth that is always to be found in a prosperous mining district, the college graduate and the most illiterate, the multi-millionaire and the poorest rub shoulders in one common chase—the chase for the almighty dollar. And yet, all is not democratic, either. Socially, were to be found all the adornments of the most select society, snobs that would make royalty seem like pikers and 'Our Crowd' everywhere, distinctly separating the sheep from the goats. The General Manager and Superintendents and their crowd are the Four Hundred, and I assure you there is no stepping over!"[35]

There was a little song about the "Red Light Ladies" popular among the miners in the 1880s. The first verse went like this:

First came the miners
To work in the mines,
Then came the ladies
Who lived on the line.

The town had really grown by the year 1883. The *Business Directory of Arizona* listed 34 businesses, including 5 general stores, a confectionery, a brewery, 2 livery and feed stables, a lumber store, a stove and tinware store, a ladies notions store, and of course 8 saloons along with the saloons in the two hotels. Included in the prominent men of the city were Aaron Mason, Superintendent of the Silver King Mine, George L. Miller, Postmaster, and R. E. Kennedy, Deputy Sheriff of Pinal County. With the opening of the Seventy-Six Mill on June 28, 1878 a party was held where the crowd consumed 30 gallons of Bell's best beer. And on July 9th of that year, when M. A. Baldwin, Boss of the Pinal Mill and Mining Company, began pounding ore at his mill under contract for the Wanna Whata Silver Mining Company, the town had 20 stamps in operation. Depending on the source, the town had between 2500 and 3000 occupants during its peak. Some estimates put it as large as 5000. Pinal actually was one of the largest towns in Arizona, but when the ore at the Silver King Mine began "pinching out" in the late 1880s, the town began to "fade away" and almost as fast as they came, the inhabitants began moving away.[36]

The Pinal County Bank opened its doors in 1881. During this time, Pinal had about 200 buildings, with many made of adobe bricks. The Pinal Bank Building was one of the exceptions as it was made of stone. There were 4 doctors, 6 lawyers and a dancing school. In 1881, there was much excitement with talk of the railroad coming to Pinal and Silver King. Additionally, there was a Wells Fargo office, a post office, newspaper, (*Pinal Drill),* assay offices, two large hotels, half a dozen restaurants, two drug stores, a watchmaker, photograph gallery, brickyard, lumberyard, two blacksmith and wagoners' shops, livery stables, general stores, two barber shops, bath rooms, six lawyers, four doctors and a preacher. There were also two breweries. The United States Brewery was in a building 143x38 feet. It and the Pinal City Brewery were both located on Pearl Street. Whole streets have been built and suburbs are becoming covered with houses. "All is life, energy and progress."[37] There is much evidence in the way of numerous old broken beer bottles left behind at all the building sites at both of the mining towns. Everywhere I looked around the building ruins I found the thick, broken brown glass bottles, remnants of the early miners drinking days.

Pinal Bank cir. 1890s. — Gladys Walker Collection

Fatty Perkins—-The Last Resident of Pinal City

After the demise of the Silver King Mine Pinal City soon became a ghost town, and it's buildings became part of other boomtowns or settlements. Only a single man remained at Pinal, Fatty Perkins, who was partner to a rancher named Mike McGrew in the cattle business. Fatty was commonly referred to as the "Mayor of Pinal." He lived in the old bank building and corralled his horses in the old dance hall. It was said that in the ravine behind his place of abode there was a pile of champagne bottles great enough to completely fill-up a large saloon and the empty beer bottles were too many to count. Pinal had certainly been a great boom town in it's day, but 16 or 17 years after the boom was over (which was about 1888) McGrew was supposed to have hauled most of the town away. First went the doors and windows, then went the timbers, then all the other usable parts of the buildings. Some of the bricks from Pinal were said to have found their way to Tempe, where they became other brick houses. It was said to be a country on hard times and Fatty Perkins earned his keep just by staying in Pinal to keep the buildings he occupied from disappearing. Fatty cooked for the cowboys at round-up time and kept them well fed. The rest of the time he just looked after things at old Pinal. For Fatty's efforts, he was allowed to brand two hundred calves.[38]

Tales from Pinal City

From the recollections of Perry Wildman, comes the following story

"In 1885 I bought the business of George Miller at Pinal. I had my hands full of business and lived at Pinal. I remember once a drunken fellow had taken an overcoat belonging to someone else was arrested and brought to trial. He stated he had simply borrowed the coat as it was cold and he wanted to go up to the King. The jury was evenly decided as to guilty or not and in order to settle it two of the jurors were selected to play a game of 'seven' to decide the question. Not guilty won the game, but Judge Benson nevertheless had the fellow sent to the Florence Jail for 60 days and he left the country. No undesirable citizens were wanted in Pinal as well at the King." [39]

The Cost of Living in 1881 at Pinal

George L. Miller & Co., a General Merchandise store in Pinal, advertised these prices for their wares in the August 16, 1881 *Pinal Drill:*

Having reduced all their prices for Cash to
The lowest figures. From amongst their immense
Stock the following are quoted as
Sample prices

Haydens Family Flour----------$0.04 cts.
Y. Powder -------------------59 cents lb.
White sugar----------------6 lbs. for $1.00
C. R. Coffee---------------4 lbs. for $1.00
Climax Tobacco---------------75 cents lb.
Lard---------------------------15 cents lb.
Bacon-------------------------18 cents lb.
Ex. Tea-----------------------50 cents lb.
H & I Axel Grease-----------50 cents lb.
Rice----------------------8 lbs. for $1.00
Coal Oil-------------------$6.75 per case

Daily Life at Pinal

In order to give a flavor of the life at Pinal in the 1880s, the following quips were taken from the *Pinal Drill* September 9, 1882:

"-Go to the Tunnel for a game of pool
-William Binkley's baby died this week
-Mrs. Summers is visiting this mountain sanitarium for the benefit of her health

-The Gila is rising
-The Court Calendar for October is remarkably small
-Sheriff Gabriel paid a flying visit to Pinal, this week, on a tax collecting tour
-The Territorial and county taxes of Pinal County have been fixed at $2.60 on the $100
-Another arrival in town this week, it was a boy, and the mother is Mrs. Rogers
-The public school will open on the 18th, under charge of Mrs. W. A. Hall
-Fresh fruit received daily at the Pomona Fruit Store
-Our Florence friends don't like the idea of the Pinal Const. Co building the road direct to the railroad
-The Grand Hotel has been leased to David Whitehead, who will have it renovated
-We see from the Phoenix papers that a large quantity of fruit is rotting in that neighborhood for want of a market, they should send it up here"

The Smallpox Scare of 1883

The *Pinal Drill* on September 15, 1883 describes the Smallpox scare that came to Pinal and Silver King:

"This malady is prevalent in various places around us. At Tempe it is severe, they have it at Phoenix, and we are told at Florence. Pinal, Silver King, Hastings and our part of the country is yet free from it, and we hope our citizens will take prompt precautions and preventative measures." The paper even prescribed a homeopathic remedy for prevention. "As the smallpox, being so near, might take a fancy to visit up here, we will urge upon all the importance of keeping off the unwelcome visitor. In the first place, let all eat they can of the best and most relished food. Beware of constipation. Also, let *all* buy half a pound of cream of tartar, none but the best. It is acid, has no other taste whatever, and will all dissolve, when put in sufficient water. Put it in a pitcher, cover it with water, stir, let water settle and drink of it whenever thirsty. Continue it. It cleanses the blood, cools it, and is a deadly enemy of smallpox."[40]

Another smallpox preventative of the humorous variety was published in the *Pinal Drill* on September 22, 1883:

"The following *recipe* for this dangerous malady was found amongst the papers of the late Chief Physician of the King of Siam at Bangkok. "Keep your feet warm, your head cool, your bowels open, your conscience clear, and drink three times a day a glass of Boca Beer. This medicine may be had at the Tunnel Dispensary."

The Quarantine Squad

The *Pinal Drill* also had this article about a quarantine squad.

> "Information having been sent from Florence that a person with smallpox had gone to Pinal, and in view of the fact that smallpox is also at Mesa City our citizens elected Geo. L. Mill and Jake Suter to the permanent committee on Sanitary Affairs—the committee reported on a suitable place for a hospital—measures were taken to *quarantine any sick person, on the roads, before reaching Pinal.*"[41]

Boys on burros cir. 1880s. — *Gladys Walker Collection*

These measures of quarantine may seem severe, but consider the medical experience and expertise of the 1880s and it was reasonable of prudent men who wanted to keep a deadly epidemic from getting started in their community.

Pinal and Silver King get Electric Lights

We take our electricity and electric lights for granted, but living in Pinal and Silver King in the 1880s was almost like living in the dark. The pioneer settlers of Pinal used coal oil lamps for light before electricity. The *Pinal Drill* on June 16, 1883 stated:

> "T. F. Crovise, representing the Brush Electric Light Co., Cleveland, Ohio, will arrive in Pinal about the time of our going to press. He comes to arrange and estimate the cost of erecting electric lights at Pinal and Silver King for the Silver King Company's mill and mine, as well as for other folks." Silver King was the *first* place in Arizona to get electric lights. The Silver King also established a direct line to the San Fransisco stock exchange in order to keep stockholders and potential stockholders aware of the progress and production of the mine. Another article in the same date's paper discusses portable electric lamps. "E. Reymert has ordered a lot of portable electric lamps; they will arrive next week. They are used exactly as other lamps, without pipes,

the chemicals will last three months and can then be renewed as easily as oil; the lamps never explode. They are cheap also."[42]

Boca Beer Saves Owner's Favorite Dog

On a lighter note in that day's paper, the following amusing article was found: "What might have easily been a tragedy happened at the Tunnel Saloon this week. Erich's handsome dog incautiously approached too near to the refrigerator and was frozen solid by the intense cold. Half a bottle of Boca Beer was forced down its throat, and it immediately revived, but a few minutes more and it would have been too late." Apparently the Tunnel Saloon was a favorite of the editor, as well a many townsfolk, and Boca Beer, a strong English beer, was a favorite of the miners as both made the papers on many occasions.

The Orientals at Pinal

The Orientals were said to be a thrifty and enterprising people. However,

Tunnel Saloon Pinal City cir. 1880s. — Pinal County Drill

they were not allowed to work in the mines, so they opened up restaurants and laundries. They did very well at these enterprises and saved their money. They did not believe in banks and were said to have buried their silver coins until they had an opportunity to return to their homeland.

A Search for Chinese Coins Turns Deadly

During the depression years of the 1930s at Superior many people were unemployed, and only men with seniority were kept on the job. Therefore, it was a hopeful Christobal Rivera who went in search of the old tale of buried Chinese coins one day. He began digging in a sandbank in Old Pinal City one afternoon. With him was his friend Concha Frias. The two searchers located a place where they thought was a likely place to look. Suddenly, the dirt embankment collapsed, and Christobel was buried alive. Concha tried to keep his face out of the sand but could not do so. She frantically went in search of help. But, by the time she located help and got back to the cave-in, Christobel was dead. Christobel left a son named Chris, who was raised by Mrs. Ramona Rivera along with the rest of her children. He grew up in Superior and raised a large family of his own.[43]

Chinese Chop House ad 1886
— Pinal County Record

A Shooting in the Chinese Block

There were Chinese sections at both Pinal and Silver King and the *Pinal Drill* of July 16, 1881, tells the following story regarding a shooting:

"Qui Gee shot Sue Gee, all in a general row at the Chinese Gambling House on the 13th. Dang Fook and Wang Wy ran away, and the police are after them. Sue Gee was shot in the mouth and will probably die. Go Ghu is dead. Qui Gee is under arrest. Dang Fook and Wang Wy, are likely to be caught, and will be returned to Florence, where they came from a short time ago. Dr. Bluett says that Sue Gee will not die. He is fixing a wire, ironclad, fire proof jaw on him, so as to secure the safety of Sue's masticating machinery hereafter." Apparently, the fireproof jaw did not work, because the *Pinal Drill* on July 23, 1881 stated that Sue Gee had died. The August 6, 1881 *Pinal Drill* has a certain "Qua Kee (Qui Gee) is held for the murder of Ah Chew (Go Ghu) and was admitted to bail for $1,000, upon a writ of habeas Corpus before Probate Judge Wratten. P.A. Brown Esq., Counsel for Qua Kee (Qui Gee), has proven himself a second Burlingame."[44]

On August 27, 1881, the *Pinal Drill* reported another murder on the Chinese Block.

"Last Thursday morning about 10 o'clock, the town was again thrown into great excitement by the report of another Chinese murder. Two persons, a China woman and a girl lay dead in pools of their own blood." Apparently what happened in this lengthy article was that a certain Jim Mason killed the mother and daughter out of revenge. Jim had bought the daughter for $400, and the family decided not to turn her over to him. Jim went to their hut with a revolver, and finding the two women smoking opium with a white man, shot the women. Jim was then apparently shot and killed by Wong Kong in a fight for the revolver."

I was unable to ascertain from the papers if any arrest took place, or if Wong Kong was turned loose for self- defense reasons. Since there was no mention of this incident in subsequent papers this may have been the outcome. It must also be noted that all of the 1881 papers were not in the archives.[45] Many Orientals worked and lived in Pinal City. In one section of the ruins during the 1960s, large quantities of opium bottles and clay pipes were found, as well as some Chinese coins.[46]

The Saga of Shoot-Em-Up Dick

One of the prominent citizens of Pinal's Chinese colony was Jim Sam, the owner of the Bank Exchange Restaurant and Chop House. One day a self asserted bad man, known as Shoot-em-up Dick, wearing two six-guns and a Bowie knife, swaggered into Jim's restaurant and ordered the best meal in the house. After eating, he demanded the best cigar available and started walking out the door without paying. Withdrawing a big six-shooter from under the counter, Jim asked the bad man if he forgot something. "No, you damned yellow heathen, I didn't forget nawthing, I was your guest. I am Shoot-em-up Dick!" "Oh so?" said Jim, meanwhile pointing his gun at the other man's belt buckle. "So-you Shoot-em-up Dick, I Shoot-em-down Sam! You pay pletty damn soon, or Shoot-em-up Dick be pletty damn dead." According to the often-told tale, Dick paid.

Shoot-em-up Dick's real name was Richard Barnes and he was arrested in April 1882, for being a horse thief. His accomplice at that time was the infamous Billy the Kid No. 2, an Arizona outlaw, who likened himself to the original Billy the Kid. Dick was released on bail, but Billy was jailed. He was in jail in Globe when a lynch mob stormed the jail and hanged two killers named Grimes and Hawley. Billy thought the mob was going to take him too. He told the jailer later that he thought he was a goner! Dick did not like what was happening, jumped bail, and official records reveal that he was never apprehended again. Perhaps things were getting too hot for him in Arizona Territory then! Billy served time in the Yuma Prison, escaped, was caught and served the rest of his time without incident. He was released on October 24, 1887. Billy's real name was never known.[47]

Shoot-em-up Dick did surface again during my research. *Recollections of an early resident of Superior* by a Mr. McPherson, relates this story about Dick.

> "In the fall of 1913, I worked in the store for E.F. Kellner. Many amusing things happened. One night a man posing as a 'hard boiled hombre,' with the title of Shoot-em-up Dick, rushed into the store moaning and said, 'for God's sake boys I've been shot and am dying.' The boys in the store ran for Doctor Homes, who was at that time physician for the Magma Copper Company. After pulling his shirt open, the doctor found one small bird shot on the fleshy part of Dick's arm. Dick was wounded by some man whom he had sent word that he (Dick) was going to kill. The shooting happened on the road opposite the Ice Plant."[48]

Some Tough Guy, Huh? Not all the old timers were tough; some just had or tried to have tough reputations.

Shoot-em-up Dick met his end in the Chiricahua Mountains, year unknown. It seems that he stole a pair of work mules from a certain Major Downing who lived in Pinery Canyon. Major Downing, then 77 years old, took off after the thief. He stated that he followed the trail for several miles and at last caught up to his mules tied up to a tree. He said that nearby, fast asleep, lay Shoot-em-up Dick. The Major stated he then took possession of Dick's guns, woke him up, and then meted out swift justice to the offender. The sheriff followed Major Downing back to the scene of the capture, and there was Dick hanging by the neck from a tree. The sheriff's posse buried Dick at this site. The Major was commended for ridding the country of an undesirable. No arrest was ever made by the law.[49]

Attempt Shooting at Shamp's Kitchen

This quip comes from the *Pinal Drill*, February 26, 1881: "At Pinal, Walter Harvey tried to shoot Frank Valenzuela in Shamp's Kitchen on Sunday last, but the trigger would not work, and Frank's legs helped him to get away from his marksman. Harvey was arrested and held in jail." It seems that Frank was a better runner than Walter was at taking care of his six-shooters.[50]

The Pinal Cemetery

After locating the Silver King Cemetery, I located the Pinal City Cemetery, after some difficulty, as it is about mid-way between the Silver King Mine and Pinal City site. I located about 75-85 gravesites there. Most of the headstones are long gone, or the writing is so weather worn that the names are not readable. Several of the miners and millworkers are buried there, including their family members. The childrens graves are recognizable by the small pile of stones that mark the graves. Typhoid and other fevers of the 1880s took the lives of many of the Pinal residents, as did some 19th century ailment referred to as consumption. Consumption was probably tuberculosis. Some victims of violent deaths are also buried there. Besides Mattie Blaylock's grave, few are identifiable. See the story of Mattie Blaylock (Wyatt Earp's second wife) in the chapter on Pioneer Women. The Mexican population used this cemetery until about 1916. Maneulita Guzman, who died in 1914, is buried here, and her descendants still maintain her grave. Elueteria Lujan is there. She died in Superior at age 32 in 1916. All of the readable headstones were of women who died in their 30s moot testimony to the rigors of a frontier life, when women died before their time, many in childbirth.

Mrs. H. E. Reese Tombstone. Pinal Cemetery — *Author's photo*

From the annals of the *Pinal Drill* comes this story about a strange death at Silver King. When an unknown man died in a hotel room at Silver King on February12, 1881, his body was taken to the Pinal Cemetery for burial. The remains were "decently placed in a coffin." Before being placed in the grave, the body "showned signs of life." Dr. H. H. Davis was immediately summoned from Pinal to examine the body. Doctor Davis, after examination at the gravesite pronounced the man "oficially dead." The coffin was again closed, and "consigned with its contents to the grave."[51] I was unable to locate any follow-up information in subsequent papers, and I wonder if this uknown man was *really* dead or maybe perhaps in some type of coma unknown to the doctors of the 1800s.

As I walked along in silence at the long deserted cemetery, I came across the headstone of Mrs. H. E. Reese, who was born in Copenhagen, Denmark, and died at Pinal in 1880. Her husband had a headstone placed there that reads, " Blessed are the dead that die in the Lord." There does not appear to be any "Boot Hill" for Pinal or Silver King. The old cemetery is on a hill, off County Route Number 8, and overlooks State Route 60 and Picket Post Mountain.

The Demise of Pinal City

Where it was once a common site to see the huge ore wagons rumbling down the mountain road to Pinal and hear the pounding away, day and night, of the stamp mills and the boisterous talk and laughter of the miners, today it is silent and empty. Now only jackrabbits and rattlesnakes inhabit where one of Arizona's most prosperous towns once stood. All the buildings are gone. Even the stone building of the Pinal Bank is gone, where Fatty Perkins once lived. Of the stamp mills, only a few iron support rods and one stone retaining wall remain. You can still see some remnants of the house foundations and some tin remains from the stove and tinsmith shop. The large wagon ruts remain that lead to the mills, but all else is gone.

All is hushed and still now at Pinal, only the wind makes noise where once the mills pounded away and huge ore wagons rumbled. What was thought to be the beginning of a great Arizona town is truly gone with the wind. Pinal City remains only in the scattered writings and memoirs of those who have gone before us!

The Little Known Town of Queen

During the year 1880, some very rich silver ore was taken out of the Silver Queen Mine. But the deeper the miners went into the hillside, the more copper they hit and the less silver. It was called "rebellious." Copper in those days was considered a liability, not an asset. Horses and mules provided the only transportation and it was necessary for the miners to live close to their work. The towns of Pinal and Silver King, which had grown up for that very reason, were each about five miles by road from the Silver Queen. This was too far for the miners to travel twice a day in those days, so the natural thing to do was to establish a mining camp near the mine. Apparently the little settlement near the Silver Queen was called "Queen." The *Pinal Drill* on March 12, 1881 stated:

"Queen City is growing fast. Restaurants, Creveau's and Deutsches' boarding houses, Charley Miller's store, Faylor and Parker's saloon, F. Wentworth's new store, numerous tents, the new Gem Mill, private residences, Mr. Deutsches' new gardens. All betokens life and prosperity. There are the ladies and children, and the fresh flowers coming out after the spring rains, to make Queen City a lovely place for the hardy industrious miners, and a charming home for them all."[52]

The mining camp of Queen was located at the present site of Superior almost completely surrounded by mountains. The Queen post office was established there on April 21, 1881. Queen had a population of about 100 people, a general store, of course saloons, a boarding house, a restaurant and two hotels, which amounted to a total of about 20 buildings, plus the traditional miners shacks and tents. At that time, in the 1880s the creek ran year round providing water for the residents.[53] The *Pinal Drill* on April 9, 1881 writes this about Queen:

"Noticeable here is Charles Miller's store of general merchandise, Faylor & Parker's Saloon, William Deutch's boarding house, also Creveau's, and a restaurant that would be a credit to any town in the way of cookery. The large new store building of F. Wentworth will be soon filled with goods. They expect soon to have a post office, and to secure the appointment of a Notary Public. We found W. H. Merrit of Florence, with his assistants, surveying the Lewis mine. Some of the adjacent mines are the Gem gold mine, Supriser Gold Mine, Silver Queen, Windsor, Lewis, Wana Whata, Copper Top, Webfoot, Erie, Two Brother, Hayes, Eureka, Silver Chariot, Belcher, Geo. Wahinton, Babe, Black Prince, Manhatten, and others too numerous to mention as these hills contain a vast network of minerals, veins, lodes and ledges."[54]

Queen, anticipating the future importance of its location, predicted that Queen Creek Canyon would some day become a main thoroughfare. Today State Highway 60 winds up along the rim of the canyon, linking Globe and Superior, but Queen has long since vanished.[55] Will C. Barnes in *Arizona Place Names*, listed Queen Canyon as:

"Located in section T. 1.N., R. 11 E. After Silver Queen, first mine located in Pioneer District near Globe. Stream flows southwest and is lost in desert near Higley, Maricopa Co. Thompson Arboretum is located on this creek, four miles below Superior. The listing by Barnes for Queen states: "Pinal Co.-Station on Queen Canyon, East of Florence, near Silver King, probably the old Silver Queen mine. Queen was listedin the *Arizona Gazetteer* of 1881, as Queen City Post Office, 31 miles northeast of Florence and 3 miles east of Pinal, at the mouth of Queen Creek Canyon. The post office was established April 21, 1881 with Charles Miller, P.M."[56]

According to *Ghost Towns of Arizona,* by James E. & Barbara Sherman,

"Queen contained about a hundred inhabitants, a general store, saloons, a boardinghouse, a restaurant, and two hotels—a total of twenty buildings and a number of tents. An important feature was the Gem Mill. The historic bluff, over which a band of Apaches leaped to their death while being pursued by Captain J. D. Walker's militia and Indian allies in the 1870, towers above the old site of Queen. Residents of the camp stated that they found human skulls and bones resulting from this mass annihilation in the early 1880s."

The Town of Hastings

In 1882 the present day town of Superior was called Hastings. Hastings was an early miner there and had a 20-stamp amalgamation mill established alongside Queen Creek where the Magma Clubhouse was later built. Thus the Pioneer District in 1882 had two major mines, three thriving little towns, and hundreds of prospectors trying to imitate the big bonanzas.[57] The Hastings mill would not operate because it was set up to handle gold, and gold was not the metal that would be the bonanza of this area. Silver was the precious metal that would make this area a mining giant during its time.[58] *Arizona Place Names* by Will Barnes in 1935, listed Hastings as follows: "Pinal Co. Found on old map, 1882, of the Pioneer Mining District on Queen Creek. First name of Superior q.v."

Directory of Businesses
Silver King & Picket Post/Pinal
(1879-1892)

PICKET POST (Pinal)		**SILVER KING**
	(1879)	
Guindani J. & Co. - Gen'l Store Miller, G. I. & Co. - Gen'l Store		
	(Jan. 1880)	
Guindani, J. & Co. - Gen'l Store Miller, G. L. & C0. - Gen'l Store		
	(Jan. 1881)	
Pinal (Formerly - Picket Post) Allen, T. F. - Blacksmith Binkley, W. T. - Gen'l Store Bluett, W. H. - Drugs Brooks, J. - Gen'l Store Caveness, Matt & C. - Blacksmiths Gardner, Hiram - Barber Goldman, & Co. Gen'l Store		**Silver King** Buckalew & Ochoa- Gen'l Store Thompson & Brown - Gen'l Store Williams R. - Hotel & Saloon

Graham, & Wright - Liquor & Varieties
Henry & Johnsee - Saloon
Hunt, J. B. - Saloon
Miller, G. L. & Co. - Gen'l Store
Murray, H. B. - Saloon
Nichols & Benson - Butchers
Passama, A. - Baker
Reymert & Co. - Drill
Schmidt, Henry - Shoemaker
Shamp, Gus - Hotel
Willey, W. H. - Hotel

(April 1882)

PINAL
Arnett, W. L. - Hotel
Becher, Gus - Brewer
Berthier & Barter - Saloon
Binkley, W. T. - Gen'l Store
Bluett, W. H. - Drugs
Brinkman, Mrs. D. - Notions
Brook, J. - Gen'l Store
Brunckant & Hilge - Bakers

SILVER KING
Bowen & Jones - Saloon
Bucklew & Ochoa - Gen'l Store
Ellis, Aron & Co. - Gen'l Store
Ludekn J. - Fruits & C.
Mayer, T. - Butcher
O' Boyle, W. C. - Hotel & Saloon
Williams, R. - Hotel & Saloon
Young, R. - News Depot

(April 1882 Cont.)

PINAL
Cavaness, Matt & Co. - Blacksmiths
Champion, Josiah - Lumber
Ellis, Aaron & Co. - Gen'l Store
Everhart E. - Drugs
Goldman & Co. - Gen'l Store
Gomez, F. - Grocer
Graham, P. B. - Saloon
Hall, & Hurley - Livery
Harter & Lewis - Saloon
Hunt, J. B. - Saloon
Hutchinson, W. T. - Blacksmith
Kimball, S. F. - Livery
Loeffler, & Co. - Gen'l Store
Lutger, J. H. - Saloon
Miller, G. L. & Co. - Gen'l Store
Murray, C. F. - Saloon
Palmer & Rowe - Hotel
Passamoua - Baker
Pinal Co.- Bank, E. W. Hopkins, Pres.
C. M. Gilmore, Cash.
Preaso, Carlo - Saloon
Reymert, J. D. & Co. - Drill
Schmidt, Henry - Shoemaker

SILVER KING

Suter, J. - Stoves & Tinware
Warnke, E. F. - Brewer
Wentworth, F. G. - Saloon
Wittrock, Henry - Grocer
Wurch, Michael - Saloon

(July, 1883)

PINAL
Arnett, W. L. - Hotel
Bocher, Gus - Brewer
Brinkman, Mrs. D. - Notions
Brooks, J. - Gen'l Store
Brunenkant & Hilge - Bakers
Cavaness Bros. - Saloon
Lumer- Saloon
Champion, Josiah -
Ellis, Aaron & Co. - Gen'l Store
Everhart, W. - Drugs
Goldman & Co. - Gen'l Store
Gomez, F. - Grocer
Hall, B. W. - Livery
Harter, M. W. - Saloon
Harter & Lewis - Saloon
Hunt, J. B. - Saloon
Hutchinson, W. T. - Blacksmith
Jensen, F. - Fruits, Confec. & C
Jensen & Werner - Brewery
Kucht, E. - Saloon
Martin & Goldman - Blacksmiths
Miller, G. L. & Co. - Gen'l Store
Murray, C. F. - Saloon
Pinal Co. Bank, E. W. Hopkins, Pres.
A. Venton, Cash.
Preaso, Carlo - Saloon
Reymert, J. D. & Co. - Drill
Schmidt, Henry - Shoemaker
Stanfield, Thomas - Feedstable
Suter, J. - Stoves & Tinware
Warnke, E. F. - Brewer
Wartelle & Walworth - Gen'l Store
Whitehead, David - Grand Hotel
Wittrock, Henry - Grocer
Wurch, Michael - Saloon

SILVER KING
Mills, Aaron & Co. - Gen'l Store
Jones & Jones - Saloon
Luedka, Bros. - Gro. Fruits & O
Mavor, F. - Butcher
O'Boyle, W. C. - Hotel & Saloon
Thompson & McQueen - Hotel &

Wildman, Perry - Gen'l Store

(July 1887)

PINAL
Brunnenkraut, C. - Baker
Champion, Josiah - Lumber
Drais, L. K. - Hotel

SILVER KING
Bennett, Charles F. - Saloon
Brown, J. H. - Cigs. Tob. & Saloon
Dryden, A. H. & Co. - Hotel

Early, W. R. - Paints, Oils & O
Jensen, F. - Brewery & Saloon
Notions,
Kuohl, Erich - Saloon
Luedke Bros. - Gen'l Store & Hotel
Marlow & Co. - Butchers
Martin, R. H. - Freighter
Reymert, Mrs. Eliza - D & F
Schmidt, Henry - Shoemaker
Silver King Mining Co., (Incorp.)
Steffy, William - Saloon
Suter, J. - Stoves & Tinware
White, Fredrick - Furniture
Wildman, Perry - Gen'l Store

Gurtisen, Maz - Saloon
Havestrom, J. - Shoemaker,

Hawley, Louisa A. - Gen'l Mdse.
Marlow & Co. - Butchers
Smith, Charles A. - Saloon
Wildman, Perry - Gen'l Store

(July 1889)

PINAL
Brunenkrant, C. - Baker
Culver & Roemer - Freighters
Howser, F. - Gen'l Mdse.
Ludeke Bros. - Gen'l Store
Marlow & Co. - Butchers
Nicholas, Theophile - Butcher & Cattle
Reymert, Eliza - Notions & Confec.
Schmidt, Henry - Shoemaker
Silver King Mining Co. (Incorp.)
Steffy, William - Saloon
Suter, J. - Stoves & Tinware
Werner, Aug. - Brewer
White, Frederick E. - Furniture

SILVER KING
Bennett, Charles F. - Saloon
Haverstrom, J. - Shoemaker
Hawley, Louisa A. - Gen'l Mdse.
Knight & Curry - Gen'l Mdse.
McNeil & Hammond - Gen'l Mdse.
Marlow & Co. - Butchers
Smith, Charles A. - Saloon

(July 1890)

PINAL
Brunenkrant, C. - Baker
Howser, F. G. - Gen'l Mdse.
Luedke, J. - Jewler
Nicholas, Theophile - Butcher & Cattle
Schmidt, Henry - Shoemaker
Silver King Mining Co. (Incorp.)
Werner, Aug. - Brewer
White, Frederick - Furniture

SILVER KING
Hawley, Louisa A. - Gen'l Store
Knight & Curry - Gen'l Mdse.
McNeil & Co. - Gen'l Mdse.
(see Wildman & Co., Florence)

(July 1891)

PINAL
Nicholas, Theophile - Butcher & Cattle
Schmidt, Henry - Shoemaker
Werner, Aug. - Brewer

SILVER KING
Kincaid, Maggie - Hotel
Knight, John - Gen'l Mdse.
McNeil & Co. - Gen'l Mdse.
Williams, R. - Hotel & Saloon

(July 1892)

PINAL	**SILVER KING**
(None Listed)	Knight, John - Gen'l Store &
Saloon	Williams,R. - Hotel & Saloon

Frank Major, Silver King merchant 1880s
—Bowen Family Collection

Chapter Eight
Pioneer Women of Silver King and Pinal

The picture that appears in my mind when I think of those pioneer women of the mining camps is one of a sturdy yet gracious woman. She follows her man into the wilderness of the old west with their brood of children, fear ridden with the unknown dangers that surround them. The charm and radiance of pure womanhood shines and her benign influence affects all that come into contact with her. They were brave, these mothers and daughters of our pioneers, brave in venture and brave in their conduct and devout in their Christian religion.

The women of the old west were deeply respected by the men of the "wild and woolly" west. Lest anyone attempt to defame, by word or deed, the character of these pioneer mothers and daughters, a dozen men were eager to beat him down. It was part of the *Code of the West*. For these men were very protective of all women, as if they were their own mothers and daughters. This fact alone was a guarantee for their protection from insult or injury.

The following poem, reflects the deep reverence and esteem held for the pioneer mothers:

Like the rose in its blooming, shy bashful, and innocent,
When full blown its sweet fragrance and charm
Reflect a character of love, honor and purity
And, when its velvet petals fall to mingle with the earth,
It's beautiful life leaves pleasant and cherished memories.
Author unknown[1]

Valiant Women

No one has written the full story of the many valiant women who followed their men into the west. They came and made new homes and raised families under intolerable conditions. Children were born regularly into large families, and one or more children, it seems, would die by the second summer. But no matter how many children were born, each was precious to the mother, and none more than those they lost all too soon. From the Atlantic states and the Mid-west states they came, across the endless prairies, beyond the Rockies, on their way to Oregon or California. More than a few even made their way to Arizona Territory.

The West owes much to these women who endured cruel hardships that now seem intolerable. The crawling ox teams, the buffalo chips for fuel, a child that must be held to keep it safe, sleeping out in the prairies while keeping watch for Indians and outlaws. They did not boast of courage, as they scarcely knew they had it. They only knew they must do what it took to survive. It took real courage for these pioneer wives, mothers and sisters to face child-birth in the wilderness, after each long day of travel, and make do with what there was available to feed their families. These women trekked with ox teams that sometimes moved only 10 miles a day. They spent endless weeks sleeping in the wagons, cooking over an open fire, washing their clothes by streams when available and walking all day beside the wagon when the roads were rough.
While the men built homes, mines, saloons, stores and gambling halls, the women sought and prepared food for the table, demanded schools for their children and attended to their religious upbringing. They made their own clothes by weaving their own cloth or obtained some from a traveling peddler. They made their own soap, and washed their family's clothes, as available water would permit it. The bravery with which they faced childbirth in the wilderness, often without medical aid of any kind, speaks of their bravery and endurance. It was the women who usually brought books and civilization into western mining camps. The Bible always, and a few other books, some poetry and often a medical or doctor's book were carried. They nursed their sick and injured as best as they could with few remedies. Always came the fevers or smallpox, gunshots, and sometimes arrow wounds, broken bones, snake bites and unknown ailments.

Size had little to do with courage. Endurance and calmness in time of trouble was a more endearing quality of these women. They accepted their role and were proud of it. They had to endure the loss of babies who either died during childbirth or within the first years of life. In mining camps, the cemeteries usually had more tiny graves than those of grown people. The long forgotten cemetery of Pinal above State Route 60 and Picket Post Mountain contain many of these small graves. Even though the markers are long gone, you can easily tell the children's graves by the small ring of stones.

Mines Named for Women

The esteem for which the early miners held their women was the fact that they named many of their mines after their wives, sweethearts or other favorite women. Some of these are as follows:

Agnes	Alice	Annie	Augusta	Cleopatra
Cora	Dianna	Emma	Emmaline	Florence
Hattie	Irene	Isabella	Jenny	June
Kitty	Lizzie	Lulu	Maggie	Nellie

And many, many more.

Women Arrive at the Silver King

Silver King dance card cir. 1880s — *Sam Michael collection*

The following article describes how the men felt about the arrival of women in the mining camps. The *Pinal Drill,* October 16, 1880, stated:

"The Boys Hop at Silver King. Not long ago, and the memory of man runneth not to the contrary, we had stag dances. Nothing but tramping on the dirty ground, with hats on, spiked boots, and horney hands clinched in the embrace of men, circling in a whirlpool of excitement to the rasping screech of a hoarse old fiddle. But now a days what have we come to, or rather what has come to us? Ladies - those sweet temporizers of men's passions, those finer creatures, at whose look the rude become polite, the aggressive passive, and for whose pleasure men assume the best attire, and ornament their mien with bland decorum. We have ladies to charm the dance, the festive board, the gala day, and to crown the pleasures of our gatherings. The boys of Silver King did themselves honor—Over 50 dancing couples were present and a large audience besides, all in elegant attire. The music was excellent. We might describe the ladies; but dare not attempt it. Splendid dresses, excellent dancers. The magic of their presence filled the hall with delight. All were happy."[2]

Dancing—A Favorite Activity at Pinal and Silver King

Dancing was a favorite activity at both places. When there was no wooden floor available, the people would stretch a canvas tent across the hard dirt floor, and wax and polish to such a degree that it would equal the wooden floors of today. After the schools were built, there were many dances and "hops," as they were often called, written about in the papers of the day. A short quip in the *Pinal County Record*, on October 16, 1885, stated: "The dancing school at Silver King, under the charge of Prof. Niles is getting along nicely. He has about fifteen pupils." The *Pinal County Record* on February 19, 1886 states: "The Calico Hop at Silver King, Tuesday evening was largely attended, and the neckties created considerable amusement."

The School Marm

In a short quip about the number of school children, the *Pinal Drill* on October 7, 1882, describes the Pinal schoolmarm. "There are now 64 children in the public school. The number will increase as their efficient teacher becomes more appreciated. Mrs. Hall confirms the saying that teachers are born, not made. She has the talent, the knowledge and the proper disposition, kind but with a due appreciation of the rod."[3]

Gossip From Pinal Surprise Mine, Pinal, A.T. September 21st, 1875 The Wedding

The following is a copy of a letter to the editor of the *Arizona Miner* in 1875. The letter is signed simply, Pop.

> "Editor Miner: I have just returned from a trip to Florence, and have had the pleasure of recording another "hop." On Thursday the 16th inst., at the residence of the bridegroom, on the Gila River, near Florence, Judge Levi Ruggles married Mr. William Long to Miss Mollie Whitlow, of Marysville, Salt River Valley. This is I believe, the first American wedding that ever took place in this part of the Territory. With the bride, who looked very pretty, I have not had the pleasure of an acquaintance; with the bridegroom I have been acquainted for several years, and I must say that there is not a more worthy and honorable young man in the Territory, and I think Miss Mollie has been fortunate in getting a splendid young man for a husband, of whom I have no doubt she is entirely worthy, for she appears to be a really nice young lady. May the happy pair enjoy long life, prosperity, honor, happiness—everything; may he prove a noble husband, and she a noble wife

who will love him long and well, and never think he is too Long. Mr. Long's wealth goes far into the thousands; and now may the mine of love which the fair young bride bestows on him have no end. Love in a silver casket is a very nice thing. What more is necessary to complete this picture?

"The wedding supper was splendid. I am told by those who participated that it was the best table they ever sat down to in Arizona. Wish somebody would get married every day and ask me to the dinner. The festivities ended with a grand ball, given by the bridegroom, in the evening at the Florence schoolhouse and from candle lighting to 4 o'clock in the morning everything went lovely as a marriage bell. The mine is looking well. Lyinan, Nash & Co. are about to erect their new 10-stamp quartz mill if they can find a suitable mill-site. Success to them. They certainly will do well here."

The Unsinkable "Molly" Long-Bailey—Reminiscences of Mary E. Bailey

Mary Bailey was born Mary Elizabeth Whitlow, of the Whitlow family cattle ranchers. She was called "Molly" by her family and the town of Marysville, Arizona was named after her. She was reputed to be the first white baby born in the Phoenix area. Her first husband was William Long, one of the founders of the Silver King Mine. Long died of smallpox shortly thereafter. She then married in succession Walter Bailey, who died, then his brother William Bailey, who also died then Jules Harte, who also died not long after their marriage. Molly lived a long life, however, and died at the age of 81 in 1939.

Mary E. Whitlow Bailey cir. 1880s — Whitlow Family/Greg Davis Collection

According to Molly: "The music had your modern swing (dance music of the 1930s and 40s) beat a hundred miles. We had harps, guitars, fiddles and piano, and the lovely Mexican and American waltzes, fox trots and other numbers—were beautiful." She said attendance at these dances was made up of the towns, and the Silver King miners and cowboys would ride as far as 30 miles to be present.[4]

Pinalino Women go-a-Hunting

Women even went on sightseeing and hunting expeditions with the men on occasion. The *Pinal Drill* of September 8, 1883 writes of such an outing.

> "On Thursday of last week, many Pinalinos, namely Matt Caveness, Miss Likes, M. Brown, Mrs. Doanes, W. A. Hall, and Miss Maggie Hall, F. Czarnowski, L. Hall and Mr. Manlove, started for a trip through the mountains. They visited Roger's District, Reavis Ranch, the Salt River, Globe, the Pinal Mountains, Pinal Ranch, and on last Tuesday they returned home sunburnt but happy. They ate quail, venison, pigeons and wild turkeys killed by the Nimrods of the party, and on the Salt River they found that feasting on fresh fish was not an Arizona fiction. The gentlemen of the party said that the ladies were 'bricks,' for they never complained of fatigue, thirst or hunger or sunburn, but enjoyed themselves to the utmost, and traveled on the steepest trails. The ladies declared that the gentlemen were the perfection of kindness, did all the cooking, dishwashing, hunting and general managing—Mr. Reavis accompanied them to Salt River through places unknown but to himself, in order to show them the scenery. Mr. Reavis spent the night in their camp and alternately delighted and made them shiver with accounts of old times, bear, and Indian adventures. Altogether the days of that jaunt will long be remembered by them all."[5]

Matt Caveness, in his reminiscence years later, gives a vivid description of this excursion, with accounts that substantiate the story above.

Elisha Reavis the Vegetable Peddler

Elisha Reavis moved into his remote area in the early 1870s after being a packer for the army at Ft. McDowell. He was the first Anglo-American settler in the vast Superstition Mountains. He was one of the Superstition Mountain's first Anglo settlers, arriving about 1865, and had a farm high in the mountains of eastern Superstitions. His farm was located at about the 4800 foot level, about 4 miles north of Iron Mountain and one and one-half miles north of Mound Mountain, near a running stream. He homesteaded about 200 acres, which consisted of 35 cultivated acres. Bearskins carpeted his cabin and deer antlers covered his walls, testimony of his prowess as a hunter. He made his living cultivating vegetables in a high, desert mountain garden. Once or twice a month he hauled his produce to the markets of Central Arizona where it commanded a premium price. He took his vegetables to the towns of Silver King, Pinal and Florence, especially when the Silver King mine was in boom. In fact, he registered as a citizen of Silver King on the Great Arizona Census. It was said that he could make about $5,000 per growing season.

Elisha Reavis cir. 1886
— Superstition Mountain Historical Society

Elisha Reavis was truly a wild a woolly looking character. He had long unkempt hair, a long scraggly beard and wore the same clothes until they practically disintegrated. Regardless of his looks, the residents at Silver King and Pinal welcomed his arrival, and women watched the trail in search of the old hermit, his dogs and burro train, for they knew fresh vegetables were soon to be had. His cabbages were said to be the size of small barrels. Reavis loaded his vegetables on his burros and traveled over the trails from his farm high in the mountains to sell his welcome goods at Pinal City and Silver King. Since fresh vegetables were difficult to obtain he was always welcome at the mining camps. He was considered an eccentric backwoods hermit by most that knew him. He traveled over a route towards the Silver King from Iron Mountain. This old trail goes past Rogers Canyon and is still called the Reavis Trail.

The Blonde Blue Eyed Lady

Mrs. Thomas Thompson arrived in Silver King as a new bride in the late 1880s. Her husband was a millwright for the mine. She had blue eyes and blonde hair. Many of the friendly Indians who came to the mine came to her house to peek in the window at the Blonde Blue Eyed Lady. Blondes and blue eyed women were rare in the early west. On one occasion an Indian who could speak good English came by her house to see her and her baby, who had red hair. The Indian knocked at the door and said he wanted to see the baby. Mrs. Thompson said it was time to feed the baby and he could not come in then. He said, "I have already been in to see the baby. Look at the back of the chair where the baby was. I put a *mark* on the back of the chair." Sure enough, the mark was there! [6]

The Last Stage from Silver King

Another story from Mrs. Thompson tells of the mine closing. The mine manager called her husband in and told him that the mine was shutting down, and the last stage was leaving in two hours. He went home and told his wife, who hurriedly packed what belongings she could. The Thompson's and their three boys made that last stage just in time. She stated that they left in such a hurry that they packed only their clothes and a few things and even left the dirty dishes from their last meal on the table. The Thompsons left Silver King behind them and took many various stagecoaches until they arrived at Jackson, California where her husband Thomas worked in the mines.[7]

Tales of Two Pioneer Women

The next two stories are about two women who came to the Silver King-Pinal Area. The first story is about a woman who traveled west with a wagon train and lived near Pinal and Silver King for a long time. A nearby mountain is named for her family. The other story is about a woman who came to Pinal to live out what would be the last years of a rather short life. She was the second wife of one of the most famous lawmen in the old west. No mountain or valley or stream is named for her. For many years she lay in an obscure grave in the old Pinal Cemetery.

Mrs. Robert A. Irion—Elizabeth Irvine

Elizabeth Irvine was born November 30, 1842 in Allegheny County, Pennsylvania and was the only daughter, with five brothers, in the family of John and Mary Boyd Irvine. Her parents were Scotch-Irish, who as soon as they were married set sail for America, arriving in New York on May 24, 1834. Thus the spirit of the pioneers was born in Elizabeth. The family moved to Sandusky, Ohio and from there began their westward journey. The country was in a fearful depression preceding the Civil War, and the family started out for Colorado in search of a new life in the west. Elizabeth was left behind at a boarding school but did not remain there long. She married Archibald Craig on September 3, 1863 and went with him to Kansas. On October 3, 1865, her young husband died, leaving her with a seven-month old son, Dudley Craig. Her brother John came to her rescue, and in the spring they bought a wagon and joined a wagon train going to Colorado. She then married Robert A. Irion on March 7, 1868.

The next ten years in Colorado were hard times. But in 1877, Aaron Mason, Superintendent of the Silver King Mine, who was a friend of the Irion's, induced them to come to Arizona with a brother of theirs. Mason told them of a beautiful little mountain dell on the trail between Silver King and Globe that they might take possession of. This place was to become Pinal Ranch. Knowing they could start afresh in a new land, and Robert could get immediate employment at the Silver King Mine, they came to Arizona Territory to seek their fortune. The following paragraphs are some of the recollections of Elizabeth Irion during her wagon trip from Colorado to the Silver King, as taken from her unpublished journal in the Geraldine Craig file at the Arizona Historical Society in Tucson.

> "Whenever it was possible the members of the wagon train observed the Sabbath Day, but many times when water was so far distant, they had to travel, for with the cattle, fourteen miles a day was about as far as they could go. With the possible exception of two or three nights, the men had to take turns guarding the horses and cattle to keep them from being stolen. With their cows they had milk for their coffee. They also had a small coop containing chickens. When the travelers were camped, the coop was opened and the chickens were allowed to roam close by. The fowl gave them occasional eggs and a treat in the form of a chicken dinner, from time to time.

Map of Iron's trek 1877 — Helen B. Craig Collection

On June 8th, they were snowed in on the Raton Pass, the division line between Colorado and New Mexico. She mentions on June 25th that it is 40 miles to the next water, and when we realize that they could only average 14 miles a day with the cattle, we can appreciate what it means. They arrived at Puenta de Agua on June 28, and she recalls that a woman and child there had smallpox, but they were not too concerned since they had been vaccinated earlier on the trip as a precaution. "We found a pretty place to camp but suffered terribly with the dust. The wind has been in the right direction to blow the dust in our faces ever since we started from home. It is worse than the heat, cold or rain or anything we have encountered. It is perfectly awful."

They arrive at Ft. Craig, New Mexico by July 15, where the soldiers had an encounter with the Indians and killed 6 of them. By August 7th, they are at the San Carlos Reservation and the journal continues:

> "We are within 30 miles of Globe City and I am anxious to get off this reservation. These Indians are miserable looking creatures; very filthy. They dress in the height of Apache fashion." She describe the dress of the Apache male as a breechcloth and moccasins. The Apache female she describes as wearing a skirt and nude from the waist up, with a great amount of beads on the arms, neck, and hair. She continues: "We eat in the wagons to try to keep away from them, but they crowd around us so, it makes me sick. I always feel something clutching around my heart, although I know they are harmless. You cannot help but feel sorry for the poor, dirty, ignorant, degraded creatures. I imagine though, how terribly a person would feel to hear 10 or 20 give the warwhoop unexpectedly in some canyon."

The next day August 8th they arrived at Globe City, and she says: "I must say this is not much of a place." Elizabeth and her sister-in-law Anna stayed at Globe City, while Robert Irion and Solon Mason rode over to the Silver King. On the 13th they started again on the last lap of their tedious journey. She recalls that the Silver King Mine was only 28 miles by trail on horse or mule, but 120 miles by wagon. This was by way of El Capitan Mountain and the Gila River, then across the hills to Florence, then back toward Globe for about 25 miles to Pinal where the mill of the Silver King was located. They arrived there on August 24th. The last page of her journal reads:

> "The Superintendent Mason kindly offered us a company house; it is not very fine but we are thankful to get under a roof again. We have been 12 weeks

> and two days on the road. We have been unwell for a few days and we have lost 4 head of cattle but otherwise we have great reason to thank His Holy Name for the many blessings."

The cattle arrived soon after, and the men all got employment at the mill or mine. Robert Irion went to work as night foreman at the mill, while Elizabeth kept a boarding house with Robert's sister Anna. On January 23, they moved to take charge of Mason's ranch near Florence. While there they secured a half interest in the Pinal Ranch and moved there on August 14, 1878. Elizabeth described Pinal Ranch as a "temple in the mountains. Surrounded by high peaks, this little valley of about 40 acres of fertile soil is covered with heavy oak timber, making it a striking contrast to the barren and boulder strewn cliffs surrounding it."

Elizabeth was a devout Christian and the Bible was read daily in her home. This is an example of her character, when we realize that at that time the frontiersmen seldom observed Sabbath worship. She was a pioneer in the true sense of the word, refined, and educated, yet with that wonderful strength of character. She was able to bravely face the perils and dangers that confronted her and give her best towards the settlement of the west. Iron Mountain in Pinal County was named after the Irion Family.[8] The Irion's built a comfortable house and many a weary traveler would speak of the hospitality enjoyed there. They acquired the other half interest in Pinal Ranch and raised a large herd of cattle, residing there until 1896, when they moved to Tempe. The *Arizona Silver Belt* on October 11, 1884, stated: "Thomas Buchanan has sold his Pinal Ranch to Robert Irion and will hereafter reside at Silver King."

The Tidiest of Housewives

This tale comes from the *Engineering and Mining Journal* of 1882.

> "Beyond the Silver King, a trail leads over the mountain. It would be more rewarding if viewed from the back of a horse than one of Mr. Schultz's wretched mules. Up the steep sides, and over the jagged edge of the basin, the trail (Stoneman Trail) winds into a shallow ravine. A steep ridge again separates this from Devils Canyon. Suddenly, the path emerges from this gloomy defile into a very paradise, where spreading oaks shade the meadows, and corn, whose feathery tops you can hardly reach from your mules back, grows in profusion. In this oasis we found a ranch, and in the ranch the 'tidiest of housewives,' who set before us on immaculate table linen, such savory soup and meat-pasty, rice pudding and fresh milk. And in her parlor,

we find as excellent food for our higher faculties; for on the shelves were books of stern Calvinistic theology, but beside them in a most judicious assortment of history, essays, and poetry. Evidently there has penetrated into these recesses, other refinement than the rough and ready sort, which we can not recognize in the miner and herdsman, bred of fellowship in hardship and danger, and freedom and fresh air, a refinement which makes these rough men hospitable and sensitive to your wants and feelings, however uncouth in their modes of expression. But here, in this farmer's home, we see the contrast in the influence between mining and agricultural pursuits."[9]

This place was Pinal Ranch and the lady was Mrs. Elizabeth Irion.

Celia Ann "Mattie" Blaylock

Many people probably have read of Mattie Blaylock, second wife of the West's most famous lawman, Wyatt Earp. But, those same people probably don't know that she came to live in Pinal City after leaving Tombstone. As I was searching for the old Pinal Cemetery several years ago, I located the old cemetery, and came across a wooden monument with a plaque and photo of Mattie. The Guzman family, an old mining family, whose mother is also buried there, placed the memorial at the cemetery. This is a tribute to Mattie, as a pioneer woman of the west who fell upon hard times after her man left her. This poem was on the plaque:

Celia Blaylock Earp

From within the swirling dust devil
of the elf owl's cactus home
for abandoned dreams at the Silver King
first shattered in Tombstone
she does not haunt the desert
in judgment nor in blame
Celia Blaylock Mattie Earp
seeks justice to her name

From her sixteen-year old bright star reach
to the laudanum whisky glass
she will tell you how a life can be
reduced to a tintyped past
depicting almost moments of
a wife who failed to lend
credence to the legendary
"honor" among men

If there be a time hereafter
when troubled souls might rest
free from careless consequence
of weakness of the flesh
I do not think it out of place
to invoke to the dry rock earth
God's rest to the yet departing soul of
Celia Blaylock Earp

Michael Papaianni

On the side of the 4x4 holding the memorial to Mattie was another small plaque. It read: "Special thanks to Manual Guzman, III, whose devotion to his grandmother resting nearby, helps maintain the awareness of the burial site of Celia Blaylock Earp, and allows us to honor the memory of 'Mattie,' Michael Papaianni, July 1995." Since that date vandals have removed or otherwise damaged the memorial, and this poem at the Old Pinal Cemetery was re-dedicated on a brass plaque on the Main Street of Superior, in front of the old Magma Hotel.

Mattie Blaylock's grave. Pinal Cemetery — *Author's photo*

Celia Ann " Mattie " Blaylock Earp came to Tombstone as Wyatt Earp's wife on December 1, 1879. She met Wyatt at Ft. Scott, Kansas, sometime after 1870. Wyatt and Mattie had a stormy relationship according to historians, and he left Mattie for a young, seductive Josephine Sarah "Sadie" Marcus. "Sadie" had been Sheriff Johnny Behan's girl, but they had a falling out. The fact that Wyatt left her was a great cause of distress to Mattie. Allie Earp thought a lot of Mattie and said this about her. " She was as fine a girl as ever lived. She worked like a slave. Stuck with him (Wyatt) through thick and thin and was there every minute."[10]

After the famous O.K. Corral shoot-out in Tombstone, the Earps all left town. Mattie was sent by Wyatt to his family home in Colton, California, with Allie and Virgil Earp. Mattie could not stay in Tombstone, as she did not fit in with the upper crust of the mining camp ladies. In fact, she was snubbed by them, as they did to all the wives of gamblers and people who worked in saloons. Mattie could not stay in California either, as she pined away for Wyatt. She soon left Colton and went back to Arizona. She came to Globe where she lived for a time with her old Tombstone friend "Big Nose Kate," Doc Holiday's former girlfriend. When Wyatt visited Globe with Sadie it was just too much for her. Unable to stand seeing her former husband who left her for another, albeit younger woman, Mattie moved on to Pinal City where she lived and worked as a prostitute.

Mary K. "Big Nose" Harony cir. 1880s. — Gila County Historical Society

At Pinal, she maintained her friendship with "*Big Nose Kate*" of nearby Globe. Kate's real name was Mary Katherine Harony. Kate was also known as Mrs. "Doc" Holiday, Kate Fisher, Kate Elder, Kate Melvin and Mary K. Cummings. In Tombstone, Kate supposedly set up a house of ill repute in a tent. Kate ran a boarding house in Globe and married George Cummings in 1888, after Doc Holiday's death. Kate managed hotels and brothels throughout the various mining camps of Arizona from 1888 to 1897. Kate was a housekeeper for several years, then went to live in the Arizona Pioneer's Home in Prescott where she died at 89 years old in 1940, a week short of her 90th birthday. Her tombstone reads Mary K. Cummings. Kate detested Wyatt Earp and Doc's relationship with Wyatt, as well as for abandoning Mattie.

There is a story that Kate was buried at the Silver King Cemetery. However this story was conjured up by her friends, to bury Big Nose Kate Harony, while Mary K. Cummings was appealing to Governor George W.P. Hunt for admission in the Pioneer's home. It seems that there was resistance by some respectable women of Arizona, who did not want a former brothel keeper to have the benefits of living with honorable women. A ceremony was held at the Silver King Cemetery and a grave was dug and filled in and a headstone placed there. However, someone has long since taken Kate's headstone, as well as many others that were there.

Mattie had been a social outcast at Tombstone because of her husband's affiliation with the gamblers and the Oriental Saloon. Mattie had been hurt by the snubs, so she knew that she had to leave Tombstone. While at Pinal, she was visited by Phin Clanton at Pinal. The Clantons were archenemies of the Earps. These mining and milling towns were all alike and Mattie found Pinal very much like Tombstone. At all times of the day the huge ore wagons could come into town making clouds of dust. She could hear the mill stamps at Pinal pounding away day and night. Nightlife at Pinal was similar to Tombstone, with the sounds of music blaring from the saloons, and women and men laughing and dancing into the night.

Pinal apparently became a nightmare for Mattie compared to the social life with all the other Earp women. It was said she was always half drunk from opium and laudanum (which contains opium), trying to forget the man who left her. Six years after she arrived at Pinal, in the spring of 1888, she told acquaintances that she could take it no longer and wanted to end her life. She was drinking heavily all the time now, and could not face the tintypes of her and Wyatt that she always kept on her dresser. On Monday July 2, 1888, her friend 65 year old Frank Beeler looked in on her. "I wish you'd go to Werner's and get some whiskey for me," she said. "And I'd like to have you go to Ludke's and get me a little more laudanum so I could sleep." Beeler went out and bought her some whiskey. "Go get the laudanum" she begged, "I want to try and get some sleep." Beeler went back and bought some laudanum and gave her about 16 drops and some whiskey. When a friend called on her about 8:30 P. M. that night, she had no pulse. Dr. Thomas A. Kenniard, of Silver King and Pinal, stated on the death report that cause of death was suicide. Judge W. H. Benson, who took deposition and ruled her death a suicide, held a Coroner's Inquest on July 4th. At the Inquest a witness testified that Mattie said "Earp had wrecked her life by deserting her and she didn't want to live."

On Saturday, July 7, the *Arizona Enterprise* reported her death as follows:

"Mattie Earp a frail denizen of Pinal, culminated a big spree by taking a big dose of laudanum, on Tuesday, and died from its effects. She was buried on the 4th."

Mattie's belongings were sent back to Mrs. Sarah Blaylock in Fairfax, Iowa. In the accompanying letter written by Judge Benson, he wrote: "Mattie had been deserted by her husband, and killed herself."

Stuart Lake, Wyatt's official biographer, *never* acknowledged the existence of Mattie, but portrayed Wyatt Earp as the personification of a frontier Marshall.

It almost seems fitting that Mattie Earp is resting at the old Pinal Cemetery, the ghost graveyard of an old ghost mining town. So ends the sordid tale of a legendary hero who deserted his wife in favor of a younger woman and caused the demise of an otherwise faithful good wife and person. This real life story reads like a Greek tragedy, with a sad ending in a lonely forlorn place like an a long abandoned cemetery on an old dirt road, where the picket fences are designed to keep out the intruding coyotes, and whose headstones have either been vandalized or removed.

Celia Ann "Mattie" Blaylock deserves to be remembered. Perhaps some Hollywood writer or author will do her story as a sequel to the books or movies about Wyatt Earp and Tombstone. The title could read *Mattie -A Frontier Wife Hidden by History* or *The Resurrection Of Mattie!*[11]

Dr. Thomas Kenniard, Silver King doctor cir. 1880s —Bureau of Mines photo

Chapter Nine
A Pioneer Family—the Bowens of Silver King

During the time I was writing this book I was searching for decendants of people who had lived at Silver King and could relate their stories. During the summer of 2001, my friend Greg Davis called and said that he had located a descendant of Robert Bowen, the foreman and last superintendent of the mine during the 1880s. I contacted Velma Bowen Tucker, granddaughter of Robert Bowen, who lived in nearby Mesa. She was eighty years old, and had previously done a marvelous job of collecting the Bowen family history, along with many original old photographs taken at the Silver King Mine. Velma had collected the history for a long time and had assembled it in the form of a family history book. I called Velma and set up a meeting, whereby I was able to interview her, and she graciously allowed me to make a copy of her family history book. The Bowen family history contains personal glimpses of the past and life at Silver King during the 1880s. It also includes stories about the Bowen's and their relatives, and their impact on the Silver King history.

Robert Bowen Sr. was the last superintendent of the Silver King Mine during the glory years of the 1880s. He was born March 29, 1849 at New Buckenham, Norfolk, England. Robert was the second son of Stephen (Ben) Bowen and Susannah Quantrel or Quantril. They were married in New Buckenham Parish in October 1845. There were six children born from 1846 to 1866, George, Robert, Arthur, Eliza, Harriet and Elijah. Stephen and Susannah lived on Snailgate Road with their children. There were many Bowen families living side by side in the area of nearby Banham. This was a quiet little village in the 1800s, and still is today according to Velma, who visited Banham in 1980. The Bowens labored on the nearby farms and attended church at the nearest parish. Life was hard in old England for the Bowens, and they sought to improve their lot in life, like many families of that era.

The Bowens learned about America from family relatives who had come across the ocean earlier. As soon as Robert went to work, he sent word back to England, and other brothers and sisters soon came to the states. Family members came to America in search of their dreams and scattered throughout the land, from Pennsylvania to California. Robert

Bowen came to America about 1873, via the Canada route, drifting out west through Oregon and Utah, and then down to California and into Arizona Territory. The 1880 Pinal County Census shows Robert Bowen at the Silver King Mine. He gave his age as 32 and single. He had been unemployed for about three months, but soon gained employment as a bartender at the Williams Hotel at Silver King, as he and the owner Col. Robert Williams had become friends. In fact, they soon became relatives by marriage, and life-long friends.

The advent of the silver and gold boom years brought many newcomers to Arizona, including the Bowens and their relatives. As previously stated, Robert Bowen got acquainted with an influential man named Col. Robert Williams. Williams was a pioneer of the West and arrived in Arizona Territory about 1860. He drove a stagecoach from Yuma, Arizona Territory to Los Angeles, California. He was known for his wild tales of Indians and outlaws. Williams had a cheerful disposition and was well liked. His father was Captain of a sailing ship, and he was born at sea. Col. Williams owned saloons and a chain of hotels, which made him financially secure. Robert Williams, who was English, had married Lena Roth from Germany. Velma said it is not clear just how they met, but nevertheless they were well-known as a successful couple, with prestige and money at Silver King.

Robert Bowen takes a German Wife

Lena Roth Williams had kept in touch with her relatives in Germany, and the news of the mining town and miners interested her two beautiful nieces, the Oehrlein sisters. So the sisters, Elizabeth and Barbara, crossed the ocean and made the long frightening journey to the Silver King Mine. They left the quaint little village of Unterdurrback, Germany, arriving in New York about 1885. Since the girls only spoke German, getting along in a new country was difficult. They went by boat to San Francisco and boarded a train, which took them as far as the little village of Maricopa, Arizona Territory. From there, the girls took a stagecoach to Florence and took another stagecoach for a long dusty ride to Pinal City. They took yet another stagecoach up to the town of Silver King. They were overcome with happiness after seeing their Aunt Lena, their mother's half-sister, and knew they would be welcome guests. Later in 1888, the sisters' parents would follow them to America and settle in the farming community of nearby Tempe. The small mining town of Silver King was an exciting place, booming with high-grade silver and plenty of single miners. Many other newcomers were arriving all the time, and the community was growing fast with rows of little lumber houses. One can understand how the young German girls felt, leaving a family in Germany and trying to get used to a dry Arizona desert town where almost all of the people spoke English.

Robert & Elizabeth Oehrlein Bowen 1886 — Bowen Family Collecton

Velma states that very little time passed before Robert Bowen was introduced to Elizabeth Oehrlein. Even with difference in age of several years, a wedding was soon being planned. Velma stated that Aunt Lena apparently made all the arrangements, as she spoke both German and English. Robert Bowen married Elizabeth Oehrlein on April 16, 1886. They were married at Silver King by Justice of the Peace,W. H. Benson. Velma writes that Elizabeth (her grandmother) "was beautiful with her sparkling brown eyes, dark brown, upswept hair and diamond earrings, a gift from the groom." The bride wore a beautiful brocaded purple velvet and satin gown, laced up the front with shiny brass buttons and a bussel on the back skirt, which was the fashion of the day. Elizabeth carried the wild, sweet, waxy, yucca blossoms from the nearby hills as her bridal bouquet. Velma states that "Robert Bowen (her grandfather) was handsome with blue eyes and brown hair and dressed as no other miner." The *Pinal County Record* newspaper reported that many prominent people witnessed the fashionable wedding of the day at the rich Silver King Mine. According to Velma, it was the custom of the day, and a German tradition, to cut a piece from the wedding dress to present to close friends and relatives to remember that special day. Velma wrote that the wedding dress and wedding portrait are on display in the Bowen, Morris and Phelps Room at the Heritage Museum located on Horne and Lehi Road in Mesa.

The Bowen Relatives come to Silver King

Eliza Bowen Brown, the daughter of Stephen and Susannah Bowen, was born on November 14, 1856 in Banham, Norfolk, England. Eliza married William (Bill) Thomas Brown also of England. Their first daughter

Elizabeth Brown was born on July 6, 1879 in Manchester, Moulton, England. Bill and Eliza and their daughter Elizabeth came to America in 1884, then traveled to Arizona Territory where Bill went to work at the Silver King Mine as a bookkeeper. A son named Harry was born at Silver King, date unknown. Their second daughter Stella was born at Silver King, date unknown. In 1888, their third daughter Bess was also born at Silver King. After the Silver King Mine closed down, the Browns moved to Phoenix. Eliza lived until 1920 and Bill until 1936. They had moved to Long Beach, California and both died there.

William T. & Eliza Bowen Brown cir. 1886
— Bowen Family Coillection

Harriet Bowen Tite, daughter of Susannnah and Stephen Bowen, was born June 12, 1861 (or 1863) in Banham, Norfolk, England. She married William Tite whose family lived nearby. When Robert Bowen became established at Silver King, he sent for his sister Harriet and brother-in-law William Tite. Tite left England in 1885, and in 1886 Harriet joined him at the Silver King Mine where he worked as a bookkeeper. They had one daughter, Jesse Eliza Tite, born in August 1897, at Silver King.

William & Harriet Brown Tite cir. 1886 — Bowen Family Collection

The last and sixth child born to Stephen and Susannah Bowen was Elijah Bowen, born January 26, 1866 at Banham, Norfolk, England. He was the little brother to the rest of the children, and also came to Silver King, Arizona Territory. He married Jennie Corbell, of the Corbell family, well known farmers of Tempe. They had two daughters. The first daughter Elsie was born at Silver King, September 4, 1889. Their other daughter, Virginia, was born in Phoenix on May 27, 1892. Elijah died December 31, 1942 at Corona, California. After Silver King closed down, the Tites and Elijah Bowen families lived in the mining town of

Ureka, Utah in the early 1900s, prior to moving to California.

George Bowen, the first son of Stephan and Susannah Bowen, and Robert's older brother stayed in England. He was born the 3rd or 5th of August 1846, at New Buckenham. He left his family home at age 14 to work at a nearby farm, as was the custom, since a father couldn't earn enough to support a large family. George married Lydia Carter in 1861 and lived in a small village named Grantchester, just outside of Cambridge. They had six children: Robert, George, Arthur, William, Sue and Herbert called Binnie.

Robert and Elizabeth Bowen start a Family at Silver King

When their first baby was soon to be born, Robert wanted his wife to have the best of care and thought that California would have good doctors. So Elizabeth took the train to California to stay with some of the family. On April 27, 1887 their first child, a girl, was born in San Diego, California. She was named Barbara, and called Betty after Elizabeth's only sister. Coming back to Arizona on the train the baby caught a cold and got very sick. At about six months old little Barbara died and was buried at Florence, A.T. Florence was the county seat where records were kept and where the nearest Catholic Church was located. Very soon another daughter named Susanne was born on March 7, 1888, at Silver King. She was named for Robert's mother Susannah. Little Susie was dressed like a little doll, which all the miners enjoyed, since most of them were not married and did not have children. Elizabeth had to depend heavily on her Aunt Lena during this time, as she was the one who sponsored her coming to Arizona.

Robert P. & Susie Bowen 1890
— Bowen Family Collection

On the last days of the Silver King Mine a boy was born to the Bowens. He was named Robert Phillip, after his father. This made the family very happy to have a boy, and at about the same time plans were being made which would change their way of life. I wrote about the closing of the mine in another section and about the Sherman Silver Purchase Act of 1890. Well, the demonetizing of silver affected the Bowen family directly. This, along with the fact that the easily located very high-grade ores of the Silver King Mine were now hard to find, brought about an end to profitable mining at Silver King. This affected the Bowens and their friends and relatives. It was a sad time for the prosperous history-making mine, and the residents of Silver King and Pinal City. On April 15, 1902 the Bowens had another daughter named Ruth. As Susie grew her mother wanted her to have the better things in life and have an opportunity to grow in a cultured environment. So the Bowens sent her to California to live with Aunt Lena. Susie grew into a beautiful young lady, but took suddenly ill and passed away on January 13, 1909 at Los Angeles.

Susie, Robert P. & Tom Bowen
— Bowen Family Collection

The Bowens go into the Cattle Business

Robert Bowen had been foreman and superintendent at the Silver King Mine for some time and was well liked. He was a good businessman and kept a keen eye on the growth of Arizona. Robert Bowen moved his family to Tempe, A.T. Bowen had met many influential people during his tenure at the mine, and one of these was Carl Hayden. Hayden and Bowen were friends, and they discussed Bowen's future now that the mine was closing. It is not known exactly why or how it happened, but the Bowen family got into the cattle business. Robert started raising cattle

Unknown cowboy at Bowen Ranch cir. 1890s
— Bowen Family Collection

on a ranch he had purchased east of Mesa in the desert. They shipped their cattle to California.

Robert Bowen and the Pleasant Valley War Connection

Some very interesting facts surfaced during my research on Roberts Bowen. He was one of the ranchers indirectly involved in Arizona Territory's Pleasant Valley War of the 1880s. (This was also called the Graham-Tewksbury War, named after the two main families in the cattle/sheep war.) The Hashknife Cowboys also were participants in this war.

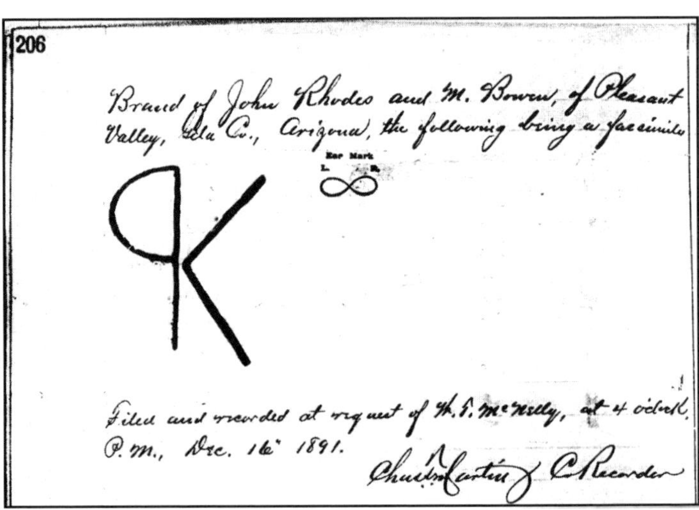

206

Brand of John Rhodes and M. Bowen, of Pleasant Valley, Gila Co., Arizona, the following being a facsimile

Ear Mark
L. R.

Filed and recorded at request of W. F. McNelly, at 4 o'clock P.m., Dec. 16th 1891.

Chas. Martin Co Recorder

Bowen-Rhodes PK Brand 1891 — *Bowen Family Collection*

Bowen was also a key defense witness in the Tewksbury Murder Trail of TomGraham. In the late 1880s Robert Bowen settled on the old Blaine-Daggs Ranch at the head of Walnut Creek. Bowen ran cattle on this ranch using the PK brand. This brand was registered to both John Rhodes and Robert Bowen in 1891 in the Mark and Brands Book #1, page 206 at the Gila County Recorder's Office in Globe. Bowen also was the owner of the Tempe Hotel in Tempe during

Bowen's Tempe Hotel cir. 1890s — *Bowen Family Collection*

the time that Tom Graham was shot at the Buttes. Bowen testified for Tewksbury at his first murder trial in 1892. Bowen stated at the trial that at 7:20 AM on the day Graham was killed John Rhodes was at his hotel in Tempe. He provided an alibi that Rhodes was with him (Bowen) at his hotel at the time of the killing, therefore could not have been with Ed Tewksbury when Graham was shot. The prosecution stated that Rhodes and Tewksbury were together at the time Graham was killed. Tewksbury was found guilty at the first trial but exonerated at the second trial. Bowen also testified that he went to Pleasant Valley in search of gold as well as raising cattle. He spoke of his ranch in Pleasant Valley as being "about the mines."

Several books have been published detailing the events of this conflict including: *Arizona's Dark and Bloody Ground* by Earle R. Forrest; *A Little War Of Our Own,* by Don Dedra; *Arizona's Graham-Tewksbury Feud, by* Leland J. Hanchett, Jr.;*Young, Arizona* in Pleasant Valley by Barbara Zachariae; *The Bloody Pleasant Valley Feud,* by Harold D. Jenkerson; *To the Last Man*, by Zane Grey (also a movie by that name) and a host of news articles and short stories. These books depict the fascinating and sanguinary times during the conflict of cattle and sheep days in Arizona Territory's Tonto Basin. I traveled to Pleasant Valley (It is properly the town of Young, Arizona north of Globe.) on many occaisions to gather information about this era and visit where many of the violent incidents actually took place during the 1880s.

The Bowen's Search for a New Home

While looking for a new home they lived at the Casa Loma Hotel on the edge of the Salt River crossing at Tempe. This is where their second son was born. Thomas Arthur Bowen was born on January 2, 1892. The Bowen family was growing, and Elizabeth was happy because her family had come from Germany and had settled in Tempe. The farming community was a welcome sight from the dry desert. The same year Susie died in California, Robert Bowen had an accident on the cattle ranch and never recovered, passing away May 18, 1909. The three remaining children devoted their lives to their mother. Her son, Robert Phillip Bowen, took over the family business and became a successful farmer and cattle rancher.

Velma's father, Thomas Bowen, left school in the fifth grade to help around the farm, and brother Bob left in the sixth grade. The boys were needed in place of hired men as the family was going through hard times. Tom Bowen married Genevieve Morris on September 23, 1916 and was the only Bowen child to have children of his own. The first child, a girl named Rheta, died at one year of age. They had four more children, Thomas Joseph, Jack, Katherine and Velma. Velma is the one who provided the information on the Bowen family. Velma married Steven Tucker June 12, 1944, who was in the army at the time.

Robert Bowen & unk. children cir. 1910 — *Bowen Family Collection*

It is with deep regret that I have to report that Velma Bowen Tucker passed away before this book was published. She died suddenly of a heart attack on January 31, 2003, ironically while working on her family genealogy and the Silver King connection. Velma was a great person and a living testimony of a good Christian. She was born on January 8, 1921 and lived a very full life, including growing up on a farm and working as an elementary school teacher for many years until she retired. Even after retirement she volunteered in the elementary school system. Velma was a quilter and made many quilts for people over the years, including me. She was the daughter of Thomas Arthur Bowen and Eliza G. Morris. Velma had six brothers and sisters, Rheta, Thomas, Katherine, Jack, Richard and Robert. Velma was the granddaughter of Robert Bowen, superintendent of the Silver King

Thomas, Tom, Jack, Velma, Genevive & Katherine Brown cir. 1940s — *Bowen Family Collection*

Mine in 1888. Velma was married to the late Steven Tucker, and had four children, Janet Hall, Steven Tucker, Barbara Dumont and Tommy Tucker. She had 14 grandchildren and many great-grandchildren. Velma was a great assistance to me in completing the collection of photos and information on life at Silver King during the 1880s. She will be missed.[1]

Velma Bowen Tucker 1950
— Bowen Family Collection

Harry Brown's Tales of Silver King

Before we leave the Bowen family, I would like to include the tales of Harry Brown about mining days at Silver King. Harry Brown was the son of Eliza Bowen Brown and William Thomas Brown. He was born at Silver King, lived there until 1889, and recalls the stories told by his father and Uncle Robert about Silver King. He stated that the memories he has about those days are in part from memory, and in part from parents and relatives who lived at the mining town. Harry recalls that "Silver King was a veritable beehive of activity and the Mecca of mining men from all parts of the world, from the captains of mining industry to the lowly desert rat with his faithful burros." Harry stated that his father worked as a miner at the Silver King Mine at one time, but then was put in charge of the Pump Station above the Stoneman Grade, at General Stoneman's old Camp Supply. The Pump Station supplied fresh water for the mine and townfolks. It was several hundred feet higher in the mountains than Silver King. He writes that there was a sufficient supply of spring water as the well was dug to a level to provide plenty of water. The water was carried via a pipeline, 3 miles down the mountain to wooden storage tanks, located at the back of the mine, and at the eastern end on town. While hiking the Stoneman Grade in 1997, I located many reminants of the old cast iron 1inch pipe on places leading down the Grade.

Harry writes that life at the Pump Station was lonely as they were the only family living there. The only way between the Pump Station and Silver King was by horseback, so he said the trips back and forth were few and far between. Also, at that time Harry writes, there were bands of Apaches still roving the country. He does go on to say that the Indians never bothered them, but sometimes came to gather the acorns which they used for food. Harry relates a humorous story how his father used blasting powder to remove a stump, but that he used too much powder and the stump flew through the screen door and landed on top of the kitchen stove, narrowly missing his wife who was preparing dinner. He writes that the dinner was badly damaged, but fortunately no one was hurt!

Harry describes the homes that most of the miners lived in at Silver King. He stated that they were mostly humble, built of up and down rough pine boards with battens. Several houses were made of a framework of wood with canvas covers and were called tent houses. The only exception was the mine superintendent's house, which was a commodious two-story structure with a veranda surrounding both levels. Harry says that it was well furnished and contained one of the two pianos in the community. The other solid house belonged to a family by the name of Taft.

He tells of going to school at the little white schoolhouse, perched high up on the hill at the west-end of town. On one school day, he recalled a lad named Raymond Owens was riding a burro up the trail to the school and was thrown and broke a leg. Years later he still remembered the cries of pain as the doctor put splints on the broken limb. Harry also tells of the John Knight mercantile store at the town, and that when the mine shut down Knight moved the store to east Tempe, on the bank of the Hayden Canal. He tells of Charles Corbell and his dairy herd on his ranch, a few miles west of Silver King, on the road to Pinal City. Corbell got his start in life by furnishing milk to the folks at Silver King Town, but later moved his dairy to a site south of Tempe. (Harry's Uncle Elijah Bowen married Jennie Corbell.) He writes of Dr. Kenniard being the physician at Silver King and attending his mother at the time of his birth. The *Pinal Drill* on September 29, 1883 stated: "Born, At Silver King on the 23rd inst., a son, to Mr. and Mrs. W. Brown. The boy is a big fellow for a little one, and he is apparently quite happy in his new situation. The mother is doing well and the nurse, Mrs. Kuhn is as proud of the boy as if he were a nugget bigger than the Silver King."[2]

Harry related this story about a character at Silver King called "Old Tutaine":

> "I suppose that every mining camp has at least one eccentric character and so it was with Silver King. 'Old Tutaine' was the name he went by. He was

a hermit and made his home in an old prospect hole driven horizontally into the hillside. I don't know that he ever did any useful or gainful work. He just roamed at will around the countryside. He never harmed anyone. His subsistence, for the most part, consisted of handouts from the miners' lunch pails. They seemed to regard him as a sort of charge of theirs. What became of him I never heard. To me he was a part of the Silver King—the picture would not be complete without him."

He recalled one other character that I will include. His name was Bob Wright, who had a tent cottage back of the orehouse where he spent part of each year.

"He was a small man, his hair almost white, and he wore a long grey beard—a typical desert rat. Now why should I remember him apart from all others? Well, I'll tell you, Bob had a hobby. He collected all kinds of fancy cuff buttons—those old fashioned ones with beautiful colored tops. My two younger sisters and I quite often paid him a social call, and of course, the girls usually wore their best bibs and tuckers. Bob enjoyed having us call on him, and when we were ready to leave he would decorate the girl's dresses with his prize buttons, as a memento of the visit. Mother didn't take to the idea very well, and our visits grew less frequent. I last saw Bob Wright in 1893 in Tempe. He made a special trip out to our ranch south of town, and spent a couple of days with us, after which he headed back to the hills and never returned. I often wondered what finally became of him. He was a good man and as a boy I admired him. The great open spaces seem to be builders of good men."[3]

Chapter Ten
Stagecoaches *Bullion*and Bullets *

During the halcyon days of the Old West, mining era, stagecoach robbery became a way of life for some of the desperate men. Mining men were taking the wealth out of the ground and outlaws wanted to take that wealth

STAGE ROBBED AGAIN.

This morning when the stage was about a mile and a half below town, between here and Florence, the driver was "held up" and told to throw out 2 bars of bullion belonging to the Silver King Mining Company. The robber was covered by some light material thrown over his head and shoulders and the driver was unable to give any description of him. The silver was estimated at about 3,500 ounces. As soon as the stage returned, and the driver told of the robbery, several of our citizens immediately armed themselves and started out in pursuit of the robber, and up to the time of going to press they have not yet returned. It is to be hoped that they will get him, and keep up the good name of our town and vicinity by speedily giving him his just deserts.

Wells Fargo & Co. have a standing offer for the arrest and conviction of any robbery committed on their line of $300.

"Stage Robbed Again," Nov. 20, 1885
— Pinal County Record

for themselves without doing the backbreaking labor of mining. The following tale depicts one of those episodes that took place in the 1880s when holdup men preyed upon the bullion that was shipped from the mills.

The stage is only a few miles outside of Pinal and Silver King, on the road to Florence, when two armed men ride out of the brush wearing bandannas over their faces, and one shouts, "Hold! "Throw down the box!" The driver says: "Thet's the same fellers thet robbed me last week, and I'll be danged if they're goin ta do it again. Hi-ya! Git-up," he yells to his team.

The bandits fire as the stage rushes by and the guard raises his double barrel shotgun and fires both barrels at the bandits. They both miss. The bandits give chase and the driver yells at the team of horses. The driver knew now that there was only one thing to do, try and outrun the bandits to the next stage station. He kept up the pace until they came upon some riders he knew coming up the road from Florence. The bandits then gave up the chase and fled into the desert mountains.

Wells Fargo agent Larkin and Silver King silver bars — *Arizona Historical Society*

Sounds like an old Roy Rogers western, doesn't it? However, this is a reasonable description of what actually took place on many occasions during the 1880s when silver and gold bullion were carried on the stages, and many desperate men tried to "take the stage." When the Silver King Mine was booming the mail and express was carried by stage to Florence, where connections were made onto Riverside and Globe. Jack Keating, one of the stage drivers, had been held up so many times that when he saw strangers coming up the road he would put the lines between his knees and hold up his hands straight. If the stranger did not happen to be a holdup man, Jack would say, "Well, by gad, the next one will ask for the box," and frequently his prophecy would come true. Jack would rather be a live stage driver than a dead hero.

Robberies became so frequent that businessmen instead of sending money by stage would seal it in cans marked corned beef or some other appropriate name and ship it by freight wagons mixed up with a lot of other freight.[1] (See story on Perry Wildman—Silver King and Pinal merchant, who had his silver dollars sent to him in kegs of nails.) The stage robberies became so frequent that they no longer became front page news, and unless a life was taken, received only a few lines on the back pages. The box they were referring to was a strongbox filled with heavy bullion. If it was on a Wells Fargo Express Route Stage, then the Wells Fargo Box was what they wanted, but all stages carried a strong box of some sort.

Bullion Shipments Announced in the Papers

Throughout the time period when the Silver King Mine was in full operation, during the 1880s, both the *Pinal Drill* and later the *Pinal County Record,* announced the shipments of bullion shipped, or sometimes *bullion to be shipped.* The *Pinal Drill* in the early 1880s put a value on the shipments, while during the latter part of the 1880s the *Pinal County Record* usually announced only the number of bars shipped. The Silver King Mine during 1885 shipped $800,000 in bullion, according to the *Pinal County Record.* These announcements could only have stirred the imagination of many a would be rich hold-up man, and perhaps this is why the paper stopped announcing when shipments would be made including the values.

The Silver King Mine was only one of many mines in Arizona shipping bullion by stage, and Wells Fargo & Co. had their hands full. They tried many ways to thwart the hold ups, including not putting bullion or cash in the strong boxes. They also hired private detectives and extra guards to ride the stages. They used undercover agents to learn of hold ups planned in advance. Sometimes these precautions worked, and sometimes their accomplices tipped off the holdup men. These accomplices included some of the drivers or guards, it is reported.[2] The mine owners soon tired of these hold-ups and in the late 1880s began to make the bullion, or bricks of silver, weigh 75-100 pounds to discourage the taking of their bullion.

Announcements of Bullion Shipments

The *Pinal Drill* listed the following bullion shipments:

June 8,1881-12 bars valued at $25,464.
July 16, 1881-16 bars valued at $25,646.02
March 4, 1882-19 bars valued at $33,247
July 15,1882- 8 bars valued at $12,875
July 7, 1883-2 bars of bullion, no value given
August 4, 1883-15 bars valued at $ 1500-$2000 each.

I was unable to find amounts of bars of bullion shipped in 1884, however, the *Pinal Drill* stated that the Wells Fargo Company report for 1883 listed

$8,183,743 in bullion produced in Arizona.

The *Pinal County Record* listed the following shipments:

September 25,1885-one bar weighing 113 lbs. and 2 bars scheduled for Wednesday at 108 lbs. apiece.
October 9, 1885-shipped 5 bars of bullion and will ship 3 bars on Monday.
December 4, 1885-The Silver King Mine shipped 4 bars of bullion and the Pinal 76 Mill will ship bullion.

The *Pinal County Record* stated on January 15, 1886 that the Silver King shipped $2.6 million in bullion during 1885.

The *Pinal County Record* on October 23, 1885 stated that two million dollars were recently sent east from San Francisco. They made a carload.[3] It must be noted that these shipment notices are only a sampling on the bullion shipments listed during the 1880s. The mine and mill superintendents were encouraged to give information on shipments as part of the daily news, and to keep the mining camps informed that their mines were still doing well. However, this information also whet the appetites of many a would be stage robber, as the bullion went from the Silver King Mine and Pinal mills by stage to Casa Grande, the nearest railroad depot. From there it was shipped to San Francisco to be minted into silver dollars. The stage robbers also knew that the payrolls for the mines and mills had to be shipped by stage, and in fact, preferred holding up the payroll shipments instead of the heavy bullion loads.

Increased Mining Activity Brings Stage Holdups

Increasing mining activity in eastern Arizona, centered on the fabulous Silver King Mine, and discovery of rich deposits in the Globe area brought a shift of holdups to that part of Arizona in 1880 and 1881. Several times stages running between Florence and Globe were robbed. Some stages evidently carried bars of bullion from the rich mines, while those headed toward the mines were believed to be carrying payrolls of gold and silver coins and greenbacks. On three or four occasions, the highwaymen were frustrated by empty boxes, but had to run for it when posses scoured the mountains after each episode.[4]

Robbery of Florence/Globe Stage

The *Pinal Drill* reported on March 10, 1883, that the Globe and Florence stage was robbed on March 1st. "The robbers took the treasure box without attacking the passengers. The box contained very little of value, however, so the gentlemen must have been disappointed." [5]

Robbery of Pinal Stage

Robbery of the Pinal Stage was the headline story of the *Arizona Silver Belt* on February 28, 1885

"The northbound stage from Florence was again halted by two highwaymen, on Monday night, about a mile and a half or two miles below Pinal, at the same spot where it was held up a few weeks ago. From W.E. Spence, who returned from the (Silver) King on Wednesday, we gather the following particulars of the robbery. When the stage reached a point mentioned, two men appeared in the road, and bringing their guns down ordered the driver to stop. Wells Fargo & Company's box was first broken open. The four passengers were then relieved of their valuables and the mail sacks also rifled. The driver was then compelled to cut the sacks open. The driver, it seems, recognized the voices of the robbers, and upon reaching Pinal, gave the names of Andy Dorsey and Henry Girard as the suspected parties. The facts of the robbery were also *telephoned* to Silver King. Deputy Pete Bousche, Jeff Bramett and Jack McCoy started immediately for the house of the suspected parties, on the road just below the Silver King.

"Once arriving there Bramett went up the door of the cabin and knocked. The door was thrown open and Bramett was confronted by the muzzle of a gun in the hands of Dorsey, who demanded what was wanted. Bramett remonstrated, because of this hostile action, and Dorsey being reassured by his manner invited him in. Bousche and McCoy soon followed, and the arrest of Dorsey and Girard was easily effected. Most of the stolen property was found secreted in the house, but Mr. W. N. Griffith, representing the King Morse Canning Company of San Francisco, failed to recover $115 in currency taken from him. Mr. Griffith had a large sum in gold, which he managed to drop in the bottom of the stage and thus saved it. After a preliminary examination before a Justice of the Peace at Silver King, the prisoners were taken to Florence and incarcerated.

"Charles Miller, the stage driver, was also arrested as an accomplice, but as far as we can learn, very unjustly, as there is no evidence to warrant his retention for the crime. Dorsey is supposed to have committed a robbery at the same place several weeks ago." [6]

$600 Awarded for Capture of Hold up men

The *Pinal County Record* on December 25, 1885 wrote of the $600 reward issued by Wells Fargo & Company, for the capture and conviction of Dorsey and Girard. It stated that "the court awarded the following apportionment of the reward: Jack McCoy, $200; Jeff Bramett, $200; J.P. Gabriel and Peter Bosche each $100." [7]

The Silver King bullion stole last Friday morning from the Florence bound stage was recovered Sunday by Mr. Henry Pruett. he being a strong believer in Christianity, said that on Saturday night, Christ told him where to find it, and on Sunday morning went immediately to the spot. The robber had thrown it into a large bush by the side of the trail about one mile from the road. The officers and citizens have been very energetic in the pursuit, although as yet unsuccessful. All the indications favor the suspicion that the reobbery was committed by one Wm. Taylor and that he had for an accomplice, a man by the name of James Shears. Shears was arrested, on suspicion Friday but succeeded in making his escape from his Keeper at Silver King Saturday night and has not since been apprehended. He made his appearance in the outskirts of town on Wednesday night, he was pursued by a possy of citizens and shot at three times, owning to the darkness he made good his escape. He then went direct to Mr. Wheeler's on Queen Creek, stole a horse, saddle and bridle. Messrs. Boschea and Preston took his track the next morning, and started in persuit. Up to the this writing they have not returned.

Arizona Stage robbery story — Pinal Drill 1883

Stage Robbed Again

This was the story title on November 20, 1885, in the *Pinal County Record.*

"This morning when the stage was about a mile and a half below town, between here and Florence, the driver was 'held up,' and told to throw out 2 bars of bullion belonging to the Silver King Mining Company. Some light material, thrown over his head and shoulders covered the robber, and the driver was unable to give any description of him. The silver was estimated at 3,500 ounces. As soon as the stage returned, the driver told of the robbery, several of our citizens immediately armed themselves, and started out in pursuit of the robber, and up to the time of going to press they have not yet returned. It is hoped that they will get him, and keep the good name of our town and vicinity by speedily giving him his just deserts." [8]

Bullion Taken in Stage Robbery

The *Pinal County Record* reported on November 27, 1885, that:

> "The silver bullion stolen last Friday morning was recovered Sunday by Mr. Henry Pruett. He being a strong believer in Christianity, said that on Saturday night, Christ told him where to find it, and on Sunday morning he went immediately to the spot." The story continues that one Wm. Taylor was a suspect along with James Shears. It says that Shears was arrested on Friday but that he escaped. A subsequent story on December 4, tells of Shears pursuit and capture, and that during his pursuit shots were fired without effect, that Shears then stole a horse and made it to the Salt River before he was finally captured. The story ended with statement that; "so far Taylor has not been arrested, but will probably will be in a short time." [9]

Attempt Stage Robbery

The *Pinal County Record* wrote of an attempted stage robbery on December 25,1885. It said:

> "Some parties under the influence of liquor, stopped the stage at the usual place Tuesday night and wanted the valuables thrown out. The driver and passengers did not agree to this, and scared the fellow off, by threatening to demolish his cranium with a rock if he did not let loose of the horse's bridle." He let go and the stage went on. It says there were three or four in the party. The article went on to say: "It is no funny thing to stop private citizens and a stage carrying U. S. Mail and Wells Fargo & Co. express. The penalty is nearly the same, whether they get anything or not."[10]

Wells Fargo Standing Reward Offer

Published in the *Pinal Drill and Pinal County Record* papers on many occasions, was the statement that the Wells Fargo. & Company had a standing offer for the arrest and conviction of any robbery committed on their line of $300 per hold-up man. This was an inducement to search for the hold-up men and thwart future robberies.

Stage Robbery at Usual Place

The *Pinal County Record* on January 22, 1886 reported this stage robbery as follows:

> "Last Saturday night as the stage was passing the usual place a man stepped out and demanded a halt, which command was obeyed. He then commanded the driver to throw down the express box, which was done, and the stage was allowed to come directly into town." The story stated that the express box contained nothing of value and a party was organized to search for the robber.

It also stated that no trace of the robber was found, although a certain party was suspected.[11]

Stage Robber Shoots at Driver

This was the story told on July 9, 1886, by the *Pinal County Record.* The story continues that the stage robber ordered the stage driver to halt and to "put out silver." The driver H. C. McDonald, owner of McDonald's stage line, was then ordered to "put out Bretta," meaning the mail sack, which the driver threw out. At this time the robber either accidentially or intentionally shot at the driver, coming very close to his head. The story then says that "Mac then struck his team with the whip and left without further ceremony." The robber got between $12,000 and $15,000 in checks, which the story said were impossible to negotiate anywhere, and that the robber who got away was probably a Mexican.[12]

"Red Wagon Brown" and the Dynamite

Stage holdups weren't the only good stories about stagecoach days. This amusing tale comes out of that era. When a driver named "Red Wagon Brown" was running the stage line to the country south of Casa Grande his load occasionally consisted of dynamite, which was generally stored under the seats. One day on the out trip, Brown happened to look back and noticed a passenger lighting his pipe. Brown immediately said, "my goodness don't smoke back there, if that powder should explode, I would not have anything left to make a living with." The stranger kept on smoking and said, "if that powder explodes you won't need anything to make a living with." [13]

Arizona Stage Lines and Routes

There were several stage lines operating in Arizona in the 1880s, although many of them were local stage operations. We are used to seeing the large Concord Stage, pulled by 4 to 6 frothing horses on TV westerns. These were used, but there were also several smaller stagecoaches used, pulled by either 2 or 4 horses.

Some of the stage lines that operated to Silver King and Pinal were: the Texas and California Stage Line, owned by the firm of Kearns and Griffith, which began at Casa Grande at the railroad station and ran through Florence to Silver King and Pinal. It also ran from Florence by way of Riverside, on the Gila River near current day Kearney, to Globe. McDonald's Stage Line, owned by Mr. H. C. McDonald, ran between Florence and Silver King and Pinal. Mr. D. W. Art also ran a stage line between Pinal and the railroad at Casa Grande. The Picket Post, Silver King and Globe Stage Line was run by H. B. Montgomery. The Arizona Stage Company, owned by W. H. Sutherland, ran a stage operation which served Central and Southern Arizona. The Southern Pacific Stage Line ran from Florence to Picacho. The Gilmer, Salisbury & Company Stage Line ran a stage line out of Pinal. The Butterfield Overland Stage Company, which carried the mail, also operated in Arizona during this time period and established a stage station at Florence in 1879. Wells Fargo & Company established a station with an agent for handling their express at Pinal in 1881.

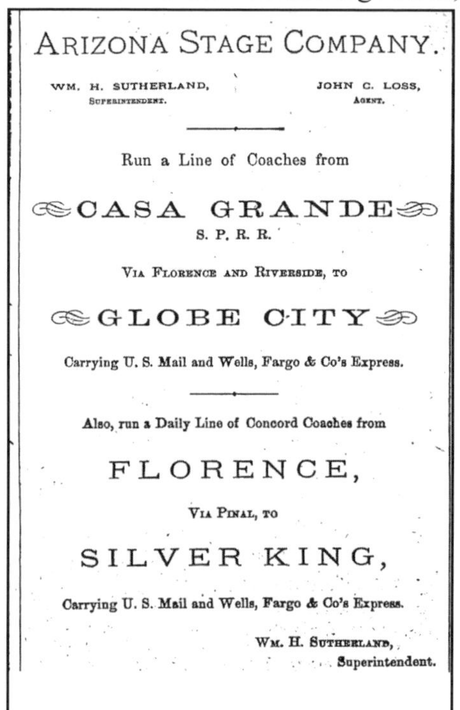

Arizona Stage Company ad cir. 1880s — Pinal County Record

Prior to 1866 there was not one local stagecoach running in all of Arizona, however, after the toll road act was passed in 1871, a number of toll road companies were licensed. These toll roads were eventually connected to one another. In 1878, there were established regular stage services and connections. In order to increase the practicality of transportation, the Tenth Legislative Session of Arizona, in 1879, authorized the sale of bonds for the construction of a road in Pinal County. It was to be built along the most practical route from Florence to Globe City, via Riverside and the

Gila River crossing. This road was conceived to give the mining districts railroad connections to the Casa Grande railroad depot.
From Florence, there were roads to Silver King and Pinal. Saxe's Toll Road began at Bloody Tanks and connected to the Silver King Trail. Saxe ran a hotel at Bloody Tanks where travelers could rest after crossing the

Pinal and Silver King stage route cir. 1880s
— *Pinal County Record*

mountains by muleback, before they were transferred to stagecoach, then onto Globe and viceversa, on the trip to Silver King. In 1882, the Howard A. Reduction Toll Road ran from Pioneer Pass to Riverside. Another toll road was Kellner's Toll Road which ran up Kellner Canyon to Russell Gulch and then connecting with the Pinal Summit Toll Road. This was the first wagon road between Florence and Globe.[14]

The Kerrin and Griffith Stage Company ran a daily stage line from Casa Grande to Florence, a distance of about 25 miles. The Pinal and Mineral Hill Express, owned by H. J. Ehlers, advertised his stage company in the *Pinal Drill,* and went from Pinal via Willow Springs and Seven

Cottonwoods. Also listed was the Mineral Hill Stage Line, owned by Wilson and LeBlanc, which ran a semi-weekly stage between Florence and C. D. Henry's Station at Seven Cottonwoods, fare $3.00 one way.[15]

Hewitt Station, on what is now the Old Magma Road, was one of the old stage stations between Florence and Silver King, named after the Hewitt family, who had a ranch there. Nearby Hewitt Canyon is also named after the family. Whitlow Station, with Lew Bailey owner, was another stage station on the Pinal and Silver King stage route, ten miles west from Pinal and eighteen miles from Florence, on the south side of Queen Creek. Whitlow Station was part of an Old Spanish ranch until Charlie Whitlow took it over in 1875. Whitlow Station was located in what is now Queen Valley in Whitlow Canyon. This stage route ran along the edge of Queen Creek, from Florence Junction into the eastern end of the Superstition Mountains. The road passed through the site of the present Queen Valley community, and the main stage stop was that town. The old road passes by where the Whitlow Dam is now located, and follows the north side of the Old Magma Railroad tracks. The road then passes around a few bends, passes Queen Creek again and goes northward into the deeper areas of the Superstition Mountains. The road goes through high beautiful passes and winds along the ridges until it reaches a rough country filled with grotesque formations of volcanic glass known as perlite. At the end of this road, you will find an entire mountain of Perlite mineral, on the east side of Picket Post Mountain. This old stagecoach road winds up at the old ghost town of Pinal City.

Silver King stage cir. 1880s — Arizona Historical Foundation

Table of Stage Rates
Texas and California Stage Line

	Departs	*Miles*	*Fare*
Casa Grande to:			
Florence	Daily	25	$4.00
Pinal	Daily	52	$8.00
Silver King	Daily	57	$9.00 [16]

Although the Wells Fargo & Express Company ran stage lines in many states, it did not run a stage line in Arizona Territory in the 1880s. It did however, ship express by the local stage lines, and set up agencies to handle their business in many communities of Arizona. Wells Fargo established agencies in Florence in 1877, Casa Grande in 1879, Picket Post and Globe in 1880 and Pinal in 1881.[17]

There are probably a thousand untold exciting stories about stagecoach days, and I have only told a few of them, centering on the shipments of silver and the Silver King Mine. The stagecoach had a long and useful life. It was first established as a mode of travel west of the Mississippi about 1817 and was to travel the western roads for about 100 years. Because of the stagecoach, early settlers could receive mail, ship bullion from the mines and other supplies. Wells Fargo with their standing reward of $300 per holdup man, if any of their express shipments were held up, had their own detectives and guards when shipping bullion or payrolls. Bonus hunters among lawmen could also expect a bounty of one-fourth of any booty recovered. Small wonder that posses eagerly pursued stage robbers, especially when the Wells Fargo express-box was involved.[18]

Many of the old stagecoaches have stood the test of time. A majority of the old stagecoaches that we see on TV westerns are survivors from the past. (That is a testimony to the quality of workmanship and material that went into them.) Without the stagecoach, the most colorful era in American history might never have come about.[19] I can still see vivid images of John Wayne riding on top of the stagecoach, shooting at attacking Indians, in John Ford's classic western movie "Stagecoach."

Chapter Eleven
Frontier Justice and the Outlaws

Arizona Territory in the 1870s and 1880s was the Wild West. Life in the early mining camps and towns was based on the *Code of the West.* Lawmen and police forces were not readily available. While in the towns there might be a Town Marshall, and in the County at large there would be a County Sheriff, the mining camps that sprung up overnight had no organized law enforcement.

A shooting match came off at Reymert's camp Wednesday afternoon in which one of the participants lost his life. It appears that two men named Louis Bromberg and Dave Robinson got into a wrangle over a game of cards, which ended in a rough and tumble fight, Blomberg getting the worst of it. After the fight Blomberg left the place to procure a gun, on his way back he met Robinson, who was standing in his doorway, and commenced firing. He fired three shots, first shot passing between the neck and collar bone, ranging downward and coming out of his back, the next shot missed him and the third passed entirely through him, going in on the left side, coming out on the right and lodging in his wrist. Robinson had an old, worthless gun in his hand but did not attempt to use it. Robinson was killed at the first fire. Blomberg surrendered himself and was taken to Florence, and Robinson's remains brought to this place. We know nothing of the antecedents of these men, and give the report as stated to us.

The coroner's jury rendered a verdict according to the above facts.

"Shooting at Reymert's Camp," 1885 —Pinal County Record

Gunplay and death were twin factors in settling many hotheaded disputes. But it was the code of the day, and a man was "*yallar*" indeed, if he failed to draw and shoot when insulted by another who was armed. As nearly everyone carried firearms, it was an act of self-preservation to beat the other man to the draw. The person who beat his man to the draw and shot or killed him was usually exonerated in a verdict rendered to the victor. Unless of course the victim was popular and had friends, as we shall see below.[1] During the period of organizing mining districts and counties being laid out, the miners and settlers usually took the law into their own hands by organizing vigilante committees ,or "posses," to pursue and capture thieves and murderers. Since there were no jails these men imposed punishment in the form of fines, whippings, ordering men to leave town, chaining them to a post and the ultimate sanction of hanging. This mentality carried over, even after the Sheriff or Marshall system was instituted, especially when someone was murdered, and doubly so if the person was well liked. The cries *Get a Rope* and *Hang em High* sometimes sounded throughout the mining camps and towns. Most mining camps had their "hanging tree" and stories to tell about men who dangled there at the end of a rope. Silver King and Pinal and the surrounding mining camps had their own stories of "frontier justice". I have included some of these stories from Silver King and Pinal, as well as, some of the surrounding area stories that somehow directly or indirectly affected Silver King and Pinal.

Rampant Ore Thievery at the Silver King Mine

This story comes from the *Pinal Drill,* July 16, 1881.

"Some time ago, Superintendent Mason of the Silver King Mine, Arizona, discovered that considerable quantities of the rich ore with which that mine abounds were being stolen by the miners. A careful watch was begun, but the opportunities for taking it were so great, and constantly increasing as the mine was opened, that little could be done to check it. On one occasion, nearly a ton of the richest ore was discovered, that had been shipped to St. Louis. This was recovered after some difficulty and expenses. That disaster to the thieves put a stop to the thievery for a while, but it soon recommenced. Nearly a thousand pounds of rich stuff was found secreted in a secluded canyon some miles from the mine, evidently the result of several weeks operations of the robbers, taking little by little each day. At the head of Pine Canyon, near Pine Creek, north of Silver King, high-graders reportedly took the Silver King Mine high-grade ore, to work in an arrastra built for this purpose. This led Mr. Mason to believe that the thievery was conducted by a number of persons in a systematic way, and he resolved to break it up. Of course, everybody proclaimed his innocence. Mr. Mason then discharged everybody at the mine and closed up the shaft, saying that he would not resume operations until he found out the names of the thieves. The act aroused considerable excitement for several days, but Mason was firm. Matters were finally settled when the miners consented to be searched, hereafter, when they came out of the mine, and under this agreement those who were not believed to have been guilty, were put back to work. Those who were guilty, were said to have commited a double crime, that of stealing, and that of throwing a cloud over honest men, who are in the same avocation. The thieves will undoubtedly be brought to justice in due time,"[2]

The *Pinal Drill* reported in subsequent issues that several arrests were made for stealing Silver King ore. However, ironically the same day, on the same page was this article. "Shipment by Silver King Today, 16 bars, 19,931 ounces valued at $25,646.02." [3]. It was about this time that the Silver King Mine put guards on full time to keep the high grading to a minimum. They were said to have also walked around the Captain's Walk on top of the Main Office Building at Silver King where they had a clear view of the mine operations above ground. There were also search rooms when the miners came up from underground.

Silver Found at Barkley Ranch from the Silver King Mine

Walt Gassler, a long time prospector and Lost Dutchman hunter relates this tale of lost silver as told by Augustus "Tex" Barkley, early 1900s owner of the Quarter Circle U Ranch, in the south central part of the Superstition Mountains. ("Tex" was called "Gus" by his family and close friends but referred to as "Tex" by most authors). "Tex" came riding in from the range one day when he saw two men digging in the hillside above his ranch house by the creek. He rode over to them and asked them what they were digging there for. They said that they were brothers and that their father used to work for the Silver King Mine as a wagon driver. He left them a map after he had died, and had asked Bill Barkley, "Tex's" son, for permission to dig there. He laughed and said go ahead.

Barkley looked at their map and then told them they were digging in the wrong spot. He directed them to dig at another location nearby, where an old cottonwood tree used to be. Pretty soon they dug up an old wooden box that contained seven small bars of silver. "Tex" kept one bar and the boys took off, slapping each other on the back. They had struck it rich, if only for a little while. Apparently these silver bars were the result of high grading from the Silver King Mine during the 1880s.[4]

Regarding Claim Jumpers

The mining camps disdained claim jumpers and would as soon as hang them as set them on their way. Claim jumping was serious business, as listed in the following article from the *Pinal Drill,* on January 8, 1881.

"The jumping fiends were lively on the first of January. They are a curious looking lot, something like a grasshopper, but with long donkey ears, and sort of a groundhog snoot. They are generally found behind large stones, hidden, and they crawl out at midnight when they try to steal locations, immediately after the stealing they wag their ears lively. They are harmless beasts, generally die out shortly after break of day, only hurt themselves, and die trying to eat themselves." [5]

Hunkydory Holmes while working the Daisy Dean mining claim about 1875 next to the Ramboz mining camp, encountered a claim jumper at his claim. The claimjumper was named Marco Banjevich. Holmes settled the claimjumper's fate with a bullet from his six-shooter. It was stated that Holmes spent most of his earnings from the Daisy Dean on the defense of his case, as he was charged with the killing. Claim-jumping in those days was considered by most miners as a justification for killing. Holmes was declared innocent by a jury of his peers.

When Pinal Had No Jail

The *Pinal Drill,* March 4, 1882, stated this about the lack of a jail.

"An effort was made, some time since, to construct a jail building at Pinal, and a large sum of money was subscribed by our citizens, for that purpose. The matter seems to have been forgotten. We think it is not right, that persons, whether innocent or guilty, who may be arrested on any complaint, should be dragged around by our constable and deputy sheriff, and kept under guard at the hotels, to the great inconvenience and exposure of the prisoners, and expense to the county. There certainly ought to be some more convenient place of confinement, to secure those under arrest, without making them public spectacles and objects of public disgrace. Many men are arrested, and discharged as innocent of crime, but they have, nevertheless, been paraded around town, which we think is a greater disgrace to the community than to the prisoners. Where is the subscription list? Why is nothing done in the matter? " [6]

When a jail was finally constructed after 1883 it was surrounded by a high adobe wall with pieces of broken glass embedded in the top. This was done to cut any rope that might be thrown over the wall, or deter any hands that might be trying to get a handhold. In those days, it was not uncommon to see different colored glass embedded in the walls around private homes also, both for protection and decoration. Glass from liquor bottles was more plentiful in those days than any other kind of ornament. [7]

The Hanging of Tom Kerr

The *Pinal Drill* of January 26, 1882, relates this hanging story.

> "We had a fine Christmas here last night. The notorious cutthroat Tom Kerr shot and killed a young fellow here last night. He shot him down in cold blood, then emptied his pistol into the body, and then beat the corpse on the head with the empty pistol. But the renowned *Judge Lynch,* called a jury and gave him a trial, and then took him out and swung him to a limb the same night. The Coroner at Globe was notified and the body was swinging in the tree all day, but tonight it was cut down by some of his friends. There is some talk of stirring up the lynchers for hanging him, but no jury will ever convict them." [8]

Tom Kerr at one time was a special deputy sheriff under Sheriff Pete Gabriel. In fact, Kerr at one time was instructed by Gabriel to adjust the rope and spring the trap in a legal hanging that took place in the Old Ghost Town of Adamsville (4 miles in the desert, west of Florence). According to some sources in the early news accounts, Kerr had a record as a cold-blooded killer, from Idaho to Arizona, but there seemed to be some contradictadory information to this portrayal.[9]

Since Tom Kerr was a Captain on the Miner Expedition of 1870-71, and a noted mining pioneer in Central Arizona, I wondered just what precipitated his lynching. It had to be a dastardly act, because Kerr seemed to be a well liked man by his peers. My research located the following newspaper article in the archives of the Gila County Historical Society. It was written by Wade Cavanaugh in November 1961, newspaper unknown.

> "Tom Kerr ran a successful saloon in Globe and was thriving from numerous investments and was a constable, deputy sheriff, and member of the school board. Well liked by his acquaintances, he had one main failing; a love for the refreshments he sold in his saloon. It is known that he spent Christmas Day of 1882 drinking in a saloon in Pioneer. By that night he had become morose and belligerent, setting the stage for the incident that followed.
>
> "Shortly before midnight, a young stable hand, William Hartnett, entered the tavern to warm himself from the cold winter wind. Hartnett, an equally well liked youth of 20, had arrived in Pioneer from his home in Michigan a few weeks before. Kerr invited Hartnett to have a drink. The youth refused and Kerr, taking it as a personal insult, pulled out his revolver and shot Hartnett several times. Hartnett died on the barroom floor.

"The startled crowd of miners present grabbed Kerr as one of them ran to sound the alarm. As the several hundred miners in town learned of the shooting a wave of anger swept over them. A hasty jury was impounded, the verdict rendered, and Kerr was given one hour to write any letters he wished to compose. A rope was produced and soon Kerr's body was hanging from the sycamore tree. The following day Sheriff William W. Louther and Justice of the Peace J. P. Allen rode into Pioneer (Mining Camp) after hearing of the shooting and hanging. They discovered the frozen body of Kerr lying on a bench outside the saloon, the rope still around his neck. They could find no witnesses who would talk. That afternoon, in Boot Hill, a half mile from town, graveside services were held jointly for the killer and his victim. In the dusty files of the courthouse, the coroner's jury still shows the cause of death, *hanged by parties unknown*." (Pioneer was a mining camp named after the Pioneer Mine It was located in the mountains southwest of Globe and southeast of Superior, during the 1880s. The Pioneer camp was a stage stop for the stagecoach after the Howard and Reduction Toll Road was built in the mountains, between Globe and Riverside, in 1882. It is now just a memory and another one of the many small ghost towns of Arizona Territory.)

A Dead Man's Boots

This tale comes from the manuscript of Charles P. Mason, nephew of Charles G. Mason, one of the original owners of the Silver King Mine.

"As an illustration of how lightly things like that were thought of in those days, several of the prisoners that night, ran tick-tacks along the walls of the various cells and whenever things were still, there would be a gentle rapping first in one cell and then another. To be sure a Chinaman we 'had in,' became so frightened that he almost went crazy, and we had to move him to other quarters. But this only added to the joke. The next day the prisoners began gambling for the *dead man's boots,* and the excitement became so great that we had to go in and separate them."

The Death Watch and a Chief's Head

Mason's manuscript relates another tale about five Indians under sentence to be hanged.

"At another time, we had five Indians in the jail under sentence to be hanged, and with the death watch over them. One was a chief. On the morning before the day they were to string, the jailer entered the cell to find three of them dead. They had put blankets over their heads and then wound their gee strings about their necks and pulled. The other two were probably too cowardly. They got theirs next day when a rope with jars of quicksilver, weighing slightly more than thc Indian, fastened to one end, were passed

> over pulleys and around the Indian's necks. The jars were sprung instead of the man, but the effect was the same, as the sudden jerk up broke their necks."

Mason continued the tale by relating the details of a practical joke played on a certain Dr. Adler, who for some reason wanted the chief's head before he was buried.

> "I overheard Dr. Adler talking to Wood Porter about getting the chief's head. The chief was one of the Indian's who had committed suicide, and the three bodies lay in a grave, lightly covered, awaiting the bodies of the other two Indians. The Doctor and Porter had planned to drive out that night and get the head. So Charlie Rapp and I got a couple of shotguns and went out and laid for them. They had gotten the head in a sack and were just climbing into the buggy, when we let go the triggers. Then we let out a whoop like Indians, and we sure could hollar when we had a few drinks in us. Well, the horse ran away and smashed up the buggy, but the two men escaped with nothing more than a few bruises and the loss of the head. The next day, a Mexican boy found the sack with the head and was so frightened when he opened it that he ran all the way to town and for a long time was afraid to tell anyone. I know that the doctor afterwards got hold of the skull some way, for I saw it later in his office. It was a long time before he found out who did the shooting. 'Doc' was a young man with a long black beard, bushy hair and thick lensed glasses. He looked more like a Russian anarchist than a doctor. He is now practicing in San Francisco. Has a small office and lots of women patients." [10]

The Lynching of Ned Fales

One of the most notorious incidents at Silver King took place on or about August 9, 1884, and was the Fales-O'Boyle feud which ended fatally for both men. The following is a brief outline of that tale of tragedy and frontier justice. Billy O'Boyle was a well-liked owner of hotels and saloons at both Silver King and Pinal. In his eating place at Silver King, O'Boyle had a young pretty Irish waitress whom he was quite fond of. A frequent visitor to Silver King was Ned Fales, a handsome colorful cowboy, typical of the kind found in the wild-west stories. Ned made it a common practice to ride into saloons on horseback and shoot out the lights. He liked the waitress, and she also thought highly of him, which was said to have caused some of the animosity between the two. At any rate, harsh words were exchanged between Fales and O'Boyle. Fales invited O'Boyle to drink with him, but he refused, and the cowboy threw a glass of whiskey in O'Boyle's face and walked out of the saloon. O'Boyle got his six-gun out and started shooting at Fales causing a flesh wound. Fales turned quickley and fired one shot, which struck O'Boyle in the heart, and he dropped dead instantly.

Following the shooting, Fales was arrested and taken into custody by a deputy sheriff and kept at the Lewis Corral. Jack McCoy, a brother-in-law of O'Boyle, was also desperately in love with the waitress, and saw the opportunity to get rid of his rival, Ned Fales. After the shooting, it was said that he stirred up the miners and finally aroused them to take the law into their own hands. McCoy was said to have led the lynching party of about 25 men, and Fales was hanged that night from a pole behind the Lewis Corral, opposite Hank Murray's saloon. The two companions of Fales, although they were not participants in the shoot-out, were considered as undesirable citizens and were duly notified that they must leave the country at once, which they did.

It was later ascertained that Fales was an assumed name, and that the cowboy's name was Goode, and that he was from a wealthy, influential St. Louis, Missouri family.[11] There is no documentation that McCoy, or the miners who hanged Fales, were charged with a crime, or that McCoy won the hand of the fair waitress. Later, after the Silver King Mine closed, the Lamb Brothers of Mesa bought O'Boyle's Silver King Hotel at Silver King. They took it down board by board, every board numbered, and transferred it to Goldfield City in 1893, just east of the Apache Trail, where it was completely reassembled and put up exactly as it was at Silver King.

String-Em Up

There were several times during the 1880s when newspapers wrote about stage robberies and the statements printed would say something like this. These stage robberies are getting to be quite frequent again and it would be a good idea to *string some of them up again* and put a stop to this business, for a while at least. The sentiment towards stage robbers and other criminals was to "hang em high," when they were caught as a deterrent to other would be hold-up men, or at least make sure the ones they had in custody would not be repeat offenders.

A Lesson on How to Hang-Em High

Methods of Hanging, was the subject of a September 5, 1885 news article in the *Pinal County Recorder.*

"The mode of carrying out the sentence of the law, 'be hanged by the neck until you are dead,' has usually been left to the discretion of the hangman, the law taking no cognizance as to what be the proximate cause of death. Culcraft invariably adopted the short drop of about two feet and a half, and if I may judge from some specimens of his ropes, which are still to be seen at Kirkdale. A slow process of asphyxia must have produced death. Marwood adopted what is generally known as the long drop, of which he was supposed by many to be the originator, though it was used long before his time, both in Paris and in Ireland.

"To Professor Haughton, we are indebted for a scientific exposition of the rationale of the long drop, and of the mode in which death takes place. Dr. Haughton also gives an elaborate explanation of the American method, which is a scientific modification of the old naval method of running the culprit up the yardarm.

"Having now briefly referred to the different modes of hanging, which have been adopted in executing criminals, we will be better able to judge which is the best and most practical method, when we have considered the various causes of death. Professor Tidy says, 'in hanging, as in drowning, death does not always take place in exactly the same way.' Thus, it may result from (1) asphyxia, (2) cerebral hyperemia, (3) a combination of asphyxia with apoplexy, (4) syncope, (5) injury to the spinal cord and pneumogastries, (neuroparalytic death)." [12]

Pete Gabriel—-One of the West's Toughest Lawmen

The next four tales all involve John Peter "Pete" Gabriel, who was at one time Pinal County Sheriff and later Deputy U. S. Marshall. Gabriel was one of the most feared lawmen in Arizona Territory. He was involved in many of the investigations surrounding stage robberies and shootings that took place in and around Silver King and Pinal. The *Pinal Drill* on September 30, 1882 describes Gabriel as *a terror to thieves*. He never received the notoriety of a Wyatt Earp or Wild Bill Hickock because many of his accomplishments were never highly publicized. Many of the old manuscripts and local news accounts depict him as one of the toughest lawmen to ever pin on a badge. Although the following tales do not directly involve either Silver King or Pinal, they indirectly affected the citizens who lived there, because of their notoriety, and because the participants were well known by the citizens of both towns.

Death of a Horsethief

This tale comes from the *Pinal County Record*, September 17, 1886.

> "In attempting to arrest a horsethief and desperado at Florence Wednesday morning, Sheriff J. P. Gabriel was obliged to shoot the man. The circumstances as near as we can learn, are as follows: The horsethief was on Bailey Street, back of J. M. Ochoa's store, and the sheriff went there and demanded his surrender. The man refused and at the same time put spurs to the horse and started off, and also making a break to draw his gun. Sheriff Gabriel fired at the man and the ball struck him in the lower part of the back, passing through him and striking the horse in the back of the head killing it. The man lived until yesterday morning, when he died. So far, we have heard of no one blaming Sheriff Gabriel for the course he pursued in making the arrest." [13]

A Lynching Next to St. Elmo Saloon

The most dramatic robbery in the Globe area took place on a hot August day in 1882. It was the 20th of the month, and Andy Hall and Frank Porter were on the mule express, in lieu of a stage, transporting the cash payroll between $5000 and $10,000 dollars for the Mac Morris Mine and the mail.(Mac Morris was also referred to as Mack Moris.) The Mac Morris Mine was one of the mines that listed its bullion shipments in the papers. Hall was the Wells Fargo express carrier. Porter was the mail carrier. They had six mules loaded and were taking the usual route through Pioneer Pass to Globe. The stage route ended at the top of Pioneer Pass, and from there the mule express took the pack-trail down the mountain into the town. As the mule train entered the gully a few miles from town, gunfire suddenly blazed from the rocks above them. The robbery was supposed to look like an Indian attack, thereby covering up all traces of a robbery. The plan went like this. While the victims fled, the robbers would make off with the money. Porter was unharmed and spurred his horse toward Globe seeking help, while Hall was slightly wounded in the thigh and hid in the brush beside the road out of sight of the robbers.

As they approached Globe, the robbers met Dr. F. W. Vail, a druggist and prospector. When Vail asked what the shooting was about, the big man stated that it was the Indians again. Vail agreed to accompany the robbers back to Globe, not knowing that they had just held-up the mule express. As Vail dismounted to hide his pack and lighten his load, the big man, who then told his smaller companion to shoot twice into the doctor, shot him in the back. A mile or so from where they had left Vail they were accosted by Hall, who at first thought they were Indians. Hall had not seen either man during the robbery but became suspicious during the trip back to Globe, when he saw the "burden" they were carrying. When almost to Globe, he was suddenly shot in the back by the big man. After Hall fell, he turned and fired his revolver at the big man, apparently without effect. He was shot eight times in the chest around the heart. Andy Hall had been the youngest member of John Wesley Powell's expedition down the Colorado River (through the Grand Canyon) some years earlier, and had been a real stalwart on the trip during it's trying times. He was well liked in the community.

Meanwhile, Porter had arrived in Globe and sounded the alarm. A posse was quickly formed and gave pursuit. They found Hall dead, and later located Vail who was still barely alive. He described his attackers as two men from Globe, one a big dark complected man and the other a smaller man. After checking the scene, it was readily apparent that one of the footprints came from a man with very small feet, first thought to be that of a woman. Dr. Vail died before they got him to Globe.

The bodies were taken to town where they were viewed by most of the men in Globe. A service was held at the Methodist Church in Globe. The Reverend D. W. Calfee "committed their souls to their Creator" and the men were buried nearby. The lawmen began their search for a man with very small feet. It did not take long for them to focus on Lafayette Grime who was the dance master in town and had very small feet. Fate, as he was called, was arrested after questioning by Pete Gabriel. Fate broke down, told the whole story and put his confession in writing. He was only 19 years old, and eager to confess to his part. He implicated Curtis E. Hawley, a man who supplied lumber for the mines, and his own brother, Cicero Grime, a photographer in town. (Cicero Grime often advertised in the *Arizona Silver Belt*, located in Globe.) Fate had been a recent member of the Globe Rangers, and saw distinguished service against the marauding Indians. Since the sheriff was unable to protect the settlements from Indian attacks, the citizens of Globe formed the Globe Rangers and the Globe Home Guard. These consisted of volunteers who banded together to assist far flung ranches and settlers from Apache attacks.

Hawley and Cicero Grime were immediately arrested and confessed their part in the crime. The citizens of Globe were aroused and wanted action, so an impromptu hearing before a Justice of the Peace was held, and the three men were held for grand jury action. That didn't satisfy the townsmen however, as they demanded a more severe form of justice. A bitter jurisdictional dispute then followed. Gabriel tried to take the men into his custody for safekeeping on mail robbery charges, and to avoid the lynch mob of 100 men who were shouting "hang the damn murderers," but Sheriff Lowther wanted to keep them in his jurisdiction for murder. Gabriel told the prisoners, "Sorry men, I can not save you. You will be dead within the hour." Almost immediately afterward that night, a lynch mob gathered before the one room adobe jailhouse and took the men by force. A plea was made for a hearing and a hasty hearing was held. Hawley and Fate Grime were to be hanged, while Cicero Grime was to be sent to prison for 21 years.

The following telegram was sent to J. J. Vosberg, the Wells Fargo Agent at Globe, after he had wired his Superintendent Valentine news of the capture of Hawley and the Grimes brothers, adding that the money had not been recovered. Valentine's answer was a classic representation of popular sentiment during that sanguinary time period in Arizona Territory.

TELEGRAM

AUGUST 20, 1882

J J VOSBERG

WELLS FARGO AGENT

GLOBE ARIZONA

DAMN THE MONEY HANG THE MURDERERS VALENTINE

Irene Vail, the widow of Dr. Vail, pleaded for the lives of the men along with others with some success, as Cicero Grime was spared the hangman's noose. Hawley and Fate were forthwith taken to a large sycamore tree on Broad Street in front of St. Elmo's Saloon and Stallo's Hall. Before the hanging, Hawley was allowed to write out a will, leaving his wife and children $5000 that he had saved. Fate requested that he be allowed to remove his betraying boots before he was hanged and promptly fainted as the noose was prepared for him. Both men were then hanged. Fate died instantly as his neck was broken, but Hawley was not so lucky. His slowly strangled breath came out in rasps, and could be heard a block away for several minutes.

The bodies hung there until late afternoon of the next day. The sheriff refused to cut them down, saying he had nothing to do with the execution. The lynch-mob took up a collection and paid to have the bodies cut down

Globe's hanging tree cir. 1880s — Gila County Historical Society

and buried. The sycamore was referred to for years as the "Hanging Tree," and the area next to it as "Dead Man's Flat," although no one was ever hanged there afterward. [14] The "Hanging Tree" was badly damaged in the disastrous fire of June 9, 1894 that also burned down many of the nearby buildings. It was cut down a short time later by Henry Mounce, under orders of Sheriff John H. Thompson. Opponents of cutting it down said it should have been allowed to stand as a reminder of *the spirit of justice* of the early years.[15]

Cicero Grime was sent to the infamous Territorial Prison at Yuma. While there he feigned insanity, and was committed to an asylum in Stockton, California. He escaped and traveled to Oregon, where his family later joined him. Yuma Prison was a terrible place to live. The sun caused the temperature to be 120 degrees during the day. The average life span of prisoners was only 10 years. It was said that the only relief a prisoner got was escape.[16]

According to the *Arizona Silver Belt,* during the 1880s and on into the 20th century, the St. Elmo Saloon was a well known establishment where many a lonely, weary miner met his "soiled doves" for a night of pleasure and companionship. It was said that eight "soiled doves" feathered their nests in St. Elmo's. The St. Elmo also boasted its collection of roses; Big

Rose, Dusky Rose, a mulatto, and Jew Rose, a very attractive Jewess from San Francisco. She had the reputation of being one of the best faro dealers

St. Elmo's Saloon, Globe cir 1880s — Gila County Historical Society

in Globe. Also at the St. Elmo were Gladys, an Italian, and Violet and Nettie. [17]Another interesting fact about St. Elmo's is that during 1906 they were showing said to be showing "moving pictures" there every night at 8 o'clock. [18]

"Red Jack" Elmer and the Riverside Stage Robbery

On the night of August 9, 1883, the stage between Florence and Globe was held up about a half mile from the Riverside Station. It was rumored that the stage was carrying a payroll of $30,000, but this was not true, as it was only carrying $1000 in gold and $2000 in silver. Suddenly shots rang out without warning as the stage was going up a grade. John Collins, the guard, was shot in the chin and throat with buckshot and was killed instantly. Sheriff A. J. Doran (who also was at one time the superintendent at the Silver King Mills), who was at Pinal at the time, was telegraphed. Doran quickly gathered a posse to search for clues as to who had committed the robbery. He had the posse meet him at the Riverside Stage Station.[19]

Someone recalled that they had seen "Red Jack" Elmer (aka Almer) on the stage before the robbery. Others also saw them heading south, towards the San Pedro area, on fast horses right after the robbery. According to Charles P. Mason, he recalled that "Red Jack" would "lay around" Silver

King or some other prosperous camp for several days waiting for them to make a shipment. He would then take passage on the same stage as the bullion or payroll. His presence on the stage was the signal to his companions to hold up that stage. Mason called "Red Jack" the "tough one of the outfit." He stated that he also recalled that Len Redfield had a ranch on the San Pedro near Reddington and posed as being a respectable rancher.[20]

It seems that "Red Jack" and a man named Hensley had been seen on the trail to Len Redfield's ranch right after the robbery. Sheriff A. J. Doran and his posse headed south, although the ranch was about 75 miles away. He said that it was a gamble to go that far to question two men and perhaps offend the wealthy Redfield with "only the barest possibility that there would be any connecting link to the robbery." [21] Sheriff Doran decided to take that chance, as it was his best lead. He had along with him on the posse ex-sheriff Pete Gabriel, Deputy Sheriff's G. D. Scanlan, Fred Adams and Special Deputy W. Harrington. Charles Mason, who worked at the *Arizona Enterprise* in Florence, also joined the posse. At Dudlyville, they learned that two men leading a packhorse were headed south at a fast clip. Riding south along the river, Pete Gabriel found some $20 dollar gold pieces, and at one point fished some masks made of gunnysacks from the river.[22]

After arriving at Redfield's house, Pete Gabriel found a Wells Fargo Box under some hay in the barn when he went to give some hay to the horses. He could not believe it, as he always thought Redfield a friend and respectable man. At Redfield's house they found some money from a previous stage robbery and a U. S. Mail sack. The posse then arrested Redfield and Joe Tuttle. Tuttle gave a complete confession and implicated the others mentioned above. The money was cached on the way to the ranch, as the silver was too heavy and inconvenient to carry. It was hidden in an arroyo about thirty miles from the scene of the robbery and later recovered, but the gold was not recovered. Meanwhile, Special Detective Casanega and Detective Thacker of Wells Fargo & Company were also involved in the investigation and pursuit at this point.

During a further search of the ranch $14,000 was found under a post and dirt floor in a small adjoining adobe building, along with a missing book from a robbery which had occurred the year before. It seems that Len Redfield's ranch on the San Pedro was a rallying point for the stage robbers, one of a series between Riverside and Benson, from which they would obtain fresh horses for flight. On the 16th of August, they arrested Frank Carpenter as an accomplice, as it seems that he gave fresh horses to the robbers during their get-away to the Redfield Ranch.

The prisoners, Tuttle, Redfield and Carpenter were jailed at Florence. The citizens were out for blood when they found out who had been behind the series of stage robberies and murders. A vigilante committee began talking of a lynching. Sheriff Doran tried to talk the leaders out of it and said that he would strongly resist any attempt to take his prisoners. They were examined in Justice Court and held for trial. The *Pinal Drill,* on August 25th reported: "The evidence is strong against them and it is said upon sufficient proof, they will be hanged without waiting for the next term of court."[23] That was a nice way of saying that they would soon be *lynched.*

U. S. Marshall Evans showed up to take the prisoners with a court order for mail robbery, but Sheriff Doran wanted to retain custody of his prisoners for murder charges. In fact, while the law enforcement officers were arguing over jurisdiction, the vigilantes broke into the rear of the jail and promptly hanged Redfield and Tuttle from the jail rafters. *Judge Lynch had again dispensed frontier justice.*

The *Pinal Drill* reported on September 8, 1883:

> "On Monday last, about 10 o'clock fore-noon, a crowd of men entered the jail and hung Tuttle and Redfield to some of the rafters. The crowd was composed of armed citizens unmasked." [24] Charlie Mason described the lynching as follows: "It is said that Redfield did a good deal of talking, but Tuttle took his medicine in silence. When the citizens were through, they went through the front door and told the U. S. Marshals that they could have their men now." [25]

Meanwhile, "Red Jack" and Henseley had fled towards the Rincon Mountains where they hid until spotted by a miner near Sulphur Springs. Posses under Sheriff Doran and Sheriff Bob Paul of Pima County were notified and gave chase. "Red Jack" had vowed not to be taken alive. Both men were pursued to a gully near Wilcox, where a gun battle ensued and they were shot and killed. It was later said that "Red Jack's" loss was not mourned.

Frank Carpenter died on November 13, 1883. The cause of death was not given, but it was of natural causes, perhaps a heart attack. The *Pinal Drill* reports on November 24, 1883

"We saw him pale-faced, standing before the stern tribunal of Justice. His long incarceration had bleached his features and shattered his frame. He was acquitted on one charge and let to $1000 bail on another. He had witnessed the hanging of his uncle (Redfield) and Joe Tuttle by the vigilantes in the jail and of all that tragedy. Not even a single whisper betrayed the terrible fact that four lives had been sent to their long rest connected with this dark episode. It seemed that the law lay dead beneath the iron heel of vengeance, and the dark veil of oblivion covered all, except Frank Carpenter. Now he is also dead. The grave has settled all." [26]

Death of Frank J. Carpenter.

The *Benson Herald* gives an interesting account of this young man's history. "He was born in Iowa 1856. He came to Arizona in 1876 with his uncles Leonard G. and Henry T. Redfield, and settled 40 miles below Benson on the San Pedro. He was industrious, and sustained an excellent character. He was amiable and obliging in his disposition. He never used tobacco, whiskey, or rude language, He died at Taylor's Ranch on the San Pedro in this county 13th inst. His funeral was the largest ever had in Benson. The People turned out "en masse," and the business places were nearly all closed."

We saw him pale faced, standing before the stern tribunal of Justice, where presided his Honor Judge Sheldon last term of Court at Florence. His long incarceration had bleached his features and shattered his frame. He had been imprisoned on two charges as accomplice connected with the Joe Tuttle stage robbery and murder. He was acquitted on one charge and let to $1000 bail on the other. He had witnessed the hanging of his uncle and Joe Tuttle by the vigilantes in the jail and of all that tragedy, his case raised the only sound that reverberated through the corridors of Justice. Not even a single whisper betrayed the terrible fact, that four lives had been sent to their long rest connected with that dark episode. It seemed as if the law lay dead beneath the iron heel of vengeance, and the dark veil of oblivion covered all, except Frank Carpenter. Now also he is dead. The grave has settled all.

"Death of Frank Carpenter," 1883

— Pinal Drill

The Deadly Duel Between Pete Gabriel and Joe Phy

On May 31, 1888, one of the west's most *classic shoot-outs* took place in Florence, Arizona Territory, at John Keating's Tunnel Saloon. It was almost as if the western writers' had written a script for this deadly duel between two former friends. Each man was considered worthy of a place among the best gunmen of the day. Phy was known to practice with his six-guns for hours, almost every day. Phy's expert marksmanship was so highly touted, that it was said he "could get on a horse and keep a tin can rolling all the way to Ray from Florence, if his ammunition held out." [27] Both men had been involved in numerous shooting scrapes and had come out winners. Pete Gabriel had been the Sheriff of Pinal County and had also been a United States Deputy Marshall. Pete Gabriel married Carrie Wratten on October 5, 1880. Judge John D. Walker perfoirmed the ceremony at Florence. It seems that Pete was very jealous of his young wife, and that Phy was very friendly towards her. This may have led to some of the animosity between the two.

Gabriel was well known to the residents of Silver King and Pinal being involved in the arrest of several lawbreakers there. Joe Phy had worked for the Silver King Mine as a wood contractor. The *Arizona Enterprise* September 5, 1885 reported: "Mr. Joe Phy arrived from Tucson this week with his teams and three wagons. He was contracted to deliver 1000 cords of wood at Pinal for the Silver King Mine." [28] Phy had also been a law enforcement officer. In fact, Phy had at one time served under Gabriel as his deputy. The two men were well-known and respected in the territory. They had mutual friends in the territory, and it was a shock when their famous gun duel took place in the Tunnel Saloon in Florence. (There also had been a Tunnel Saloon in Pinal for many years.) It made the newspapers for days, and would be the topic of conversation whenever their contemporaries gathered for the next half-century.

Carrie Wratten Gabriel cir. 1880s
— Arizona State Archives, Phoenix

On one occasion, when Gabriel and Phy had been in pursuit of an outlaw, they camped together day and night for a month. The *Arizona Enterprise* August 15, 1885 reported: "Sheriff Gabriel and Joseph Phy returned Tuesday from their protracted chase after the person who has Goldburg's horses." [29] This and other duties made the two good friends. You can't live on the trail with someone, in constant danger for a month, without making a lasting friendship.

It seems that bad blood that developed between the two was caused when Gabriel did not support Phy for Sheriff after he had said he would. Phy had embarrassed Gabriel by severely beating up one of his (Phy's) outspoken critics and this did not sit well with Gabriel, and apparently he changed his mind about supporting Phy. Phy had always claimed later that Gabriel had promised to help make him the Sheriff when he quit the office. Phy was nominated for Sheriff of Pinal County in 1886, but lost the election and blamed Gabriel for the loss. The *Arizona Enterprise* reported Phy's loss of the election in the November 20, 1886 edition. [30]

After the election, Phy never lost the opportunity to blast Gabriel behind his back and was very outspoken in his criticism of him. Gabriel did not reply with criticism of Phy, but avoided him whenever possible. Despite Gabriel's efforts to avoid the showdown, all of Florence knew it was only a matter of time. That time came a year and a half later.[31]

For several weeks during May of 1888, the community felt that a showdown was coming. The inevitable showdown occurred on Thursday, May 31, 1888. Gabriel was in John Keating's Tunnel Saloon socializing with friends when he glanced out the window and saw Phy watching

Keating's Tunnel Saloon cir. 1880s — *John Swearengin Collection*

him. Word was that Phy was "gunning" for Gabriel. Witnesses say Phy approached and entered the front doors of the saloon. They differ as to whether he had a gun in his hand or not, although most witnesses stated that he had a pistol in one hand and a knife in the other. There was a quick exchange of shots. The lights were shot out and then the gun battle moved outside. Each man emptied his revolver, and both men were seriously wounded. Phy finally went down after being shot in the abdomen, hip, thigh and wrist. Gabriel was shot in the chest which went through his one good lung and through the groin. Phy was taken to Doctor Harvey's house and treated, but he died about midnight. Gabriel would not use Doctor Harvey who treated Phy, his enemy, and waited until the next day for Doctor Sabin to come from Sacaton, about 18 miles away, to treat him.

The *Arizona Enterprise*, Florence A.T. June 2, 1888 reported the shoot-out as follows:

A DOUBLE TRAGEDY

An impromptu duel with fatal results, Joseph Phy and Ex-Sheriff Gabriel Meet in Deadly Encounter-One is Dead and the other Mortally wounded

"About 8 o'clock Thursday night the people on Main Street were suddenly startled by the loud reports of pistol shots in rapid succession proceeding from the Tunnel saloon. No previous noise of an altercation was heard. It was at first thought that someone under the influence of too much booze was emphasizing his hilarity in harmless pistol practice. When two struggling forms came through the screen door into the street, still firing at each other, and one fell on the street and the other staggered down the sidewalk towards the Florence Hotel, the fact was recognized that a fearful tragedy had been enacted in the semi-darkness that surrounded the entrance to the Tunnel Saloon.

Joe Phy cir. 1879
— *John Swearengin Collection*

"Eleven shots were fired, and the echo of the last had scarcely died away ere dozens of people had gathered about the prostrate form of Josephus Phy. Joe was a well known resident of this place, who was evidently badly wounded, but was still flourishing a keen bowie knife in a threatening manner, that for a short time repelled the friendly offices tendered him. He was carried to the stage corral where Doctor Harvey examined him—he died about 12 o'clock the same night after receiving the shots.

"The other party to the tragedy, Ex-Sheriff Gabriel, had proceeded down Main Street after the shooting, to the entrance to the O. K. livery stable where he fell. He was assisted to a room, where his wounds were stanched, until a doctor could be summoned from Sacaton, 18 miles distant. Dr. Sabin reached town about half past 1 o'clock yesterday morning and found Mr. Gabriel bearing heroically under his severe wounds."

The article went on to describe the fact that an inquest was held by Justice John Miller on June 9, 1888 and the coroner's jury of C.F.Palmer, W.J.

Bley, N.M. Hickey, Jas. Matthews, W.V. Elliot, Geo. E. Evans, L.K. Drais, C.H. Starr and Jacob Connor rendered a verdict of self-defense. It also stated that Gabriel was improving, and that he would likely be about again in a week.[32]

Pete Gabriel cir. 1870s
— *Bancroft Library, California*

Although shot in his only good lung, Gabriel lived another 10 years, then died at his mining camp near Arizona's Dripping Springs Mountain. It was said by his close friend Mike Rice, on occasions before he died, that he would wake up in his sleep and draw his six-shooter and cry, "Joe, Joe, Joe", and with each "Joe" he fired into the air. The *Florence Tribune* reported his death on August 6, 1898, stating: "The Ex-Sheriff J. P. Gabriel died last Saturday morning at 3 a.m. at his mining camp on Mineral Creek. His partner found him, a Mr. McCalister. The cause of his death is not know, but it is supposed the water he was drinking was impregnated with arsenic and other mineral poisons. The body was buried at the mine." [33]

Mike Rice stated later that he thought Pete was "done in" by friends of Joe Phy. In June of 1936, a bronze plaque was cemented in a headstone of rocks and placed at his gravesite by friends, led by Johnny Gibson of Florence, who journeyed to his isolated grave east of Ray, Arizona. Gibson was one of four men who helped bury Gabriel in 1898. The inscription reads " Pete Gabriel, Died Aug. 1898, Calif.-Ariz. Peace Officer, Pinal Co. Sheriff 1879-80-81-82-85-86." The grave is far removed from regular travel paths and can only be reached by foot or horseback. It was said that he was absent from his mine several months, and when he reached the mine he drank the water in the shaft, and that apparently the burro also drank the water. The burro was found dead at the mine site, and Gabriel was desperately ill. According to a man named McCalister who found him, Gabriel died within a few hours. [34] Pete Gabriel, the fearless lawman of many gun-battles, died with his boots off, far from the scenes of many of his famous gunfights.

Gabriel began his illustrious career as a deputy under Sheriff Billy Rowland in Los Angeles, California, but after being shot in the chest by

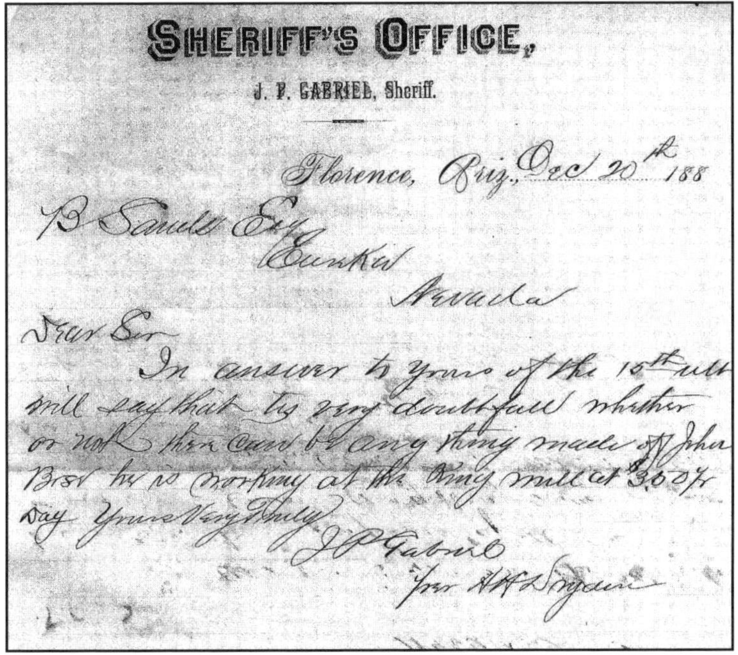
SHERIFF'S OFFICE,
J. P. GABRIEL, Sheriff.

Florence, Ariz., Dec 20th 188

B Samuels Esq
Eureka
Nevada

Dear Sir
In answer to yours of the 15th ult will say that tis very doubtfull whether or not there can be any thing made of John Brown he is working at the Army mill at $3.00 pr day Yours Very Truly
J P Gabriel
per H H Snyder

Gabriel Letter cir. 1880s —Sam Michael collection

a squatter that he knew, he resigned and came to Arizona where a stage company operating between Williamson and Ehrenberg first employed him. After killing a man in a card game in Prescott he went to Nevada and was said to have cleaned out a gang of hoodlums and claim jumpers. He returned to Arizona, and in 1878 was elected to Sheriff of Pinal County where he served for 8 years. During his career as a lawman he was also a U. S. Deputy Marshall.[35]

During January of 2003, I went to Florence, Arizona, to watch a reenactment of the famous Gabriel-Phy shootout. The reenactment was conducted on the main street of Florence, close to where the actual shootout took place at the site of the former Tunnel Saloon. The Tunnel Saloon is not there anymore, as another building was built on the site, but the flavor of the old west gunfight was very adequately portrayed. Actors and actresses from the Old West Reenactment Company put on the perfomance and followed the historical script very close to the actual shootout, as recorded by historians of the day. After the performance, a reception was held at the Florence Museum. I met the great-grandaughters of both Gabriel and Phy at the reception.

Judy Pintar & Margaret Down, Gabriel & Phy's great-granddaughters, 2003 — Author's photo

The Apache Kid and the Death of "Hunkydory" Holmes

The forever prospector, "Hunkydory" Holmes, gave up the prospector's pick and shovel at the end of the 1880s and became a deputy sheriff. On November 1, 1889, while transporting the infamous Apache Kid and seven other Apache convicts and one Mexican convict by stage to the Territorial Prison in Yuma, "Hunkydory" Holmes was killed. Deputy Sheriff Holmes and Sheriff Glen Reynolds stopped the stage at a place near Riverside Station, because of a steep grade, and let some of the prisoners out to walk to lighten the load. This occurred at a steep hill by the Ripsey Wash on the Old Florence-Kelvin Highway. After the driver Eugene Middleton and the shackled Apache Kid went ahead on the stage, the other prisoners suddenly attacked and overwhelmed the two officers. Holmes and Reynolds were both killed. Middleton was shot in the neck and was temporarily paralyzed and left for dead.

The Apache Kid and his companions stripped and mutilated Reynolds and Holmes and then fled into the hills. The Mexican, Jesus Avota, took one of the stage horses and rode to Florence for help. He was later pardoned for this deed. Thus began the largest manhunt in Arizona history. All of the escaped Apaches were either captured or killed, except the Apache Kid. He was never captured again, although some bounty hunters and old-timers claimed to have killed him at various locations in Arizona. "Hunkydory" Holmes, who had survived Indian attacks, disease, and the rough life of a prospector, and who boasted of being one of the area's earliest settler's, was dead at forty-four.[36] I have been to this site near the old Ripsey Wash many times and there is a very steep hill there. I can envision the stagecoach stopping and letting out the passengers while the horses strained to pull the stage up that steep hill. Traveling from Florence there is a great view of the Ray Mine from this hillside view point.

James Addison Peralta-Reavis —-the Infamous Baron of Arizona

Author at Apache Kid's cave, 2005 — Jack Carlson photo

I certainly would be remiss if I did not include a story on James Addison Peralta-Reavis, the scalawag who tried to *steal* most of Arizona's mineral wealth by virtue of a fraudulent Spanish Land Grant. James Addison Reavis was a former St. Joseph, Missouri street car conductor who, while working in the land claims office in San Francisco, became excited with the possibility of obtaining his own land grant. He had been scrutinizing the Old Spanish Land Grants for some time. He became obsessed with not only obtaining a land grant, but also acquiring a huge tract of land and the self-proclaimed title, Baron of Arizona.

James A. Reavis's Spanish Land Grant claimed title to over 18,000 square miles, including 3.5 million acres in prime mining country of Arizona. His claim included the cities of Phoenix, Tempe, Mesa, Casa Grande, Florence, Globe, Safford, Clifton and Morenci. This area included several hundred mines, including the rich Silver King Mine and the huge copper deposits of Globe, Miami, Ray, Clifton and Morenci. His claim also included the rich gold bearing mines of Goldfield near the Superstition Mountains. He also claimed key areas of New Mexico, and the Southern

Pacific Railroad's right of ways. Reavis claimed this land by virtue of his marriage to Sofia Peralta, the alleged last living heir of Don Miguel Peralta de la Córdoba y Sánchez, caballero de los Colorados con grandeza de España. In reality, Sofia was not royalty, but a peasant girl who Reavis transformed on paper into the heiress of a vast Spanish Land Grant. Shortly after his marriage Reavis came to the Phoenix area to look over his empire and to deal with the people who were "squatters" on his lands.

The Reavis Fraud.

"The Red Barron" was brought into court on an examination in a suit brought by Attorney General, Churchill at Phoenix to quiet title to a valuable tract of land adjoining Phoenix, and the testimony shows the character of the scheme, and who are his aids and abettors. The Barron makes a sorry show in court, he is anxious to have it understood that he is backed by the C. P. R. R. Co., the S. P. R. R, Co., and the Silver King Mining Company. He swears that Barney paid him $5000. 00, and, to quote the *Phoenix Herald* "he very adroitly fails to swear that this money was paid to him on his claim through the Peralta Grant, nor does he remember what bank paid the money, or on what bank the check was drawn." We know that he gave the Silver King Co., a quit claim deed of the Silver King property, but we venture to say that the favor of receiving that for nothing is the extent of the aid he has received from the Silver King Company and we sincerely regret that the swindle should be thus placed, apparently under the *color de rosa* of Mr. Barney's purse and name. We are not surprised at the S. P. R. R. Co., lending him a hand. Their effort to absorb our Territory is too notorious, but that they should be driven to association with a wandering tramp and use him as a pretense and an instrument to rake up mexican traps for a blinder, that is surprising, and shows to what depth of contemptible meanness rich men will sink, under the craving passion of avarice. Mr. Churchill has done the country a great service in thus laying bare the foul plot. The Red Barron and his henchmen may have money, but the country is mightier than them all, and in this country a public wrong will not succeed.

Reavis Fraud story 1884

— Pinal Drill

The vastness of the land grant not only contained enormous wealth, but the grant held the possibility of depriving thousands of innocent people of their homes and possessions. Many worried homeowners paid money to Reavis for quitclaim deeds. The Silver King Mine and the Southern Pacific Railroad was said to have paid huge sums of money for their deeds. The *Pinal Drill* June 16, 1883 edition, stated that Reavis had made his quitclaim deed to the Silver King Company for the mine and mill grounds. The *Pinal Drill* also stated on February 23, 1884 that the Silver King Mining Company and James Barney paid him $5000 and that this money was paid to him through the Peralta Grant. But no one knows just how much money the mine and the railroad paid Reavis, and it was rumored to be $50,000 apiece. At this time James and Sophia Peralta-Reavis were living in regal splendor, with homes in the United States, Mexico City and Madrid, Spain. Reavis had retained Robert Ingersol, one of the great lawyers of his time, who apparently regarded the claim as legal.

Reavis called himself the "Baron of Arizona," a title that he wore with dubious distinction; however, his house of cards soon came crashing down. It seems that his nemesis Royal Johnson was once again appointed U. S. Surveyor General. Johnson put his watchdogs on Reavis, and after much digging, document examination, and taking witness statements, they unearthed the evidence needed to discredit Reavis. Johnson publicly declared that Reavis was a fraud. Reavis could have remained the Baron and gone on living his opulent lifestyle, but he filed suit against the government for 11 million dollars. This was really his undoing. Some people just can't deal with success. Reavis lost the suit against the government and they proved him to be an elaborate scam artist.

Tom Weeden, the editor of the *Arizona Enterprise Newspaper,* and a document collector, had one of his assistants, William Truman, who was a calligraphy expert, closely examine some of the Peralta Land Grant papers. Truman found that they contained writing made by a steel tipped pen, not a quill pen. Steel tipped pens were not invented until the 19th century, so the Peralta documents from the 18th century had to be fraudulent. In 1895, the U. S. District Court in Santa Fe, New Mexico declared the Peralta-Reavis claims to 12,750,000 acres of land in Arizona and New Mexico fraudulent. A real Peralta once governed Santa Fe and this was a strange twist of fate for Reavis. Reavis then went to trial for filing false claims against the government. But, by this time Reavis was broke, and he had to represent himself. The case against him was substantial. He brought his wife and small twin sons to court each day, expecting sympathy. James Addison Peralta-Reavis was convicted of perjury and sent to the penitentiary for two years, but he was released after only twenty-two months. Reavis' wife Sofia had divorced him, and he ended up walking around the Phoenix area spouting grandiose dreams of making the Salt River Valley a garden paradise simply by taming the Salt River and using irrigation. He had spouted these same dreams about 20 years earlier, and had he pursued these concepts instead of scheming to defraud the government and the Arizona people out of millions of dollars, he well may have become famous, not infamous.

Thus ends the saga of a man who would be King, or at least the only known Baron of Arizona. Reavis died penniless in Denver, Colorado on November 20, 1914, two years after Arizona statehood. His wife Sofia lived until 1934, and continued to call herself the Baroness of Arizona. Their twin sons lived an admirable life, and served in the army during World War I in France. Several books and short stories were written about James Addison Peralta-Reavis. A Hollywood movie was made about him in 1950. Aptly enough it was dubbed the Baron of Arizona, featuring the soon to be horror movie star Vincent Price as Reavis, and the beautiful drama actress Ellen Drew as his wife.[37]

The Notorious Clantons of Tombstone

This last vignette on "Frontier Justice and Outlaws" deals with the notorious Clantons of Tombstone. Phin Clanton, son of Old Man Clanton, and brother of Ike and Billy Clanton, was one of the earliest settlers of the Globe area. "When the curfew rang for the Clantons in Tombstone, the two surviving boys, Ike and Phineas fled to Apache County. There, Ike was killed by the famous sheriff, Commodore Perry Owens, and Phineas was captured." [38] Phin drew a term in prison at the infamous Yuma Prison, and when released went back to his old stomping grounds in Gila County, where he started a goat ranch. Phin and Tom Hammond had a 160-acre homestead ranch. While living in this area, he is said to have visited Matty Blaylock Earp, who was living in Pinal at the time. Quite a coincidence wasn't it? The ex-wife of Wyatt Earp, archenemy of the Clantons, being visited by one of the Clantons! Phin was involved in more than one brush with the law while living in then Gila County. The last incident occurred in May of 1894. In Case Number 205, of the Second Judicial Court at Globe, Arizona, in and for the county of Gila, Phin was indicted for the robbery of Sam Kee on May 15, 1894. He stood trial but was found not guilty. Freed from jail once more, Phin returned to his ranch and was never in trouble again—at least not for the record. He died in 1905, and is buried in the Globe Cemetery. Thus the last of the infamous Tombstone Clantons lies in a crowded cemetery, the letters of his epitaph slowly fading away, and few left that remember him. [39]

But some still remember Phin Clanton. On August 8, 1996, while coming back from the Silver King Mine, I stopped to eat at the BuckBoard Restaurant, on Highway 60 just prior to the Town of Superior. Anyway, the menu had a real interesting story of a character named Phineas T. Booglebrand on the back, which in reality is a tale of the real Phin Clanton. I thought it would make a good addition to this book, so I met the author, Dan Wright, who gave me written permission to use this unique tale in the book. Incidentally, I have stopped there many times and the food was always good.

The Legend of Phineas T. Booglebrand by Dan Wright

"Lots o'folks ask us, 'Jest who is Phineas anywayz?' Well we done sum research o' the recurds' round hear n'found some pretty interest'n facts and a few tails. One thang iz cleer, until now people din't know thet so meny famous histarical people frum Arizona's past strolled theez hear partz.

It zeems after the famous gunfight hat O. K. Corrale in Tombstone, all the mane charakterz soon left town for partz unknown. Evrbody knowz that Wyatt Erp left hiz girlfriend Mattie Blaylock in Tombstone, whilst he ran off with a younger more tempestuous beauty to Californi. Doc Holliday wound up in a sanatorium in Colorado where he died. Only one Earp bruther got kilt in Tombstone and all the uters left town too. Course all o'thet stuff evrbody knowz frum the moviez. The question waz about PHINEAST. So we haf to look fer infermation about whut happend to the bad guyz after the gunfight!

First ov all, Wyatt's brokenhearted girlfriend Mattie moved on to a little town called Pinal, Arizona Territory. The gunfight waz in 1881 and the recurds show she died in Pinal in 1888. Don'tcha know Pinal is jest downstream from present day Superior? Yep, sum ov the foundations are still there today. Now Ike Clanton's brother Billy waz kilt in the famous gunfight, and Ike waz filled with hate n'vengence. It's figured that it waz he thet done in Morgan Earp, by shoot'n him in the back whilst he played pool. Anywayz, Ike left Tombstone and meandered fer a while n' wound up at hiz brutherz place near present day Springerville. Seemz hiz bruther owned a ranch up en the high country and hiz name was PHINEAS T. CLANTON. The recurd clearly showz that Ike Clanton waz no account, and kept gettin inta trouble n' running frum the law.

Filled with hate n' vengence in hiz hart, folks 'round thoze timez talk how he and hiz bruther came down to Pinal to vizit Wyatt's ole girlfriend Mattie. No doubt they waz upta sumthin nefarious, cuz they din't use their last name o'Clanton. No, sum folks say they uzed many namez round theez partz. One uv them waz BOOGLEBRAND? Seemz it gave'em an airo'stockracy sorta feelin. Whilst in Pinal, which waz a young mining town itself, the Booglebrand brutherz dabbled in this n' that, but nev'r got their vengence on ole Wyatt. Evr'one knowz Wyatt died o' ole age in Californi about 1929 or so. How much is fact n'how much iz legend? Like anythin who knowz fur sure. The recurds do show Mattie Blaylock kilt herself in 1888 downstream frum here in the little town o'Pinal. The recurds also show that Ike Clanton wuz shot ta death by a Marshall up'round Springerville, and died out at hiz brutherz ranch. But what happenta PHINEAS T. BOOGLEBRAND (Clanton)? We can't be sure o' him, but if ya looks up at the front o' aire building after dark, yule see him gettin gussied upta go see ole Mattie nex door ta him! Maybee they'r still plottin howta do in ole Wyatt?

Whal who knoz, but do enjoy eatin at hiz namesake restaurant. We felt it waz time more folks knew the uther side ov the famous gunfight—sumthin 'bout whut happen to the bad guyz! [40]

1880s Tombstone hearse, 1996
— *Author's photo*

Tombstone's Big Nose Kate Saloon, 1996
— *Author's photo*

Tombstone's OK Corral site of the famous Earp/Clanton shootout, 1996
— *Author's photo*

Tombstone's Boot Hill - Clanton & McLaury's markers, 1996 — *Author's photo*

Chapter Twelve
Tragic Tale of the Lost Soldiers Mine

Many of the searches for the Lost Dutchman Mine started at the Silver King Mine, or were conducted by people who worked at the mine, or were familiar with the Lost Soldiers Mine story, that originated at Silver King. This story, which has been told and retold many times, figures prominently in the history of Silver King and Pinal, even though this lost mine, thought by many to be the famous Lost Dutchman Mine, is in the Superstition Mountains. The importance of the legend of the two soldier's story cannot be overstated. The searchers of the Lost Soldiers Mine included legendary figures Joe Deering, John Chuning, Jack Fraser, "Dutch John" Pipps, Dick Holmes, Jim Bark, Sims Ely, George "Brownie" Holmes, Ernest Albert Pankin and many many others, including the author.

During the early 1880s, two young French-Canadian soldiers, who had just been discharged from Fort McDowell, showed up at the town of Pinal. They stated they wanted to work at the Silver King Mine. The ex-soldiers heard it was a good place to work, and that the wages were much higher than they could get doing any other kind of work at that time. The morning after they had arrived the soldiers went up to Mr. Aaron Mason's office to apply for work. Present at that time were Robert "Bob" Bowen, a mine foreman, and A. J. Doran, the Silver King Mill Superintendent at Pinal. Mason asked about their experience and they said they had no experience in mining, but they were strong and could do any labor job that he might have available for them. They also said they had recently completed their enlistment in the army at Fort McDowell, and instead of going home, they thought they would learn to be cowboys, and then they thought they would learn to be miners, since they heard mining paid more. Mason referred them to Bowen, who said that he could put them on as muckers (those who shovel the ore into carts after the rock wall of the mine is exploded by powder or dynamite) until they learned to be miners. Mason stated if there was no work at the mine, Mr. Doran may have work for them at the mill, referring to the Silver King Mill located on the bank of Queen Creek in Pinal City.

Mason asked if they needed money. No, they said, they'd saved their pay while soldiers, and they were okay right now. Instead of taking the stage to Pinal, they saved the money for stage travel and instead walked from Fort McDowell, taking a short cut through the Superstition Mountains.

They started at Ft. McDowell, passed by the Salt River, taking the Old Military Trail, believed to be what is now the Apache Trail, and entered the mountains at a creek crossing, perhaps Tortilla Creek. They followed the creek upstream. After that, the trail appeared to go north, and since they knew that the Silver King was south, they said they cut across country. They went up this creek a little more, came to a waterfall and could go no further. The soldiers then came back down the creek, got out of the creek towards the south side, and went up over a very rough and high mountain. At this point there was no trail, so they went in a southbound direction, and tried to work their way through this rough country towards the Silver King Mine.

They came onto somewhat of a trail and they thought it was a strange place for a trail. They decided to follow it for a while and see if it couldn't lead them out of this godforsaken, terribly rough country. They followed it for a short distance in hopes that it would lead them out, when in fact it lead them through a cave-like opening between some peaks. As they went on up the trail they came to an old mine tunnel that had been covered up. Next to the tunnel they located a mine dump which contained some type of ore. They weren't quite sure what they had found, but they collected some samples and put them in their backpacks and continued on the trail towards the Silver King Mine. After telling their story they placed their backpacks onto Mason's desk and spread out the ore. After looking it over carefully, Mason told them it was hand-sorted ore, extremely rich in value. He told them they had between six and eight hundred dollars here, and said he would have his assayer check it and would pay them the full value. Mason said that after the assay they should have plenty of money.

> "If I were you," he said, "I wouldn't worry about being a mucker or a miner here at the Silver King. I'd go back to that old mine and take away as much of the ore as you can find. It must be an abandoned mine from what you told me, so it looks like you had a right to take it. How much more of this ore do you think is there," asked Mason? "Well, Mr. Mason," the ex-soldiers said, "perhaps a whole wagonload just laying out there." "I don't think so," said Mason, "you'll probably find that there's not much more and the rest is just a waste dump."

Mason explained that the miners who sorted this ore were probably killed by the Apaches, or were scared during the latest Apache uprising, and probably never came back to the mine. Mason advised the boys to go back to where they located this ore, place some monuments and file a claim on it. He said that he would show them how to file a claim. He asked if they could find their way back to the mine. "You mentioned how difficult it was to come up across the mountains and get here when you found the mine," he said. The two soldiers said they were certain they could go back to the place where they had found the mine. They remembered that it was in a northerly direction from the towns of Silver King and Pinal, and from some point they could see a sharp peak, which they had seen in their scouting days. Mason thought that this sharp peak they had seen was probably Weavers Needle.

Norm Johnson & Author at Weaver's Needle 1990s — *Author's photo*

They said in their soldiering days, whenever they were in unknown country, they were taught by their officers to look back on their trail from time to time. They were told to remember whatever landmarks they saw from the opposite direction, and to make a point of this in their mind. They said they had done this. On their way out from the very rough place through the mountains they had struck the old trail that led them through a small gap in the mountains. This was not far from where they had seen that large peak. (Others think this was Miners Needle not Weavers Needle.) They said the valley ran east and west, and they followed the eastern part of it, until they crossed some canyons and headed in the direction of Pinal. They had passed a horse ranch where there were some corrals. From that ranch they said there was a trail that led into Pinal. They said they could go back to that old mine all right. They were sure that they could find it again.

The ex-soldiers bought two burros and packsaddles, bedding and food. Enough food to last them for about two weeks. When they went back into Pinal, they stopped at Mason's office. They asked Mason if he would keep the gold that they had not spent. When the two soldiers gave the gold to Mason, he put a slip of paper in with both their names on it, the amount of gold, put it in some jars and put it in his safe until their return. They then left for the old mine in the Superstitions.

An old miner named Wiley Holman, who sometimes lived in Pinal, was there at the time and saw Mason putting the gold into his safe. He related this incident sometime later to pioneer rancher Jim Bark, who bought the Caveness Ranch, where the two soldiers came out of the mountain. He said that Mason gave the ex-soldiers instructions on how to file a claim, and how to set up monuments. That was what they were prepared to do. Go back to the mine and set out their monuments. During that particular time there was always the possibility of danger from small groups of renegade Apaches, especially if there were only two men in the group. Even if they were former soldiers, Mason said, they ought to be well armed. They said they had bought pistols at Ft. McDowell. Mason advised them to buy two rifles also, and to keep their eyes peeled for Apaches. They told Mason they'd keep their eyes out for any stray Indians. According to Mason's instructions, they were to locate two lode claims along side of each other. Each claim was to be 1500 feet long and 600 feet wide, with the mine in the center of the claims. Mason also cautioned them to be quiet regarding what they were about to do. He warned them that many a good claim was jumped before all the necessary, routine paper work had been filed. (This 1500 by 600 foot measure was the standard for a mining claim in the 1880s, and is still the standard today.)

After listening to what Mason had said, the ex-soldiers told Mason they appreciated what he was doing for them. They said they know there is plenty of gold there, and that they wanted him to be their partner. Mason agreed, but not until he had a look on his own at the old mine. He told them that if it was worthwhile, he would put up the expenses to start with and charge the outlay to the enterprise. The two soldiers left Pinal after dark. *They were never seen alive again, nor were any of their outfits ever located.* The boys and Mason had figured it was about 25 miles from Pinal to the mine. The journey could easily be made with burros, about two days going, and about two days coming back to Pinal. Mason figured it would take six days at most to set up the monuments and to do the necessary exploratory work. Mason said he expected them back in Pinal about ten days from the time of their departure. He thought that would give them plenty of time to establish their claim and bring back some high grade ore. The two soldiers' story is corroborated by many of the unpublished manuscripts of people who were at Silver King at the time. They were called *"two Frenchmen,"* "*two Dutchmen,*" *"two foreigners"* and the *"soldiers."* There were simply too many references to this incident by independent sources not to have occurred, basically as stated.

When over two weeks had gone by, Mason felt uneasy about their failure to return. He feared that the young soldiers had either been killed by the Apaches or captured. He then sent out a posse of about twenty armed men, led by Col. Robert Williams, who was a seasoned pioneer, to look for the two soldiers. Robert Williams was the owner of the Williams Hotel at the town of Silver King. This posse was instructed to make a thorough search in the region of Weavers Needle. The posse returned without having seen a trace of the soldiers or their outfit. The time of year was hunting season for the Apaches. The Apaches might be wandering through the mountains hunting deer, and it might have been easy for them to murder the boys and take away their animals and all of their camp goods.

Mason was thinking about sending another group to the area of Tortilla Mountain, since he thought they may have come that way to the mine, when he happened to see a rancher named Whitlow in camp. Whitlow lived on Queen Creek, about ten miles downstream from Pinal. Mason asked him if he knew the Weavers Needle country, and Whitlow said he was very familiar with it. Mason asked Whitlow to take a cowboy or two and conduct a search of his own. Whitlow agreed to go with some of his men the next day and make the search. The Whitlow Ranch was not far from the area of the Caveness Ranch where the two soldiers saw the corrals on the way to Pinal. On the second day after his undertaking of the search, young Willie Whitlow found a body about one-half mile from the old ranch house of Matt Caveness. Many people believed this body to be one of the young soldiers. The young Whitlow boy went and got his father. The Caveness ranch house may have appeared vacant at the time. Matt Caveness' wife lived in the old board house in what is now the Quarter Circle U Ranch area. She had a dairy ranch and sold milk and butter to the miners at Silver King and Pinal. She was living there with her three children at the time the two soldiers were murdered. Matt Caveness was probably involved in a mining or freighting venture at the time, so he did not live at the house, in what is now known as Bark Valley.

The body was nude and in terrible shape. There was a bullet hole through the body which was plainly visible. Whitlow and a cowboy put the body into a hastily dug grave at the place where they found it. On the ground, a few yards away from the body there was a black hat, similar to what the soldiers wore. There was no other item of clothing to be found. Whitlow took the hat with him, and the next day went back to Pinal and told what he had found to Mason. Mason recognized the hat as being similar to one of the hats worn by the two soldiers. Everybody in Pinal concluded the body found by Whitlow was one of the two young soldier boys, there couldn't be any doubt about it. Bill Kimball of Mesa also found a naked body about the same time period, which some people believed to be the other soldier, near Bluff Spring Mountain. They considered this another case of Apache murder. Two bits of evidence pointed to the Apaches. It was the Apache way to remove all clothing from the bodies of their victims if they were indulging in some system of torture. Also, they made it a rule to take away all animals for eating and camp equipment for plunder. Even though the Apaches were considered the likely suspects for the murder of the two young soldiers, some new developments then occurred which put considerable doubt on the theory that the Apaches committed these murders.

Author near Soldier's Grave and Miner's Needle
— *Jack Carlson photo*

Clubfoot the Swamper

These new developments that surfaced about the murder of the two soldiers involve a man with a clubfoot, who hung around the town of Silver King. This tale goes like this. At the town of Silver King there was a fellow who had a crooked foot, and he was called Clubfoot. He was also called a *swamper,* and did odd jobs that no one else wanted to do. He was also a kind of a hanger-on at gambling games. Whenever mealtime came at the games, whoever happened to be the dealer would usually toss Clubfoot some coins, and tell him to go get something to eat. As stated before, this clubfoot fellow never had any money of his own. So, when all at once, he began gambling, with money out of his own pockets, including gold coins and silver dollars, all the local people were astonished. In fact, he was seen betting $20 gold pieces at different games. When they asked where he got his money, he stated that when he had been to Florence recently, and that he had made a killing at gambling. Such things were known to happen, and at the time his explanation seemed satisfactory.

When news had reached Pinal and Silver King that Whitlow had found the body of the soldier, a saloon keeper said to Clubfoot by way of a joke, "Hey, you've been out of camp, did you do that killing?" While both the saloon keeper and the Clubfoot had a laugh, the remark was overheard by a customer who had seen Clubfoot gambling and decided to do a little investigating on his own. Perhaps Clubfoot hadn't made any winnings at gambling after all. As it happened, someone from Pinal was going to Florence on business the next day, so he was told to make some inquiries. He brought back word that Clubfoot hadn't been seen in Florence. All the freighters and stage drivers were questioned, and none of them could remember that he had ridden with them. He was such an unusual character that his presence on a stage or on a freight wagon would surely have been known. Suspicion was running high now, but they still wanted to give this man the benefit of the doubt. The period of Clubfoot's absence coincided with the first few days of the departure of the two soldiers from Pinal. Since it was clear that Clubfoot hadn't acquired his money in Florence, perhaps there was a holdup, or a burglary, or robbery somewhere where he got his money. (The soldiers were said to have had $400 in silver and gold coins on them.) But it was soon established that nothing of that sort had happened in the time period in question. Even if a holdup had occurred, Clubfoot wasn't that type. He was too sneaky and too cowardly. Knowing his character, his style was to do murder and steal money off dead bodies. He could easily have seen the two soldiers buying goods with a large quantity of money in camp. He could have followed them out of camp, killed them and made it look like the Apaches had committed the murders.

Shortly after Clubfoot was questioned about his activities he was advised that perhaps he'd better leave town as things weren't going to go well for him at Silver King. The belief became widespread that Clubfoot the swamper had followed the two soldiers into the mountains and murdered them, not for their gold mine, but for the money in their pockets. It was felt that he turned the burros loose to join with the other wild burros, and buried their clothes and outfits, then hurried back to Silver King. In the tradition of frontier justice for murderers, lynch talk was running rampant. But some cooler heads prevailed and he was told to leave town immediately. He disappeared shortly thereafter, and was not heard from again. Jim Bark's unpublished manuscript states: "The miners and gamblers became very suspicious and they got together, held a secret meeting and agreed that at the next shift change they would hang Clubfoot, and make him tell where he got his gold. Somehow it got to Clubfoot's ears and he disappeared."

Panknin's Search

Nobody ever knew whatever happened to the man referred to as Clubfoot for several years. But about 1915, a man named Ernest Albert Panknin relayed a story to Sims Ely, who in turn told it to Jim Bark. Panknin told Ely he met a clubfoot man while he was living in Alaska, and that he had befriended him. In return, the clubfoot man told Panknin he would tell him where there was a very rich gold mine in Arizona. But he said that he could never go back to Arizona. He said that the mine was in the Superstition Mountains in Arizona. Clubfoot then drew him a map. The map that Clubfoot drew started at the Silver King Mine, went in a western direction, passing Florence, went into the Superstitions passing a green spring and went westward some five miles to the mine.

Panknin stayed in the Phoenix area, where he worked as a security guard for the Phoenix National Bank for seven years, and continued to search for the elusive gold mine. The fate of the two soldiers was often the topic of discussion at the Silver King, and was mentioned in many of the unpublished reminiscences of the old timers of Silver King and Pinal City. Of course, they weren't privy to the information about Panknin and his unnamed clubfoot friend in Alaska. This information offered an explanation as to who may have killed the two soldiers. How else would Pankinin's Alaska friend know about the rich gold mine in the Superstitions? And why would he use Silver King as a starting point on his map, unless he followed the soldiers to the mine and killed them, and then left Arizona, never to return. He also had the money to go to Alaska. Why did he run so far if there was nothing to fear?

This appearance of Panknin certainly sheds light on who may have killed the two soldiers, but does not clear up the mystery of the soldiers' lost mine that many people believed to be the Lost Dutchman Mine. Panknin's obituary was listed in the *Phoenix Gazette,* December 21, 1934, and I have a copy of his last will and testament, as well as other documents identifying Panknin. It was generally considered by many Dutch Hunters that the two soldiers had in fact found the legendary Lost Dutchman Mine. The area where they located the mine, the description of the mine's walled tunnel, and the workings above the mine, fit many of the clues associated with the Dutchman's mine.[1]

"Dutch John" Pipps and the Lost Soldiers Mine

A man called "Dutch John" Pipps was employed at the Silver King Mine about the time of the disappearance of the two soldiers, and was very familiar with the details of their finding a rich gold mine in the Superstitions. He also was well informed of the story of their disappearance and deaths. John Pipps (also spelled Phipps) real name was John Pipps Monk. Pipps had prospected in the Superstitions prior to the two soldiers' incident and continued to search the Superstitions around the area where the two soldiers' bodies were found. That area was around Miners Needle and Bluff Spring Mountain. Pipps said that on one of these prospecting trips he observed an old man working an outcropping, in a rough mountain area somewhere north of Miners Needle. After the old prospector left his diggings, packed up and departed the area, Pipps went to the outcropping and inspected it. He looked down into the hole and was surprised to see what appeared to be an eighteen-inch wide vein of quartz imbedded with gold. Pipps pulled out several samples of ore from the vein and returned to Silver King. Some contemporary lost mine hunters think that this mine was the Bull Dog Mine that is currently located near Goldfield, northeast of Apache Junction, just off the Apache Trail, State Highway 88. The Bull Dog Mine located in the 1890s, was the only gold mine in this region known to have an eighteen inch wide vein of gold ore.

Pipps showed the ore to Aaron Mason, the Silver King Superintendent, and told him that he had located the Two Soldiers Mine, or perhaps even a Dutchman Mine, since there were lost mine stories even in those days.

Mason supposedly gave Pipps a grubstake in return for part interest in the mine, should it become a paying mine. When Pipps returned to the mine he said the old man was there and confronted him. He said the old Dutchman said that if he (Pipps) came back to the mine he would kill him. He said that he had killed others before and would do anything to protect his mine. Well, Pipps went back to Silver King, gave the grubstake back to Mason and explained what had happened. He said that he would wait until the old Dutchman died before he went back. Pipps continued to work at the Silver King until 1886 when he left to start up a small ranch.

During one of the times he traveled through the state, Pipps stopped at the ranch of Dick Holmes, who had once tried following Waltz to his mine. He related his story to Holmes who asked him who the old man was.

"He said his name is Waltz, Jake Waltz, and an ornery devil he is, too. He told me he could kill me at any time he wanted to, and I believed him. He

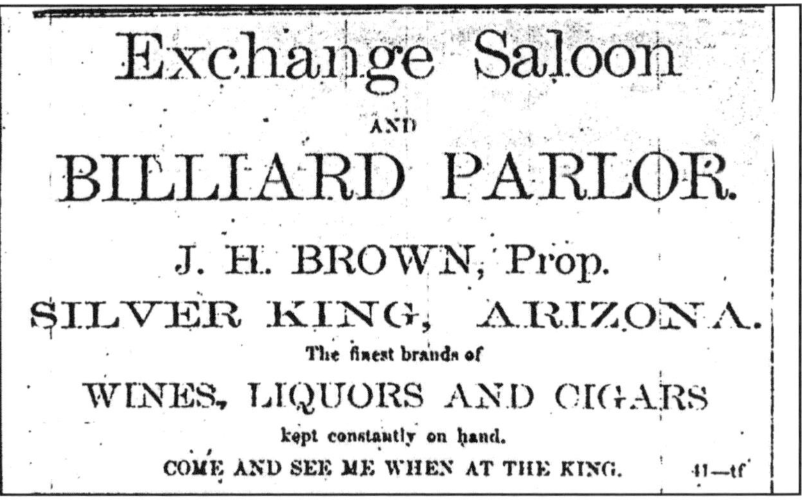

J.H. Brown's advertisement cir. 1880s — Pinal Drill

scared me and I won't go back there till he dies," said Pipps. "Why don't you locate the mine and file a claim on it?" said Holmes. "Not me," Pipps replied. "That old Dutchman has killed many a man and he'd kill me in a minute when he found out I located the thing. Nope, I'm getting old, but I've still got a right mind." Pipps said "Dick, by God I'm afraid of him, and I'm not going back until I know damn well he's six feet under the ground." "Well, I'm not afraid of him," said Holmes. "If you show me where the mine is, I'll locate the damn thing." Pipps replied, "Dick, the German is nearly eighty years old now, he can't last much longer. When he dies, I'll make you a partner, and we'll locate the thing. I guess I've said too much already."

Pipps then left Holmes' ranch without giving him the location of the mine.

Pipps ranch was located in Round Valley, in the vicinity of the Mazatzal Mountains. He was building a small house and decided to put in a well. While digging the well, which was 16-18 feet deep in sand and rocks, the well collapsed in on him. He had not used lagging to support the sides. He had been in the well for two days when four men came along and tried to get him out. Lumber was used to try and shore up the well, but it kept caving in on top of Pipps. While he was dying he pleaded with the men to get him out. He stated that he would lead them to a rich mine if they would only please get him out. Pipps died in the well without getting out, so the men just filled it in and buried him there.

Phoenix Daily Herald September 15, 1886

"Messers. Frank D. Wells and F. A. Shaefer while returning from Tonto on Monday passed a place where a man commonly known as 'Dutch John' was buried and suffocating in a well which he had been digging. He lived about forty-eight hours covered up to his head in earth and finally the earth closed over him entirely.[2]

Messers. Frank D. Wells and F. A. Shaefer while returning from Tonto on Monday passed the place where a man commonly known here as "Dutch John" was buried and suffocating in a well which he had been digging. He had already been found but those who were endeavoring to assist him had been unable to rescue him as the ground constantly caved about him. He lived about forty-eight hours covered up to his head in earth and finally the ground closed over him entirely. The man at one time worked here in the valley, a part of the time for Doc Jones. He was left buried in the hole where he died.

Dutch John's death notice 1886
— Phoenix Daily Herald

Joe Deering and the Lost Soldiers Mine

About a year after the death of the young soldiers, during the summer of 1884, a young prospector named Joseph Deering appeared at the Silver King. Deering was not at Silver King when the tragedy of the two soldiers occurred, but he had heard about it. So Deering set out to try and back track the soldiers trail from the stories he had heard. Other men had tried to find the mine but had failed.

Deering was sure he had located the old mine, but he was broke and he came to the Silver King Mine looking for work. He wanted to work for a few months, save up some money for an outfit, and wait for a friend to come to Silver King and be his partner. There was no immediate work, as the mine was temporarily shut down. Not a man to be idle long, Deering went into Brown's Saloon at Silver King and asked for any kind of work. Brown said later that he liked Deering, and after ascertaining that he didn't drink or gamble, he gave him a job. Deering and Brown soon became friends and Deering confided the mine story to Brown. He said that he had a partner and was waiting for his partner to arrive from Prescott.

On last Saturday night, at the King mine, Joseph Deering met with a terrible accident which terminated in his death Sunday morning. It appears while he was working in the 500 foot level a large boulder rolled over him literally mashing a portion of his body almost to pulp. He was immediately conveyed to his room, and Dr. Kenniard called in to attend him. Amputation was necessary, but he was so weak from the loss of blood, that he died while the physician was in the act of taking off his leg. No blame is attached to the company. Coroner's inquest rendered a verdict of accidental death. The company suspended work and attended the funeral. Mr. Deering was a good man and well liked by his friends and fellow-workmen.

Joe Deering death notice, 1885
— Phoenix Daily Herald

Deering had worked in mines in Colorado and Arizona, so he knew the mining functions. When the mine resumed operations on a full scale, Deering asked Robert Bowen for a job. Mr. Bowen, the mine foreman, seeing that Deering was familiar with mining operations, immediately put him to work. Deering kept quiet about his discovery for a while, but this secret was just too good to keep. He just had to tell someone. After working for a while, Deering struck up a friendship with his shift boss, John Chuning. Chuning was older, but he thought he could trust him. He took Chuning into his confidence, but would not give the location of the mine. He described the mine as a somewhat funnel-shaped pit, with a tunnel leading into the walled-up mine. Deering showed some ore from the mine to Chuning, which allayed the older man's skepticism and made a believer of him. Deering was anxiously waiting for his friend, and to make enough money to get up a good outfit. He wanted to claim the mine and the fortune that he knew awaited him. But "Lady Luck" was not with Joe Deering. While working on the 600 foot level of the Silver King Mine, with fellow miners James Green and Thomas Gormley, a rock wall gave way and Deering's legs were crushed. Deering was brought up out of the mine unconscious, and Dr. Kenniard, the mine doctor, amputated one of his legs while trying to save his life. Joe Deering never regained consciousness and died the same day. This accident occurred at 2 a.m. on September 27, 1885. At the coroner's inquest, James Green testified that the rock that crushed Deering's legs weighed about 500 to 600 pounds. Doctor Thomas H. Kenniard testified that Deering had a severe compound fracture to the left leg, and that he amputated the leg above the knee. *No map to the mine was found in Joe Deering's possessions.*[3]

John Chuning's Search for the Lost Soldiers Mine

After the tragic death of Joe Deering, the search for the soldiers' lost mine continued. John Chuning took up the task, based on the limited clues that Deering gave him. When the Silver King Mine shut down in the late 1880s, Chuning began the search in earnest on a limited basis. He would work at one place or another until he got a grubstake to begin the search again. In late 1892, he went to work at Jim Bark's ranch in the Superstition Mountains. Chuning discussed the lost mine with Jim Bark. He said that Deering told him he placed monuments around the mine. He also said how Deering came to find the mine. According to Deering, he was camped in the mountains when his burro got loose, and when he went to find him, he found a well worn trail, followed it six or seven miles and came to the worst place he ever saw. There he found the tunnel and the mine. Chuning said that he had seen the ore Deering brought back from the mine and said that it was very rich ore. Chuning worked off and on for several years for Bark. During his search for the lost soldiers' mine he came across a good prospect and told Bark about it. Chuning and Bark worked this prospect for about two years, when Bark said the ore was not good enough to work anymore. Chuning then set up a base camp near Tortilla Flat on the Apache Trail and searched for the lost soldiers' mine for several years. He also worked for Jack Fraser searching for the mine. Jack Fraser had also worked at the Silver King Mine, and was very familiar with the soldiers' story. There is a story told by an old prospector who knew Chuning, which states that Chuning did find an old mine about

1910, by following an old and nearly obliterated trail from Tortilla Flat. Chuning supposedly took some small Bull Durham sacks full of nuggets

News was brought to Florence this week that John Chuning, who has been prospecting for the past five years in the neighbourhood of Weaver's Needle and the Four Peaks, has at last found what he believes to be the Lost Dutchman mine—a rich gold property with a history. Chuning discovered old workings consisting of a shaft and tunnel, which he is now cleaning out, a short distance west of the Needle, which can be plainly seen due north from Florence in the Superstition mountains.

John Chuning's gold story, July 7, 1901 — Florence Tribune

from this mine. He did not file a claim according to the story but did buy an outfit and a string of burros and hired a packer.

Chuning was so overjoyed at finding a mine with gold that he sent for a good friend who showed up on the Roosevelt stage with an abundant supply of whiskey. Chuning was in ill health at the time with Yellow Jaundice (a disease of the liver), and the overindulgence of whiskey resulted in a bowel obstruction which caused his death. Well, so much for good friends. You can say Chuning died trying, because if he did find the Lost Soldiers Mine it is still waiting for some lucky soul to stumble upon it! Chuning died on November 11, 1910. John Chuning *left no map of his searches*, or a written document describing his searches. I have searched all of the places in this State where archives or unpublished manuscripts are kept. Apparently his experiences went to the grave with him.[4]

The Dutchman Jacob Waltz and the Silver King Connection

Jacob Waltz stated on his deathbed that he had *killed two soldiers* that he found working his mine. He related this story to Dick Holmes and this story was written by his son Brownie Holmes in 1944, and published by the Superstition Mountain Historical Society.

"I had been gone from the mine several months Dick. It made me mad to see anyone going in the direction of the mountain. I made up my mind to hide for two days. And the two soldiers came up to the mine and began working it. They only worked for a few hours and then got on their horses and rode away. I followed them and killed them both. I made it look like Indians had done it. After I killed the two soldiers I went back to the mine, took out some more ore and then went back to Phoenix. When I returned the next winter I saw where someone else had been there. I knew then that I was going to have do something if I intended to keep the mine. So I started concealing it, so no one else could find it. I enlarged the shaft two and one half feet all around and left a ledge six feet below the surface. I went then up the mountain and sawed timbers the right length to fit the ledge. I worked all winter sawing, dragging, and placing those timbers. They're in the shaft now…crisscrossing to a depth of six feet. I left about two feet near the top so I could fill in with dirt and rocks. No one will ever find it unless he finds the rock house down in a brushy canyon.

"Waltz was familiar with the town of Silver King and went there for supplies after the town of Adamsville closed down. Brownie Holmes wrote in his manuscript of just one such incident. "On another occasion Wolz (sic) went to the Silver King where he purchased a complete new outfit and a large supply of food. As he was making these purchases, he was accosted by an old acquaintance, Joe Guggit, who had worked with him at another mine. Joe inquired if he had made a rich strike. Wolz told him he was doing a little placering, and when Guggit asked if he could accompany him on the return trip, Wolz flatly refused to consider the proposition, explaining that there was not enough gold for both of them. Joe left, but returned shortly for something he had forgotten, and found Wolz opening a bag full of gold to pay for his supplies. Brownie stated that Joe Guggit related this to him in person." [5]

Backtracking the Soldiers Trail

Why has no one ever backtracked the soldiers trail? Well, Joe Deering said he did, but left neither map nor directions that he took. However, by using a topographical map and aerial photography, as well as the description of the soldiers' route out of the mountains, I have been able to backtrack a portion of the trail. Beginning at the old town site of Pinal, the trail runs west, along the south side of Queen Creek to Whitlow's old ranch location. Here the trail crosses to the north side of Queen Creek, and follows the old stagecoach road from the Silver King Mine. I followed the trail into the Superstitions by going in instead of coming out. I took the old trail, a few miles past the old Whitlow Ranch site, onto an old trail that continues north and then west. The trail enters Bark Valley and crosses it, until it goes northward, up past Miners Needle. This is where the soldiers' route is lost. There used to be an old trail that went past Miners Needle into the heart of the Superstitions, but it is no longer visible. The route could have taken two directions at this point. One, it could have gone west towards Bluff Spring Canyon, and then into La Barge Canyon. Or two, it could then have continued through La Barge Canyon all the way to the Salt River.

Another possibility is that the trail could have gone east towards Whiskey Springs Canyon and then northeast into Upper La Barge Canyon, back northwest into Peters Canyon, northward all the way to Tortilla Creek and to the Salt River. There are so many trails running in and out of La Barge Canyon, in all directions, that it is virtually impossible to determine which trail was taken by the soldiers. Actually, the soldiers may not have taken any of these trails. La Barge Canyon was a fairly well known trail at the time of the soldiers, and they implied that they were lost until they encountered the main trail to Pinal. This was the old stagecoach trail (previously described) that passes the Old Whitlow Ranch and goes on through the town of Queen Valley on the Old Silver King Road.

There is another trail that the soldiers could have taken. This trail goes from Ft. McDowell to Mormon Flat, crosses the Salt River and goes northward to the Apache Trail, across the high country, passing Tortilla Mountain on the westside. In this very rough country the soldiers could have lost their way, and may have taken a trail past Horse Ridge, Horse Camp Basin and Herman Mountain, entering Trap Canyon's headwaters. From here, they could have followed Trap Canyon southward, where it joins La Barge Canyon and passes Upper La Barge Box Canyon. Continuing southward, they could have gone via Whiskey Spring to just east of Miners Needle. Then they could have taken the pass just east of Miners Needle near Castle Rock, where there is a view of the entire Bark Valley and Picket Post Mountain to the east. Picket Post Mountain was familiar to most soldiers of that era, since it was an army outpost during the 1870s. From there, they could have taken the old stage route to Pinal and Silver King. An alternative route from Horse Camp Basin and the headwaters of Trap Canyon would have been to take the eastside of Herman Mountain south, via a small canyon that leads east into Red Tanks Divide. From there, they could have taken Red Tanks Canyon, to the southern rim of the Coffee Flat Mountains. They could have continued on to Randolph Canyon, and on to Whitlow Canyon and the old stagecoach route east to Pinal and Silver King. Estee Conatser called this old trail the "Old Spanish Trail," in her book *The Sterling Legend.* She said, "this trail leads to and through some of the most rugged terrain to be found in the area. This could easily have been the trail the soldiers found after leaving the old mine dump where they found the rich ore." [6]

Trying to follow the trail of the soldiers, as they entered and left the mountains, is something that I have explored over the years and it is difficult to do only from the ground trails through hikes, so I used aerial photography as well to view the topography and possible routes. I also spoke to an old cowboy familiar with the Ft. McDowell and Verde River/ Salt River area who told me about an old Indian trail, which later became the cowboy trail from Ft. McDowell, south past the Salt River. It comes out near an area called Blue Point on the Salt River, well known to the Salt River tube floaters. This trail leads up through Bull Dog Canyon and past the old ghost town of Goldfield, then onto the old Apache Trail, which in essence was the old military trail, from Ft. McDowell to the city of Florence. This trail also intersected the old stagecoach road to the towns of Pinal and Silver King. Another route the soldiers could have taken was past the Goldfield site and east into the Superstitions to the First Water Trailhead. They may have then traveled eastward past Needle Canyon, onto the Bluff Spring Canyon Trail, going southeast past Miners Needle and to the old stagecoach road near Whitlow Canyon.

As you can see, there are many routes or trails that the soldiers could have taken, but there is only one route or trail that led past known gold deposits, and that was past the gold fields at the old Ghost Town of Goldfield. But, that does not say that this was the location they found their gold. It may well have been the Dutchman's mine they found and then it was subsequently covered and lost to history, maybe to be inadvertantly found by some lucky person in the future. Even more tantalizing is the belief of some treasure hunters that the Soldiers' and Dutchman's Mines are two completely different mines. This would explain some of the different clues to both mines.

On October 6, 2001, I hiked the Dutchman's Trail to Barks Draw in search of the old mining claims and mine tunnel of Chuck Crawford. My hiking partner, Jack Carlson, accompanied me. We started at the Peralta Trailhead at the dead-end of Peralta Road. As we hiked up the first hill we had a good view back towards the parking area of the Peralta Trailhead as well as the Peralta Trail and Peralta Canyon. There are several scenic rhyolite pinnacles of volcanic rock along this route. The trail switchbacks back and forth, up and down hill. Near the bottom of the first hill is the intersection of the old road leading to Barks Ranch, later the Quarter Circle U Ranch. Just southeast of this spot is the place where one of the soldiers (who left the Silver King Mine) was killed and buried about 1883.

Anyhow, we hiked down the hill to a wash where we headed left or west, and within a short time located some prospect holes where chrysocolla was abundant at one of the holes. Then about ¼ mile further west up the canyon called Bark's Draw we located what appeared to be Chuck Crawford's claim and mine tunnel, called Casi # 1. The old hard rock tunnel had been dug about 90 feet into the hillside, in the old style, whereby it needed no timbers to shore it up. The material on the dumps and the walls of the mine were orange and red in color, similar to iron ore stain. About one third of the way into the tunnel Jack Carlson, who was leading, almost stepped on a rattlesnake. Luckily the snake gave a warning, because Carlson never saw it. I never saw Carlson move so fast. In fact, he knocked me up against the stone wall of the mine and I banged my head, which resulted in my having a headache the rest of the day. We went back to where the snake was curled up and took a photograph. It was a black rattlesnake. After exploring the tunnel, we spotted an arrow painted on the north wall of the canyon above, and just to the west of the tunnel. Jack Carlson climbed up to it, but nothing significant was located. I went back to this old mine at a later date and went back about 90 feet to the end, which doglegs to the right. Someone had apparently been trying to work the old mine because there was a plastic bucket and some clothes and signs of new digging.

We also saw some great views of Miners Needle and Castle Rock (or the Spanish Graveyard) on this trip. From the area between Castle Rock and Miners Needle you can look southwest and see the Quarter Circle U Ranch. This was the old board house and ranch the two soldiers probably saw across the broad valley. If you look southeast from this vantage-point you can see Picket Post Mountain. When in this area I easily saw Picket Post Mountain (where old Army Post was during the 1870s) and realized that the soldiers would have been very familiar with this landmark from their days in the army. This was one of the significant places familiar to all soldiers.

The following are the primary clues to the legendary Lost Dutchman Mine that writers, researchers and serious Dutch Hunters have described over the years.

In a gulch in the Superstition Mountains, the location of which is described by certain landmarks, there is a two-room house in the mouth of a cave. The cave is on the side of the slope near a brushy gulch. Just across the gulch about 200 yards, is a tunnel, well covered up and concealed by bushes. Here is the mine, the richest in the world, with a rich gold ledge where the gold is so pure it can be picked off in big flakes. Above the mine is a pit or large prospect hole. This is the famous "Lost Dutchman Mine." The mine lies within an imaginary circle whose diameter is not more than five miles and whose center is marked by Weavers Needle.

On January 29, 2002, while hiking in a well known canyon in the Superstitions with Jack Carlson we came across some interesting clues to the Lost Dutchman Mine. At the base of the canyon was a rock formation that looked like a gun-sight. We hiked up the canyon until we came upon a large pool of water and then continued hiking in the canyon wash, since there was no trail, just a route. Hiking in the wash was slow and we had to climb over the large boulders in the bed using a hiking technique called bouldering. About one mile up the canyon from where we started the canyon makes a left turn, going eastward, and at this juncture another canyon begins by going straight (south) instead of going left. This route gets somewhat brushy and there are several obstacles to overcome. You have to bypass several huge boulders, and then cross over standing waterholes (during the rainey season) by zigzagging across the creek to where there is dry land again. There were many cattails and bamboo trees growing in the wash. During the rainy season you might have to select another route or bushwhack it through the brush alongside of the canyon.

In any event, we continued up the canyon passing many potholes, then we came across a large cave on the left side, almost hidden by brush. Prospectors must have used this cave for years, based on the trash left including many old tin cans. I even located an old 1880s broken beer bottle, similar to the ones I found at the Old Silver King Mine site. In one of the old Dinty Moore Stew cans I found a small gecko lizard that was using it for a nest. After taking several photos of the cave, which incidentally had a stone arch above it, we continued our search.

After bouldering some more upcanyon, we climbed a small hill on the right, below a large butte. We had passed this location on a previous hike, and had seen some mine tailings on this hillside. We climbed a little further up and located an old mineshaft with a pit above it, where digging had begun, then had stopped. The mineshaft was about 75-80 feet deep with a 5 foot square shaft. There were signs of chrysacholla (copper stain) in the mineshaft, and some timbers used as cross braces were made out of ironwood trees, like the oldtimers made them, not square timbers. The ore of the mine and mine dump contained reddish rocks, but I could see no visible signs of metal within the ore. There was evidence that the mine had beeen worked as recent as 40-50 years ago, as some newer nails and debris indicated. The ore and mine dump contained material similar to those I have seen at other old mines that contained copper ore.

Author at Dutchman's Cave 2005 — Jack Carlson photo

The pit and mine were located about 200 yards across from the old cave. Also from above the mine, high on the opposite hillside, you could see Weavers Needle to the south. The canyon at this point is a north trending brushy canyon. Many of the facts fit the clues to the Lost Dutchman/Lost Soldiers Mine. The cave 200 yards across from the mine, the pit above the mine and the reddish ore and the location of the mine next to a hill between two hills are good clues. Then, there is the canyon wash with many potholes. The "trick in the trail," where this canyon goes left (where it is easy to miss the trail) is another clue, which also fits the Lost Dutchman/Lost Soldiers Mine. The proximity of year round water, and the fact that you can climb above the mine and see Weavers Needle towards the south, are other good clues. The presence of the 1880s beer bottle and the type of wood used as cross braces in the shaft, indicates that this may be a very old mine site dating back to the 1800s. This makes this location in my opinion worthy of more searching, thus I have not specifically given directions to the cave, lest other searchers destroy some of the landmarks and other important clues found.

While other caves, including the one in Randolph Canyon at Dripping Springs, have been called the Dutchman's Cave, this particular cave contains many of the clues left over the last 125 years that would link it to the Lost Dutchman Mine. Alas, this location only lacks the one critical ingredient to make it the long sought after Lost Dutchman Mine, gold ore. Note: The "trick in the trail," is a clue left by Joe Deering and others, who search for the 1880s Lost Soldiers Mine, and this mine is thought by many to be the Lost Dutchman Mine.

Reavis the Hermit and Jacob Waltz

Although Reavis preferred to be alone in his high, mountain retreat, people did visit his home on occasion, and Reavis also encountered various individuals on the trails winding through the Superstitions. Such was the case one-day, when according to old-timers in the area; Reavis met Jacob Waltz in Rogers Canyon. Waltz was prospecting for gold, and Reavis was packing his vegetables to market. "Jake," Reavis said, "You are searching for gold and someday you may find your bonanza, but when you do you will forever destroy the peace and solitude of this wilderness wonderland."

Reavis could not have known at the time how prophetic his statement was, as the legend of Waltz's lost mine eventually grew to the point where thousands of people have come to the Superstitions to search for the Dutchman's gold. Another possible trail for the two soldiers could have been the old Reavis Trail to Silver King via Rogers Canyon. To get to the Reavis Trail the soldiers could have taken the Old Spanish Trail to Tortilla Ranch and traveled southeast via Fish Creek to Rogers Canyon on the Reavis Trail. This trail was a rough trail during the time of the soldiers. It is only about two miles as the crow flies from the Reavis Trail Rogers Canyon. On Iron Mountain above Rogers Canyon you can see the Four Peaks to the north looking as one and Weavers Needle to the south. Both landmarks are clues to the Lost Dutchman Mine.

Also not far from the Reavis Trail is the old Indian ruin near Angel Spring, which contains a two-room house in a cave. I have been there many times and always considered this as a possibility because of the above clues.

Within Rogers Canyon James Rogers and others located and filed on about 17-20 mining claims. The Silver Chief Mine was supposed to have contained free gold or gold not found with copper and/or silver. Also From Rogers Canyon one can travel south and intersect the old stage route to Silver King which is what the soldiers said they did. There were cattle ranches located in the Hewitt Canyon area and a large wide valley to the southeast which is what the soldiers may have seen during their hike to Silver King.

Ron Feldman writes of the possibility of Rogers Canyon being the location of a lost mine and cache in his book *Crooked Mountain* and his Treasure Trove exploration took place within Iron Mountain near Rogers Canyon.

In closing this chapter, I previously stated that during 2001 I met with Velma Bowen Tucker. Velma was the granddaughter of Robert Bowen; the last superintendent of the Silver King Mine, just prior to it's closing in 1888. The story Bowen told to his family, as well as the notes in his family's files, confirm the story of the two soldiers coming to Silver King, and their tale of finding a lost mine, never to be seen again. I will continue my search for the Lost Soldiers Mine on future excursions, as well as looking for new information on the soldiers.[7]

Tintype of young Jacob Waltz
— Greg Davis Collection

Chapter Thirteen

Silver King Mine's Impact on Nearby Mining

As soon as the Silver King claim was filed, the rush began to stake out claims and locate other rich sources of ore. The Silver King Mine was only located on one 600 x 1500 foot claim. Consequently several claims were located all around the Silver King, hoping to intersect that same ore body. I have endeavored to list all the mines adjacent to the Silver King, as well as the mines that were nearby, and their development. This will show just how rich the ore body was for the Silver King. As reported in the *Pinal Drill* on August 8, 1883, "There are some 5,000 locations in the District, (Pioneer) but only one paying mine—the Silver King—that paid its dividend this month." [1] Active prospecting of adjacent ground followed the success of the Silver King. By 1883, at least fourteen groups were being worked and three mills were operating. The largest mill was the Windsor Consolidated Company's mill. During 1884 this mill was leased by the Silver King Company to treat ore not amenable to concentration.[2] The Eastland Mining Company owned the Tilden Mine, which adjoined the Silver King Mine to the east. Next to the Silver King on the westside was the Bilk (properly the Mowry) owned by Aaron Mason, Robert Bowen and Harry Jones. To the immediate north was the Northern King, and to the immediate south, was the Silver King South. The following nearby mines surrounded the Silver King. They were the California, Lewis Consolidated and Surprise. In addition, the Belcher, Eureka, Webfoot, Union East, Union West, Telegraph, Cedar Tree, James A. Garfield, Silver Queen, Athens, News Letter, Helpmate, Redeemer, London, Orphan Boy, Black Diamond, Crispin, Silverado, Emma, Pinal Chief and the Peachville Mountain Mines all surrounded the Silver King.

The Blue Bird, Silver Belle, Anna Bell and Martinez mines were located to the south in Martinez Canyon, closer to Florence, where they were big silver producers. The Santa Maria, Victoria, Silver Duke, Beebe, Columbia, New Year, Irene, Alice and many others too numerous to list, were also located nearby. The Reymert mines were located southwest of Picket Post Mountain. J. D. Reymert, owner of the *Pinal Drill Newspaper* from 1880 to 1884, established the mining camps of Reymert and DeNoon. I have endeavored to be as accurate as possible, however, the nature of early mining caused many mines to be bought and sold, and names sometimes changed as many times as the owners.

On July 1, 2000 Jack Carlson, Elizabeth Stewart, John and Cathy Matthews, my grand-sons Michael, age 11 and Tony San Felice, age 10, accompanied me in 4-wheelers to the trail to the old abandoned Reymert mines south of Picket Post mountain. We proceeded onto old Route 60 and turned right to the Reymert Wash. We then proceeded left or south up the wash. About a mile up the trail is an old windmill and corral on the right side about 120 yards from the trail. About 1-mile uphill from the old windmill there is a good viewpoint of the Superstitions and Weavers Needle. We proceeded uphill 1.9 mile to an elevation of about 3425 feet. We continued upward until we passed several abandoned mines. One was a trestle mine and two were slot mines.

Author's grandsons, Tony & Michael San Felice at Reymert Mill ruins
— *Author's photo*

On July 31, 2000 my grandsons Tony and Michael and my friend Jack Carlson took the trail to Reymert Mill Town. Walking up the trail we saw several old stone foundations, which were apparently miner's cabins from the 1880s when Reymert was a booming silver area. A little further along the trail we came to the old mill site. The old smelter ovens were still standing, or I should say most of them were still there. There were also foundations of several other mill town buildings at this site. We walked up the wash following the old pipeline about a tenth of a mile and came to the spring that still had water in it. On the right above the mill town was an abandoned mine, while below the old ovens a rancher had built a water trough for his cattle.

Mines Close to the Silver King

A news report of the *Pinal Drill,* February 4, 1882, said this about the mines in the Pioneer District.

"The Emperor can boast of a ledge that shows good mineral rock on the surface. There is a tunnel twelve feet long, a cut twenty feet long, another shaft above, and it appears as if another separate ledge runs parallel. The prospect is excellent, and strongly indicative of a good mine, but there has not been sufficient work done to show up what appears to be there. In the tunnel there are good quartz veins, showing mineral. It is speculative ground, with a fine show. The 'lay of the land,' as far as we can conclude, from the bowels of the King, is about E. S. E. and W. N. W. That would strike the Bilk to the west, and the Mountaineer, Silver Rock, Emperor, Black Diamond, and Washoe & Co. to the east." [3]

Monarch of the Sea mine cir. 1879 — *Bowen Family Collection*

The Pike Mine was located about a half-mile north of the Silver King. The Alice Bell was situated about two miles from the Silver King. The Wide Awake Mining Company owned the Gem, a gold mine that had a four-foot ledge and produced gold. They had a ten stamp gold mill on Queen Creek.

This report on the Silverlock Mine came out in 1880. "The Silverlock Mine is situated about 1/2 mile northeast of the Silver King, under the Pinal Palisades. There is a shaft 91 feet deep, well timbered. The vein is well defined, 6 feet wide, same class of ore as the Silver King, with regular walls. Assays have shown from $55 to above $1,000. As greater depth is attained, the mine improves. We should judge that there was over 50 tons of ore on the dump." [4] The Eureka mine was considered one of the best properties in the District and was purchased by eastern people. It had a five-stamp mill known as the "76."

The *Pinal Drill,* August 1, 1883, reported:

"In a mine close to Silver King, called the Monarch of the Sea, they have recently struck ore reported to be good. They have been trying to find the King body of ore, but what they have struck is reported to be of a different character. The Eureka work goes on steadily. The ore now reached is low grade, but there is plenty of it. They had some small quantity assaying $2,600. The ore is improving as the work progresses." [5]

Eyeball to Eyeball

A mineshaft just west of the Silver King called the "Bilk" was sunk 1,000 feet, but found no paying ore. They were also looking for the Silver King ore body. It appeared to be a characteristic of the mines that if they had good ore on top it did not go down. The Bilk connected at the lower levels with the Silver King. In 1883, the Silver King Company built a tramway from their shaft to connect with the Bilk shaft. A 53-foot bridge crossed the wash, and nearly the whole distance was trestlework. The tramway was built to carry timbers over to the Bilk, and waste and ore from the Bilk to the Silver King. As soon as they completed the tram, they drained

Bilk (Mowry) Mine cir. 1880 — *Bureau of Mines*

the Bilk shaft. A San Francisco firm sent a 10-inch Cornish pump for use in the Bilk. The Bilk had a double compartment shaft, and used a piece of equipment called a horse whim. [6] (A whim was also called a windlass which is a hoist worked by men. A horse whim was a windlass or hoist which used horse power, hence the name horse whim.)

While I was doing research on this book a funny story was related to me by a friend and old timer named Clay Worst. He stated that when the miners were digging the Bilk Shaft they were at the lower level of the shaft, drilling, when their drill-bit went through the tunnel wall. At the same time they saw a another drill-bit coming through their tunnel wall. One of the miners from the Bilk Shaft looked through the opening his drill-bit had made and saw another eyeball looking at him. The other eyeball came from a miner from one of the tunnels of the Silver King Mine. Apparently the miners in the Bilk Shaft were trying to hit the ore body of the Silver King, when they came *eyeball to eyeball* with the Silver King miners. Needless to say, the Apex Mining Law was invoked and the Bilk had to discontinue drilling. But there is somewhat of a happy ending to this story. The Silver King owners needed another shaft, and bought the Bilk Shaft, and used it to haul ore from the Bilk Shaft to the Silver King. The ore was taken topside from the Bilk, and then put on ore carts and sent over to the ore house of the Silver King. The owners of the Bilk Shaft just happened to be the Aaron Mason, the superintendent of the Silver King Mine, Robert Bowen the mine foreman and Harry Jones. The Silver King owners paid $250,000 for the Bilk. In this case, it was "all's well that ends well."

The *Pinal Drill,* August 1, 1883 reported this activity on the following mines:

> "The Speciepaying is sinking for water for their mill to be built. They are down 70 feet and have found water, but not in sufficient quantity yet. They are going deeper. We understand they are now getting about 800 gallons per hour. They are working for 1600 gallons per hour. The Surprisor: Ten miners are working, further developing the mine and accumulating ore and they obtain the best results. The mine will now be thoroughly developed before starting the mill again. A fair quality of ore is now obtained and the mine looks decidedly better as work progresses." [7] That same day the report on the Frankfort Mine said: "The Frankfort shaft, about one mile north of Silver King is now down over 40 feet. It is the same general character as Silver King ore. The vein is between good walls and the pay-streak 4 feet wide. They are in ore similar to the Silver King. Syenite carrying zincblende is averaging $40 per ton silver. It is a big strike and likely to prove an immense mine." [8]

The Silverado Mine was one and one-half miles from the Silver King. *The Mining and Scientific Press* of July 16, 1887 reported that the Silverado had two prospect shafts on small veins at the surface that expand to several feet. The Silverado contained ore that assayed from a few ounces of silver up to thousands per ton. [9]

Silver King Mine view from Bilk Mine cir. 1880s — *Bureau of Mines*

Mines North of the Silver King—Rogers Canyon

About 14 miles north of Pinal, on the old Happy Camp Road, the mines of the Rogers District were located. It was said that there was a spring of water that ran six inches deep year round. The nearest route for a wagon was from Hewitt's Station, which was about ten miles from the mines. There were three main ledges running parallel southeast and northwest. The names of some of the mines were the Snap, Chloride King and Monarch on one ledge, the Columbia, Silver Chief and Manhattan on the other.

Early prospecting was conducted in the area of Rogers Canyon by James Rogers (referred to as Captain Rogers) and others during 1873, or perhaps even earlier. James Rogers located the first mine in this new mining district in 1875, called the Silver Chief, hence the name Rogers District. The mines took out native silver and carbonate ore. Free gold was reported to be found in the Silver Chief tunnel during 1882.[10]

The Rogers District was described in the *Pinal Drill*, March 11, 1882.

"Altogether, the facilities for mining and milling in the district are unsurpassed in the territory. There are three main ledges running parallel, S. E. and N. W., dipping to the north, on which the chief development work has been done. The Snap, Chloride King and Monarch are on one; the Colombia, Silver Chief, Manhatten and World Beater on another; the Bluch, Dickens, Goodenough and Plato on the third."

The Silver Chief, with owners Geo. DeWalt, Chas. Ceslinger and F. F. Broerman, was the first mine discovered in the district, having been located by James Rogers in 1875. It also has more work done on it than any of the other mines. There is a 200 foot tunnel, on the ledge, which is of carbonate ore, showing native silver, from three to twelve feet in width. Free gold has also been found in the tunnel. Assays of the ore go from $120 to $9,500. At a depth of 20 feet there is an 80 foot shaft to the surface. About 300 feet west of this shaft is another 60 foot shaft, at the bottom of which the ledge is twelve feet wide, and they are going to work in the tunnel, to reach this shaft, which they will do at a depth of about 250 feet." The article then goes on at length to discuss the details of the other mines, stating that they all have good ore, along with other valuable properties nearby. [11] The Rogers District mines were located in the are now known as Rogers Trough Trailhead and Rogers Canyon, located in the Superstition Wilderness Area and closed to mining now. I have seen some of these old mines, but the Rogers millworks are no longer there. However, there are some pieces left of the old mill, near the stream above the trail through Rogers Canyon.

Iron Mountain Ore Cart Mine

On November 22, 2003 I drove out to Rogers Trough Trailhead on Iron Mountain, accompanied by Jack Carlson and Tom Lamonica. The drive to Iron Mountain takes you several miles through Hewitt Canyon on a very rough jeep trail. The jeep trail zig zags upward to a hairpin turn which requires you to stop and back up prior to again proceeding. The drop-off is very steep at this point and there are no guardrails on this precarious narrow road. If you encounter a vehicle coming down some one must either back up or find a place to move to the side so one or the other can pass. We drove on up to the Rogers Trough Trailhead where we parked. On the way up the mountain we saw 7 mule deer about 1/3 way up Iron Mountain. There was a buck, 5 doe and a fawn.. We hiked out through Rogers Canyon north towards an area where some old abandoned mines were located. We followed the trail for about ½ hour or ½ mile until we came to a side canyon that led us uphill on the left side. This canyon was supposed to lead us to a mine with an ore cart and another mine with a shaft. Well, we hiked up this very steep canyon through some very rough manzanita and catclaw and other dense brush for about 2 hours but did not come to the mines. It appears that we took the canyon on the wrong side. We did find some quartz from a blowout or prospect that was higher up, but did not find any mines.

Jack Carlson with Iron Mountain ore cart — Author's photo

Jack Carlson indicated that perhaps we had located part of the 1880s trail that Reavis the Hermit took and James Rogers used. Carlson took another GPS (Global Positing System) reading and re-checked his map, and we walked to the right or north for about another hour. We came to some of the lower mines at last, and located a mine dump, prospect and covered mine tunnel. We walked down hill and I spotted the ore cart, and we came to a tunnel and mine dump near the ore cart. We inspected the tunnel which went over 100 feet into the hillside. We found chryacholla, azurite, malachite and evidence of some metal in the quartz on the dump. The ore cart was still in pretty good shape and there was a lot of ore cart track there. We walked out to my jeep just as darkness fell, and I drove out in the black of night off Iron Mountain. We were really exhausted after bushwhacking it all day. This 4-hour hike took about 7 hours, and we were really glad to get back to the jeep in one piece. We were all scratched up from the brush and I looked like I had tangled with as mountain lion not a mountain. But all in all it was a good trip and we plan to go back to locate the other mines in the area.

Ron Feldman and the Iron Mountain Peralta Mine

In 2000, Ron Feldman of the O.K. Corrals Riding Stable in Apache Junction authored a book called *Crooked Mountain* about Quenton "Ted" Cox and his search for the Lost Dutchman Mine on Iron Mountain, in the eastern part of the Superstition Mountains. *Crooked Mountain* is a compelling novel of treasure, deceit and death. Although the book is fiction, it is based on the real life adventures of Ted Cox of Globe, Arizona. Cox searched for the Lost Dutchman for many years and kept copious notes of his searches. During 2004 Ron Feldman the co-owner of the OK Corral riding stable received a Treasure Trove permit from the US Forest Service to dig for the Peralta treasure cache, he believes is buried on Iron Mountain. This permit enabled Feldman to further Ted Cox's search on Iron Mountain. Feldman believes that the Cox story along with the Dutchman's clues lead him to an old mine in the Iron Mountain area. There may be some truth to this story. James Rogers mined this area for silver and other metals in the late 1880s and the early 1900s. (See above section Mines North of the Silver King.) Many others, including Jimmie Jinks, have also documented mineralization on Iron Mountain for precious metals. I have seen the old area where a mill was set up to process ore during the 1880s. The *Pinal Drill* stated that the mines in Rogers Canyon were taking silver sulphurets and copper pyrites out of white quartz during the 1880s.

Ron Feldman with hand-hewn mine timber, 2005 — *Author's photo*

On September 21, 2004 Ron Feldman his partner Bob Schoose and Feldman's sons Jesse and Josh, along with several members of a group, named Historical Exploration And Treasures (H.E.A.T), began a historic expedition within the Superstition Wilderness, in search of a what they believed to be a lost Spanish (Peralta) Mine. They conducted this exploration in order to prove that the Spanish had mined north of the Gila River prior to the late 1800s, when Anglos started

Author at Treasure Trove site, 2005 — *Jack Carlson photo*

mining in the area. Feldman's exploration is based on an unpublished manuscript of a man named Ted Cox, who stated he saw bars of bullion stored in the abandoned mine over 50 years ago. Cox stated he was unable to recover the bullion himself, due to various life-threatening circumstances. H.E.A.T. also hopes to recover bullion that was stored in the old mine tunnel prior to 1848. At the Treasure Trove dig site there was a 20-foot tall wooden headframe designed by Bob Schoose, Feldman's partner, along with a 15-foot shaft that had been shored up with wood planks. Old style timbers have been taken out of the tunnel including hand hewn timbers. I was last there during late May 2005, and they were still excavating the old tunnel. The final

Heat Members. L-R Duane Short, R.J. Schroeder, Eric Shervey, Jesse Feldman, Ron Feldman — Author's photo

Heat Team & volunteers, back: Joe Lalich, Ron Feldman, unk., Dave Wenham, Josh Feldman, Robert Knect. Front: Eric Shervey, Bob Schoose, Tom Riches — Author's photo

chapter of this saga will be written in a future book by Ron Feldman called *Deep Fault.*

Mines South of the Silver King

This report in the *Pinal Drill* February 4,1882 described the mining activity south of the Silver King.

"The Beebe Mine is seven miles south of Pinal, and one and a half miles north of the Pioneer Con Mine. It has a very large well-defined ledge, a vein five feet wide in the shaft, which is fifteen feet deep. Average assays run $28, 32, 41, 43, 52, 78, 89 and 108 dollars. The tunnel below is 60 feet long. There the vein is three and a half feet wide and assays average $15, 16, and 108 dollars at various points. Some assays run up into high figures. The Beebe belongs to E. A. Sarrick. It is an old location and a mine of great value. Ore is there in quantities and the formation shows a permanent ore body." [12]

The *Pinal Drill* September 8, 1883, described the Australia Mine.

"The winze in the Australia on the Continental ledge, is now about 50 feet deep. The vein is unbroken from the top, some 80 feet. In the bottom the mineral is 20 inches wide. Assays by J. D. Baker, show 17 percent copper, 44 percent galena, and $15.37 silver per ton. They will now sink on the vein, which is very wide, and then drift into the mountain on the vein. The depth of the mountain is over 45 degrees, so that depth will be gained rapidly. The Crispin Mine was located on the first cone south of the Silver King, about 900 feet south. The shaft was said to have reached considerable depth and had a fine ore body similar to the Silver King in character."[13]

Happy Camp Mines

The *Arizona Business Directory and Gazetter of 1881* described Happy Camp as follows:

"Situated three miles northeast of Pinal, in Happy Camp, is the property of Mr. William Johnson, called the Uncle Billy, which is one of his favorite mines. There is a shaft of 50 feet, heavily timbered, and a large dump showing galena ore, rich with silver. There is also a shaft 25 feet deep, also timbered, and a large dump of the same class of ore. The ledge is in avonite and granite, 4 to 6 feet wide. The highest assay was $1,025, next $630, and so down to $50. They are now working the mine. Also at Happy Camp were the Augustin, Leon, Lancing, Rockland, and Hard Scrabble Mines." The Merrimac Mine was situated 3 1/2 miles northeast of Pinal and adjoined the

> Uncle Billy. According to 1880 reports: "The ore is of the same quality as that found in the Uncle Billy. A shaft is down 20 feet, width of vein, 5 feet. The ledge is very prominent along the whole belt. Assays average $60." [14]

During April 1998 I travelled the old road to Happy Camp with Greg Davis and Jack Carlson. We found the remains of an old stone cabin near a spring identified by the large cottonwood trees in the wash nearby. We found some mine tunnels and reminants of old mining activity. The old stone corral is still there. From a knoll near Happy Camp there is a great view of the desert hills and Picket Post Mountain.

Peachville Mountain Mine

On April 1, 2005 Bob Stambach and I were 4-wheeling in search of wildflower locations to photograph, on what I thought was the old Silver King to Happy Camp road. We drove northeast up the Silver King Road from Route 60. When we reached the area past the Herron Ranch windmill, we turned left up the old jeep trail and drove one mile to a fork in the road. At this fork we drove right .7 mile upward on a very rough jeep trail that I had never been on before. There were three very difficult spots in this old road. The old road dead-ended at this spot below Peachville Mountain. As we got out of the jeep to check for a turn-around place we spotted an old mine dump. Walking up to the old dump we located a long abandoned mine tunnel. This tunnel went back about 200 feet where it ended in a large stope room and a shaft went downwards and another tunnel/drift went off in another direction. There were plenty of old timbers left in the mine and it looks like quite a bit of mining took place during the 1880s, based on the style of timbers left there. After we exited the tunnel collected some ore, which had chrysacholla and malachite (green) stains and visible native copper along with calcopyrite and some gold colored metal substance unknown. We hiked up the steep hill above the tunnel and I located a shaft that went downward 60 feet until it had apparently collapsed and closed. Well, this mine was definitely old and had been extensively worked. We took some ore samples to be assayed in order to determine the content of precious metal.

On April 6, 2005 Bob Stambach, Jack Carlson and I drove to this old Peachville Mountain mine to attempt to find out the name of this old mine. On the drive from Silver King Road on the old jeep trail we saw three mule deer and a Gila Monster. We continued uphill to the abandoned mine and entered with plenty of flashlights this time. Once inside threshhold of the mine I heard this nasty buzzing sound. I had almost stepped on a large Diamondback rattlesnake. I immediately backed out and then removed this dangerous obstacle. We proceeded to the end of the large tunnel and also saw a bat. We photgraphed the old shaft and timbers left behind. On the way out we checked the next tunnel and I saw some green eyes at the rear of the tunnel. Those eyes belonged to a ringtailed cat. After taking some photos of the cat we exited the mine. Jack Carlson plotted this mine on the topo map before we drove down the steep hill. After returning home I reviewed my files and located an article in the *Pinal Drill* dated March 17, 1883 that stated:

"The Peachville Consolidated Mining Company has incorporated; object to work the mining claims located by A. W. Blair at Peachville, Pioneer mining district, Pinal County, A. T. The Directors are A. W. Blair, James C. Plunkett, R. A. Redman, V. W. Plunkett and james Capel; all residents of Oakland."

Martinez Canyon Mines

The *Pinal Drill* September 8, 1883 reported this information on the Martinez Canyon mines.

> "The Blue Bird, the property of Thomas Farrel, near Martinez Canyon, is an immense ledge ascertained to extend 150 feet in width. It adjoins the Martinez Mine. The mineral vein on which work is done is 16 feet wide. There is a streak of spar and gray carbonate near the hanging wall, then a streak of waste rock, then a streak of galena and carbonate, then waste rock." [15]

Also located in Box Canyon, near Florence and the Buttes on the Gila, were the Silver Belle, Columbia and Anna Bell mines. They were active for years in the 1880s and revived about 1900 by Josiah Champion. The Anna Bell Mine carried copper as its principal ore. However, some of the specimens carried gold values of $15 to $50 per ton. Other mining prospects near the Anna Bell were also said to have carried gold values.[16] On May 26, 2001 Jack Carlson and I drove out Route 60 east of Florence Junction to the Martinez Canyon jeep trail. At the entrance to Martinez Canyon there are many cottonwood trees and other large trees said to be Arizona ash trees. We drove as far as we could and parked in the wash, where there was running water. We got out and walked where we encountered some majestic red canyon walls. We located what appeared to be a mine tunnel with an iron door, which I later learned was a saloon and brothel for the mining camp during the 1880s. Jack Carlson had two small flashlights and we entered the tunnel. It quickly dead-ended, and as we were about to leave I noticed some movement on the ceiling above us. We turned the lights upward and saw what appeared to be about a million daddy long-leg spiders all over the ceiling and walls. They were pulsating with a rhythm and appeared to be doing a mating ritual. That vision of the closeness of all those spiders gave me the creeps, and the hair on my neck and arms stood up. That place gave me the willies. We were in such a hurry to get out that we neglected to take a photo. We walked a hundred yards or so from this place, where we encountered some older buildings and a new small tan building. There was an old stone and adobe building there, which must have dated back to the 1880s, and a larger wooden house, painted red, said to have been built in 1947. Next, we walked up the wash where the sides of the canyon were brilliant red in color and found a large cave where plenty of campers have stayed in the past, according to the smoke stains on the ceiling. We proceeded another 1/8th mile or so up a boulder-strewn wash and came upon the Martinez Mine works. There was a large processing mine plant or mill house still there, with very old and more modern equipment in it. Nearby were some old mine tunnels and prospect holes. We photographed the mine works

Martinez Mine ruins 2001 — Author's photo

and then proceeded back to our vehicle because I wanted to get out before dark.

The first mine in the canyon is the Martinez Mine, second the Columbia Mine, and farther north up the canyon are the Silver Belle and Anna Bell Mines. The locations for these mines were made in 1879. The principal ores were lead and silver. The Pinal Consolidated Mining Company, founded by Aaron Mason of the Silver King Mine, obtained the rights to these mines in 1882. Later in the 1920s, the Silver Belle Consolidated Mining Company acquired the mining rights to these mines. Recorded production from 1926 to 1928 was $100,000. During 1937-38, the Sunbeam Gold Mining Company produced 3000 tons of ore. In 1943, the California Steel Products Company built a mill and milled about 2000 tons of ore until 1946. The difficulty of mining this area was the fact that Martinez Canyon Wash, which is a steep walled canyon, was said to have frequently flooded and was the only passage way. [17]

And flood it did! On August 16, 1883 a great deluge created a flood that devastated the mining camps of Silver King and Pinal, as reported in the *Pinal Drill* of August 18, 1883. This downpour also flooded Martinez and Box Canyons. Years later, in 1951, there was another devastating flood in the canyon. The *Florence Blade Tribune* on August 3, 1951 tells the story of the Villa Verde family who was almost swept away by the flood. The headline read: "Mother and Two Children Almost swept away when House Battered By Water." Peter Villa Verde and his wife Connie and two children, Peter and Sylvia, were living in the house when the flood hit. They recalled later that no rain fell, but that hailstones "as large as doorknobs fell for a few minutes." [18]

The Coke Ovens at Cochran

Many people have seen the famous Coke Ovens at the old town site of Butte, across the river from the ghost town of Cochran, on the Gila River, north of Florence. But few people know the real history of the ovens or huge brick kilns. In 1881, the Pinal Consolidated Company was organized and incorporated in San Francisco, California by Aaron Mason, the first superintendent of the Silver King Mine. Mason along with his associates Henry Donnelly, ex-superintendent of the Eureka Consolidated Mine, Henry Linkton, F. Logan, S. Hart and H. H. Noble, purchased two claims ten miles south of Pinal City. They were the Silver Belle Quartz and the Columbia, located in what is now called Martinez Canyon. The company built a wagon road from the mines to a mill site on the Gila River. This site was later called Butte, for the large butte nearby, now called North Butte, and a mining camp was soon established there. The company also purchased ore from nearby mines. Mason, in January of 1882, had the Pacific Iron Works build a 40-ton galena-silver plant for smelting ore at Butte. Reportedly, it was a Rankin-Brayton smelting furnace, and one of the most extensively used smelting furnaces in western mining.

In 1882, the Pinal Consolidated Company built five large beehive stone ovens at the site close to the smelter. The wood from the mesquite trees along the Gila River was harvested and burned in the ovens to create a product called coke which burns very hot in the smelter. The smelter was reported to have turned out 1,800,000 pounds of bullion during the first six months of 1883. The rich ore soon was exhausted, however, and as the price of silver dropped production was ended in 1886. The smelter was eventually dismanteled, but the stone ovens remained. They are known today as the Cochran Coke Ovens, after the nearby ghost town of Cochran. The mining camp townsite of Butte faded into obscurity. At the turn of the 20th Century, one of the coke ovens became home to a woman named Stella Lee Frakes, and her daughter Violet Frakes, who was born in one of the ovens. The five large beehive kilns loom larger than life along the

Author at Coke Ovens 2002 — Jack Carlson photo

north side of the Gila River, and is a favorite place of 4-wheelers to visit. They remain a testament to Aaron Mason's Pinal Consolidated Company.[19]

During the summer of 2002, I drove out to the ghost town ruins of Cochran several times. I parked on the bank of the Gila River bed, which was almost bone dry, and hiked across the dry riverbed to the coke ovens. This was the first time I was able to do this in ten years. Because of the drought, one could now walk across the dry riverbed, where previously there was always too much water. I was able to get some good close-up photos of the coke ovens. It was apparent that someone had been living in one of the ovens recently as the inside was partially rennovated and made liveable. Subsequent trips during 2003 revealed that someone had vandalized the living quarters.

Superior and Mineral Creek Area Mines

Two of the largest mining areas and copper producers in the area near the Silver King Mine were developed right after the Silver King. They are the Magma Mine in Superior, and the Ray Mine south of Superior, on Mineral Creek. In 1897 George Lobb left the town of Silver King, and went to the nearby town of Hastings, where he relocated the Gem and Hastings mining claims. At this time there was very little activity around Silver King or Hastings. Hastings was a small settlement named after its founder, who

Silver Queen mine (Magma) cir. 1900s — Gladys Walker Collection

went to California to become wealthy in real estate. Meanwhile George Lobb and his sons, Dick and Archie, along with Richard Trevethan (his father-in law) and Wiley Holman, worked the dumps around the Old Silver Queen Mine, founded at the same time as the Silver King Mine.

Superior was founded on Buried Money

George Lobb founded the Town of Superior on $10,000, which he had buried under his house at Silver King. After moving to Hastings he remembered the money, and sent his sons there to dig up the silver coins. Lobb soon built a general store, with living quarters at the rear. This store was located next to the present day Superior post office. George Lobb was the Town of Superior's first postmaster. In 1902, Lobb sold his Golden Eagle Group to the Lake Superior and Arizona Mining Company (L. S.

& A.). The Lake Superior and Arizona Mine started operations in 1902, with the Lobb's, Richard Trevethan and Charley Finch as the first employees. As word spread that there was work at the L. S. & A mine, in the new town of Superior, people began moving there and Superior was on its way.[20]

George Lobb, Superior Post Office— Gladys Walker Collection

The area of the present day Town of Superior had three names. It was called Hastings, but part of it was called Queen. Finally it was named Superior, after the Lake Superior and Arizona Mining Company, which was incorporated in 1902. George Lobb, the "Father of Superior," sat on the board that named the town after the mine. By 1903, Superior had about 100 residents. Mrs. Innes of Silver King owned the first boarding

Superior area 1910 — Gladys Walker Collection

house there, and liquid refreshments were soon available at Tom Kelley's tent saloon. A crisis happened in January of 1904, when the supply of "True Medicine" ran out and left the miners in a state of mutiny. Two barrels of whiskey were shipped immediately and the crisis was over!

The *Florence Arizona Blade* of April 2, 1904, stated that the town was entirely dependent on the operation of the L. S. & A. Mine. It further

Magma mine smelter 1999 — Author's photo

stated that the town was made of rough buildings, but that you could see the influence of ladies there because of the curtains on the windows. The L. S. & A. kept Superior alive until 1910 when Boyce Thompson started work on opening up the old Silver Queen Mine which, during the 20th Century, became the famous copper producer, the Magma Mine.

Magma Mine at Superior

The Magma, as stated earlier, was first called the Silver Queen, and was initially a producer of silver. The prominent outcrop of the Silver Queen vein was located about the same time as the discovery of the Silver King Mine. The Hub claim, 700 feet long and 600 feet wide, was located March 29, 1875, by W. Tuttle. In the center of this claim was the Silver Queen shaft, which became the No. 1 Shaft of the Magma Copper Company. Adjoining the Hub on the west was the Irene claim, 1500 feet long by 600 feet wide, which was located September 1, 1876. Philip S. Swain and associates organized the Silver Queen Mining Company. Mr. Swain was actively interested in the property until 1910, when it was sold to the Magma Copper Company. The Silver Queen Mining Company patented the Irene claim on October 31, 1885, and the Hub claim, November 3, 1886. All productive stopes of the mine were within these claims prior to 1920. Many fantastic tales were told of the richness of the ore in those early days.[21] In the more recent era, the Magma Copper Company used the Old Silver Queen shafts as one of its major developments. The claim was relocated after the discovery of the long-sought after Silver King. [22]

William Boyce Thompson and the Magma Mine

In 1910, Col. William Boyce Thompson, along with his partner George Gunn, bought the Silver Queen Mining interests for $110,000. They established the Magma Copper Company. The Magma Copper Company was incorporated in 1910 and essentially built the Town of Superior. It soon became one of the first major copper producers in Arizona. Boyce Thompson later founded the Boyce Thompson Arboretum on State Route 60 just west of Superior. While developing the mine for copper, they came across a small fortune in silver when they crosscut the old Silver Queen shaft during January 1911. Once again, the hills were alive with prospectors, looking for the mother lode of silver. This new activity also helped the stage line of Gene Robles.

The development of the Arizona Eastern Railroad in 1912 greatly assisted with major improvements at Magma. The Magma Company built a short line common carrier narrow gauge (36 inches between the rails) from Webster to Superior in 1914. It consisted of 31 miles of track operating on 30-pound rails. It connected with the Phoenix and Eastern Railroad at Webster (later Magma Junction) near Florence. It was a big improvement over the wagons pulled by 20 mule teams or horses. However, the ore had to be transferred from the narrow gauge railcars to the standard gauge railcars of the Phoenix and Eastern Railroad. During 1920, the Magma Copper Company decided to build a smelter at Superior to treat its ores and concentrates. This posed a problem, since the narrow gauge railroad could not carry the heavy steel needed in construction of the smelter. Therefore, the Magma Railroad Company changed its tracks to standard gauge, (70-pound rails), by April of 1923. The Magma Railroad also carried passengers and mail to Superior until 1938. From then on it carried only freight.[23]

Until the railroad came to Superior, eleven wagonloads of ore made the trip to Florence every four days bringing back coal on the return trip. The Pinal Mills that had processed the Silver King ore had long since been shut down. Phil Grigsby was the Wagon Master. Transporting ore by wagons was costly, and plans were made to develop the railroad to the Magma Mine. The Magma Company's first concentrator was built in 1914, on the hillside above the town. Once the Magma Arizona Railroad (narrow gauge) was built on May 25, 1915, the railroad began carrying ore to the smelter in Hayden. This made the ore wagons obsolete. [24] The Town of Superior began to flourish and grow around the mine. New buildings were being built and the mine employed several hundred men. World War I brought a heavy demand for copper, and Magma and Superior was on its way.

Impact of Silver King Mine on the Town of Florence

Soon after the Silver King Mine was founded, it diversified its holdings by investing in the Town of Florence on the Gila River. This set in motion a major boom in the Florence area. The Silver King Company opened the Silver King Hotel and a dry-goods store. These enterprises, coupled with the smelter north of town at the mining camp of Butte, brought prosperity to Florence. The stage and freighters ran their lines into Florence to get a share of the growing commerce. Mining at the Silver King provided an economic base for Florence. During 1877 twenty mule team wagons were passing through Florence, with ore from the nearby mines, and supplies for the miners. The Southern Pacific Mail Line began regular stagecoach service between Casa Grande and Globe, going through Florence to Silver King and Pinal. In 1879, Wells Fargo began security service to and from Florence. Florence also became the main stage stop between Silver King and Pinal, and Tucson and Yuma.[25]

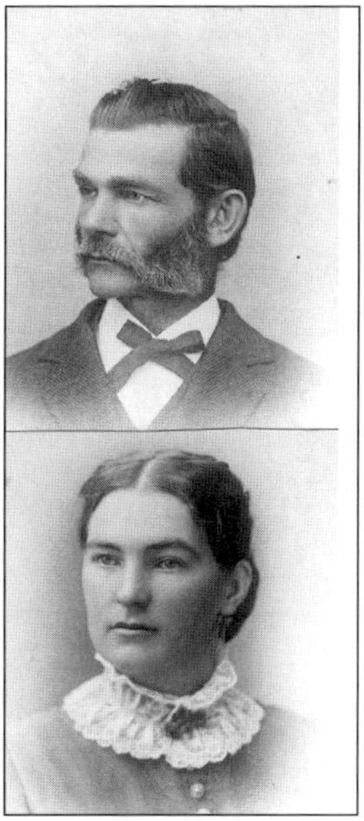

Silver King Engineer William Clark & Ella cir. 1880s— Pinal County Historical Society

The rapid growth, and incoming miners and settlers, made the town receptive to growth and prosperity. New stores opened, and the first Pinal County Courthouse and a two-story wooden schoolhouse were built. In 1877, John Clum of Tombstone fame established the *Arizona Citizen*, which was the first newspaper in Florence. A year later he took it to Tombstone and became the editor of the *Tombstone Epitaph* during the sanguinary days of the Earp-Clanton Feud, and their famous shoot out at the O. K. Corral. The *Arizona Enterprise* was started up again, with editor Thomas Weeden at the helm. With prosperity came social problems. Twenty-eight saloons soon occupied space along Florence's Main and Bailey Streets, and thirsty customers came in droves, to drink and fight. The Silver King Hotel, and its competition the Lewis

Hotel, advertised tables set with the finest food the market offered, as well as the best whiskeys, brandies and even a billiard room. The Lewis Hotel had a huge wire birdcage containing over a hundred of Arizona Gambel's Quail. There was a brewery, a Chinese bakery, flourmills, restaurants, a barbershop and even an ice cream parlor. The stages and teamster wagons ran daily between Florence and Pinal and Silver King. Cuban cigars and French cognac could be found locally. Civilization was coming to this part of the West.

Restoration of the 1880s Silver King Hotel in Florence — Author's photo

One of the most classic six gun shoot-outs in the old west took place in 1888, at the Tunnel Saloon in Florence between former Sheriff Pete Gabriel and a former deputy named Joe Phy. This famous shoot-out is discussed earlier in the book. The transfer of the land office from Florence, the closing of the Silver King Mine and other nearby mines, caused Florence to cease as a major hub of commerce. But the town recovered and renewed mining activity, this time for copper, caused a rebirth of the town's prosperity. Florence still thrives today, but with an emphasis on agriculture and the prison industry, as several federal, state and local prisons were established nearby just off State Highway 179. Prisons are not new to maintaining Florence's industry. During World War II, several thousand Italian and German prisoners were housed in a POW camp, just north of the Gila River. The large trailor parks of Florence Gardens, Casita Hermosa and Caliente De Sol now stand on the former POW site. Cotton growing, cattle ranching and other farming became the new industry of the 20th Century, as the nearby Gila River provided the nourishment for the desert farms. The most attractive feature of Florence during this time was its social life, reflecting the amicable mixture of Hispanic and English speaking residents.[26] Florence, at the current time, is once again experiencing a revitalization and historic redevelopment of the old 1800s houses. In fact, the old Silver King Hotel is one of the restoration projects and Florence holds an annual historic open house festival.

Mineral Creek Mines and Ray Mine at Kearny

After the Silver King Mine closed the first time, during the late 1880s, the Ray Mine was established south of Superior and became one of the major copper producers in Arizona. Mineral Creek, which used to flood, now flows under the Ray Mine through a viaduct. In fact, the Ray Mine is an

Town of Ray cir. 1950s— Deen Family Collection

active open pit copper mine today, called ASARCO, and owned by the the El Groupo Corporation. Ray was located on Mineral Creek, six miles from the Riverside stage station on the south side of the Gila River. In the fall of 1877, many miners were working along Mineral Creek at Monitor Camp, eight mines from Riverside, and at Republic Camp, three miles further north. The Ray Mine got its start as the Pinal Copper Company in December of 1880. William Bollinger, D. B. Spangler and other San Francisco men developed the Ray Mine and other claims. A smelter was soon erected on the Gila, and burros carried the ore between the mines and the smelter. The early settlement was called Bollingerville, after its founder. The rich ore, easily available, was soon depleted, the company had financial problems and went bankrupt by 1881.

Louis Zenkendorf, a Tucson merchant, bought out the claims of Pinal Copper, along with some New York associates, and renamed it the Ray Copper Company, then established Ray City near the mines. The mines operated with some success until 1898, when the mines were purchased

Ray open pit copper mine 1990s— Author's photo

by a British Company and called the Ray Copper Mines, Ltd. The English engineers built a huge 250-ton mill at the confluence of the Gila and Mineral Creek. Here the townsite of Kelvin was laid out, named after the renowned British mathematician and physicist, William Thompson Kelvin.

The Ray mining towns of yesterday are gone, and there is now a huge open pit mine at the Ray complex. Ray consisted of a number of separate villages, the largest being Sonora, where the Mexicans lived in their own homes leased from the company, while Anglos lived in the townsite of Ray. Barcelona was the domain of the Spanish people. The Austrians lived in Boyd Heights. The company engineers and officers lived on the hills above the Ray business district, on the eastside of the creek. The African-Americans lived in Hercules, near Sunnyside.[27] In 1912, the Ray "Mountain of Copper," contained enough high-grade ore to make underground mining profitable, but by 1948 underground mining ceased and open pit mining began. The Ray Mine, which was named after the daughter of the prospector who found the mine, still survives today. The Kennecott Copper Corporation bought out the huge Ray Mine Complex. It was called ASARCO, and in recent years it was bought out by a Mexican firm called El Groupo.

The *Pinal Drill,* September 8, 1883, reported this information on the Mineral Creek area mines in 1883:

"The Copper Depot Mine has a shaft of 15 feet and shows the same formation and ore as the Ray Mine. The road from Schoueberg's group runs westerly to the Ray Mine. The Ray Company employs 25 men and has arranged everything very conveniently, but since it is against orders to visit the mine, I can only say what I can see, viz: a large lot of native copper from which I must conclude that they take the ore out by the carload. Gus Shamp has charge of the Company's boarding house and store. Half a mile northerly from the Ray lies Tom Haley's Camp, amidst lovely shade trees, giving it more the appearance of a garden, than of a mining camp, and owing to its pure cold water and shade, is the real camp of Mineral Creek. Half a mile easterly from this place, there is seen a large mass of copper ore, that is the Copper King Mine, which belongs to Tom Haley and his partners. The shaft is 40 feet deep in carbonate and galena ore. West of this the Keystone Mine, also property of Tom Haley and partners. This mine is developed a distance of 900 feet, and shows an ore-body 150 feet in width of rich carbonate and galena. There is a shaft of 50 feet in depth, one tunnel and several cross cuts. The ore dump has about 5000 tons. The Elder Mine. They are working steadily and have a width of 5 feet milling ore carrying free gold and silver. It is about 2 miles westerly of Mineral Creek." [27]

The *Pinal Drill,* November 11, 1881 reported this on the Mineral Creek District:

"Mineral Creek also contained the following mines: The Bauman near Walnut Springs, which contained a large reef of galena and carbonate ores; the Monroe, one mile from Mineral Creek, whereby in 1881 they took out 200 tons of copper ore; and the Pennsylvania and the Lookout, about a quarter mile from Riverside on the Gila, which in 1881 reported values of 22 to 90 percent copper, and silver, \$11 per ton, and gold, \$19 per ton. The Poor Man, Boos and Copper Depot Mines, come next. The Poor Man has a cut of 80 feet in length and 20 feet in depth and shows a ledge of 10 feet in width and wonderfully rich copper ore, native copper and oxides. This is a fissure vein and the croppings are traced northerly, through the Future Mine, and southerly through the Boos Mine. The walls are in quartzite, impregnated with iron ore and native copper." [28]

Brief History of Miami/Globe Mining

The famous silver strike called the Globe Ledge in Alice Gulch was located in 1873, just prior to the discovery of the Silver King in 1875. The Anderson brothers, along with (three of the four locators of the famous Silver King Mine) Ben Regan, William Long and Isaac Copeland, and some other prospectors located the Globe Ledge. Development of this strike was delayed due to Apache attacks, but these claims proved to very rich in silver when finally developed.

The history of major copper mining in this area begins with the Old Dominion Mine, which was developed by the Old Dominion Copper Mining Company in December of 1880. This company also bought the New York and Chicago Mines, southwest of present day Miami. Two 30 ton furnaces were erected in what was called Smelter Wash, and a small settlement that grew near the mine was called Bloody Tanks. This is the location where pioneer Indian fighter King Woolsey killed several Apache chiefs in 1864 during his efforts to reduce the Apache trepidations upon the settlers of Central Arizona. The Woolsey expedition stated they found gold and some good looking quartz lodes in the Pinal Mountains.

In 1869 William A. "Hunkydory" Holmes, along with H. B. Summers and the Anderson twin brothers, made seven claims near a stone fort in Big Johnny Gulch. Holmes also had a profitable claim called the Daisy Dean in 1875, near the Ramboz mining camp. Holmes along with the Anderson brothers and others also located some rich silver claims they called the Silver Queen, near present day Superior. Holmes was quite a lucky character having survived several shootouts, but he was killed in 1889 trying to transport several renegade Apaches, including the infamous Apache Kid, to Yuma Prison.

Globe Founded on Silver Discoveries

The first silver mines to be profitable near Globe were the Ramboz, Rescue and Bluebird, located in 1872, which had rich ore on the surface. Among the richest mine in the Globe area was the Stonewall Jackson, located by two prospectors named McMillen and T. H. Harris. Charles McMillen and Doby Harris located the "Famous McMillen," or Stonewall Jackson Mine, on their way out of Globe in 1877 going from Globe to Ft. Apache. It was said that almost pure silver was cut out of the veins of the Stonewall Jackson by chisels. The locators sold it for a profit of $120,000 to some California investors who also made money on the mine, but underground water caused the mine to cease operations.

Munson's Chunk was another discovery. A prospector named Munson was the discoverer of the large almost pure silver boulder, and since he could not transport the boulder, wrote his name on a piece of paper and placed it on the boulder. It supposedly read, "This is Munsons Chunk." It was later broken into smaller pieces and transported to Globe.[29] At Richmond Basin to the north of Globe, a German prospector named Schultz found some silver nuggets. He called his claim the Nugget. A small mining camp with a post office sprung up in 1881, but the camp closed in 1884 after a brief but illustrious career. But, as with the way of mining camps, it was a ghost town by 1890.

The Mac Morris Mine in the nearby Richmond Basin was supposedly one of the largest producers of silver during it's time. It was founded by a prospector named Mac Morris and earned him a fortune before he sold it to Fisk, Stout & Company of New York for $90,000. Morris constructed a 10 stamp mill at a place called Wheatfields to process the ore. The Mac

Old Dominion miners — Gila County Historical Society

Morris payroll was the subject of a famous holdup on August 20, 1882, just outside of Globe. Andy Hall and Dr. F. W. Vail were killed during the process of this holdup. Subsequently, Lafayette Crime, Cicero Grime and Curtis Hawley were arrested for the holdup and murders. Feelings ran high and Lafayette Grime and Hawley were hung just outside of the St.

Elmo Saloon in Globe at the "hanging tree." Judge Lynch had pronounced an immediate sentence. However, a plea was made for Cicero Grime and he was spared the noose, but went to prison. He later escaped and left the territory. (This story is detailed in the chapter on Frontier Justice.)

Closer to Globe City, the Buffalo Mining and Smelting Company began operating a small furnace in 1881 to treat ores from the Buffalo and Hoosier mines. In 1882, the Old Dominion Company purchased the rich copper properties of the Old Globe Copper Company, and they moved their furnaces to Globe, below the Globe Ledge, which started the silver rush to Globe in the 1870s. Old Dominion continued copper operations at their properties for the next several years until the copper content of the ores fell, and there was a heavy flow of water into the mine. A new Old Dominion smelter was built and by 1904 it had turned out 2 million pounds of copper. The advent of the railroad made it profitable for the mines to operate with lower grade ores. The old Dominion shut down in 1931 during the Great Depression. However, the water from its mine was used in the mining industry for several years to come. In fact, water from the Old Dominion Mine is still used at the (BHP) Pinto Valley Mine today.

During the years of the development of the Globe mines and smelters, the area of Miami was also being prospected for rich copper ores. The Miami Copper Company was formed in 1907, and several mines including the Black Warrior was purchased as part of this larger organization. The Inspiration Consolidated Copper Company was organized in 1911 by consolidating with the Live Oak Development Company. It's mines were developed using the method called block caving. A prospector named Marshall first located the Live Oak Company in 1890. A large mill was built to utilize the revolutionary method of extracting copper metal, called flotation.

Miami Becomes a Town

The Town of Miami was officially established in 1909, but the name Miami had been around for years. In 1875, G. C. Nolan and Caleb M. Dodge located the Miami silver claim west of Globe. The Miami Mining Company later bought the property and built a mill and smelter at Bloody Tanks Wash. The settlement that grew up near Bloody Tanks smelter was called Miami City, and had a general store, restaurant, a hotel and the famous "Up and Down Saloon." The mill along with the "Up and Down Saloon, was shut down in 1883 and sold for taxes.[30]

Silver was the start of the mining industry at Miami and Globe, but by the end of the 19th century high grade silver ore had all but disappeared. The

Old Dominion mine 2005 — Author's photo

silver mines shut down and were soon replaced by copper mines. With the advent of electricity, and other uses for copper, Arizona then became the Copper State during the 20th century. Miami concentrated its first ores in 1911, and the ore concentrates were sent by rail to the Cananea Consolidated Copper Company in Mexico. However, the Mexican Revolution created problems, and the concentrates were then shipped to the International Smelting and Refining Company at Toole, Utah. The rising output of copper concentrates became the catalyst for the development of a large smelter at Miami. In October 1913, the International Smelting Company purchased 32 acres at Miami and built a large smelter there which was completed in 1915. The cost of the new smelter was $3,000,000. The new smelter was built without consideration of blast furnaces. The goal was to glean 97 percent of the copper values from the concentrates in the form of wire bars. Inspiration production rates reached 200,000,000 pounds of blister copper in those days. During 1927 a reverberatory furnace was built at Miami. The practice of a wet-feed charge was then continued successfully for many years at Miami. Miami continued to be a major producer of copper for the duration of the 20th Century.[31]

Pinto Creek and Pinto Valley Mines

About six miles west of the town of Miami is Pinto Creek, where prospectors filed their claims in 1881. They were filed north of the present day Route 60 bridge over Pinto Creek. Many other claims were filed from 1905 to 1921. The Miami Copper Company purchased these claims and operated an open-pit mining and concentrating complex called the Castle Dome Mine, from 1942 to 1953. Cities Service Company obtained the properties in 1969 and funded the development of the Pinto Valley open pit mine, which developed the low-grade sulfide copper porphyry deposit. Strip mining began in earnest in 1972, with concentrator production in 1974. The leaching operation began in 1981. The properties were then purchased by the Magma Mining Corporation, which in turn sold the properties to the BHP (Broken Hills Properties) Corporation in 1996. During 2002, I had the opportunity to tour the mine with the Operations Manager Jesse Gage along with Jack Carlson and Marcie Greenberg, and observe the operation of extracting copper out of acid baths which produces sheets of copper. It was a fascinating tour.

Gibson mine ore wagons 1906 — Gila County Historical Society

The Gibson Mine

On the upper Pinto Creek area in the Pinal Mountains is the Gibson Mine. Sam Gibson and William Henderson founded it about 1902 filing the Summit Claim. A small mining camp soon grew up around the mine. The town was comprised of dwellings, stores boarding houses, bunkhouses and stables. It even had a stage line that ran from the mine to Miami, during 1906. Gus Schute and Rudy Wolf owned the stage line. The mine had an incline shaft some 600 feet deep with two copper bearing veins. The Gibson Mine produced 12 million pounds of copper from 1906 to 1918. According to other reports it produced over 2 million dollars in copper by 1934. It was worked during the later years until 1959, when it still produced small amounts of copper ore.[32]

Lower Pinto Creek Gold and Al Sieber's claims

The early reports of mining in this area of Pinto Creek often mentions gold being located along with other metals like silver and copper. There are some current mining claims on Pinto Creek, by the Iron Bridge and near Gold Gulch. One of these is called the Diamond Claim. John Wilburn, author of numerous short books about Goldfield, Arizona and the Lost Dutchman, at one time had some mining claims on Pinto Creek.

Al Sieber cir. 1880s — Gila County Historical Society

It seems that Al Sieber, famous Chief of Scouts for General Crooke during the Apache Wars in Arizona, during the 1870s and 1880s, also had some mining claims on Pinto Creek during the 1890s. His claims were in lower Pinto Creek, about ½ mile or so downstream from the current Pinto Creek Bridge. They were called the Hal & Al, Lost Coon, Dan & Mac, Monroe Doctrine and others. These were promising properties according to letters Al wrote. Not only did these properties look good but Sieber and the co-owners, Bill McNelly, Hal McNelly, and Dan Williamson contracted with a person named Con Crowley for their development. Digging began, tunnels were dug and the ore looked good. The Dan & Mac, Lost Coon, Hal & Al and Monroe Doctrine are still valid claims. I located them on the current tax assessor's map for Gila County, on Section Map 204 during April 2002.

About a ½ mile down the Pinto Creek was the Yo Tambian Mine (still on Inspiration USGS map) and other mines owned by the Pinto Creek Mining and Smelting Company. When Sieber was bored with working the mines he would track some outlaw or another, with the aid of his Apache tracker friends. He reflected on his past scouting days and told Lee Middleton (brother of Gene Middleton, who survived the outlaw Apache Kid's escape at Riverside), that he once killed a couple of renegade Indians near his claims on the lower Pinto Creek and buried them under a rockslide. He suspected that their bones were still there. Sieber tired of the Pinto Creek mines after a few years and sold them for a few dollars. McNelly and Crowley retained their interests, and in January 1906, the partners sold the mines to the Arizona National Copper Company for a good price. So good, in fact, that the partners could retire. Sieber had lost out on another good deal.[33] I have explored this area several times and have located several of the old mines along Pinto Creek. This an excellent place to hike during the time there is water in the creek, but difficult to access. Al Sieber was killed February 19, 1907, by a rockslide while supervising the building of the road to the Roosevelt Dam (now known as the Apache Trail). He had survived numerous hair-raising adventures and was shot several times during his career as a scout. Stories have it that Seiber was wounded 19 times. The most serious wound was when he was shot in the foot with a large caliber rifle. Sieber was buried at the Globe Cemetery. I located his grave monument there and the inscription and name are fading out. Some small American flags were placed next to the monument. Another monument was placed near where he was killed, by the Roosevelt Lake Bridge. He was immortalized by Dan Thrapp in the book, *Al Seiber—Chief of Scouts*. This is an excellent book on the Apache Wars in Arizona Territory.

Al Sieber monument Roosevelt Lake, 2004
— *Author's photo*

Apache Kid, 1889
— *Gila County Historical Society*

Chapter Fourteen
The Death and Re-births of the Silver King Mine

The last year of profitable operation was 1887. The grade had fallen to 21.08 ounces per ton for concentrating ore, and 32.47 ounces per ton for amalgamation ore. During the first half of 1888 the Silver King Mining Company operated at a loss, and the President, H. H. Noble, reported a debt of $75,000. An assessment was levied, and operating costs were lowered from $40,000 to $5,000 per month. By December, the indebtedness had been paid, and the company had a balance of $74,000, and sufficient ore was in sight to run until January 1, 1889. Prospecting from 1889 to 1890 failed to find a better grade of commercial ore, and the mine was closed in January 1891. It was opened however, in September of that year, after 44,000 delinquent shares were called in. In October 1891, a strike was made in a new shaft, east of the old workings. This ore was developed in 1892, but the company again was in debt. Ten stamps were moved from Pinal to the mine, and a small production of concentrates was made during the remainder of the year. With the decline of the price of silver early in 1893, the mine was again closed. After the Silver King Mine had been abandoned, it was filed on at different intervals by several people. Several big mining men tried to get control of the property, among them John Hays Hammond, but they were unable to reach an agreement with all of the claimants.

Letters regarding the closing of the Silver King Mine in 1887

THE SILVER KING MINE, ARIZONA
***ENGINEERING AND MINING JOURNAL*-JANUARY 26, 1889**

Editor: Engineering and Mining Journal:

"Sir: The two paragraphs in my letter upon Mining in Arizona, in which I refer to the Silver King Mine, this journal. December 22, but did not mention the late superintendent, have drawn from him the communication, which appeared in your issue of the 29th. He says: Quoting again from Professor Blake, continuous costly preparations making, for increased capacity for working the ores with additional stamps, and a $40,000 compound condensing engine just laid down. I must say that those who know the facts as a deliberate attempt to mislead can only construe such an observation.

This charge is unjust and unnecessary. It is not refutation, nor is it argument. My statement was general and retrospective, and was a criticism upon the policy of making surface improvements while development work below was neglected. It did not address itself to the cost of such improvements, further than to illustrate the fact that costly improvements had been made. As to the cost of the engine, Mr. Macy states that it did not cost $40,000 but explains that, including the cost of oil tanks, grading, trestles and "all other incidental expenses incurred for the handling and application of the new fuel, the cost was in round figures $40,000. Again, in Mr. Macy's official circular to the shareholders, dated July 6th, 1888, he writes: We have in addition, been put to large, expense in arranging for a change of steam fuel at the mills, from wood to crude oil. To use oil, which is cheaper than coal, it was necessary to replace the four high-pressure non-condensing engines that furnished power for the mills by a single large compound-condensing engine. This engine is now in operation: and the extensive tankage, piping and general mechanism for storing, hauling and burning oil is all provided. The various necessary changes and improvements about the mills since the first of the year have cost in the neighborhood of $45,000, and this outlay with the unfortunate and unavoidable decrease in output, has depleted the treasury.

It thus appears that the engine did not cost $40,000, but the engine and its essential accessories did cost $45,000. We also have the secretary's circular of the same date explanatory of the expenditures for improvements, which caused the company to be in debt to the extent of $75,000. In this circular the cost of the improvements above ground was originally printed as $65,000, but the copy in hand has been corrected to read $45,000, and the new 175 horse-power engine is specifically mentioned as the chief item. However, if I had been aware of the details when I wrote the letter, I would probably have mentioned the accessories by writing engine and accessories at a cost of $45,000.

But the question of the cost of the engine and its appurtenances, whether it was more or less than forty thousand dollars, is not material, and should not divert attention from the main issue.

It is clear, I trust, from the official statements that the surface improvements I referred to have been made, and that they are not those which have been in use for the production of dividends for the period of five years before July, 1887, when dividends ceased. The ten stamps added early in that year could not have contributed to the dividend very long. The observations in my letter were impersonal and served to show my dissatisfaction with the past management of the property without attacking Mr. Macy. I had no desire to arraign him, or enter into a public discussion of the matter, or have I that wish now, but it is forced upon me by Mr. Macy's article. He has thrown the gauntlet, and I take it up, but I will not yet consider him as alone responsible for the management I criticize. It is not the conduct of mining operations at the mine alone, but the general policy of the company in its payment of dividends over a hollow mine; its policy of expenditures in anticipation of reserves not yet developed, to which I object. The Eastern stockholders are not entirely blameless for leaving their shares in the hands of trustees, and in that they did not from time to time make independent investigations of the property. The directors, superintendent, and the shareholders should bear their respective portions of the responsibility and odium of a policy of management, which has brought this once splendid property to bankruptcy, to be rescued only by an assessment upon its shareholders.

I consider the management as open to severe criticism, first, in adopting and carrying on the policy of making improvements above ground when it was known that no dividend paying ore-bodies had been found below the 700 level bonanza. Second, in not keeping the exploratory work in depth so far beyond the immediate requirements of the mill, that at least two or three years supply could be safely counted upon before making additions to the plant above. In other words a good decisive and intelligent policy of deep underground exploration should have been undertaken long ago, especially when management knew the speedy exhaustion of the 700 level bonanza. Failing to find new ore-bodies below, a policy of retrenchment and not expansion should have been adopted.

I do not "discredit" any statement made to me by Mr. Macy and my knowledge of the 800-level and the 900 level is derived chiefly from him. My present understanding is, that while the ore-bearing porphyry-ground and the quartz body or "vein" are still there in depth, there is practically no ore for extraction in sight.

As for the reason that prompts me to question or criticize the past management, it has its foundation in my special interest in the property, having, before Mr. Macy's administration, frequently visited it for the purpose of making a description of it. This description was published nearly six years ago, and the Silver King was one of the few properties that I could commend. My opinion of it has since been frequently asked and feeling great confidence in the intrinsic value of the mine, as also confidence in the management. I have generally commended it, and have taken the fact of large outlays for surface improvements as evidence of ore reserves below. It now appears that I have been misled by such improvements. I do not charge intentional misleading, but that has been the effect, or result, in my own experience, if not, in that of others. My interest has also been that of a shareholder, who paid the assessment on shares the value of which, to his astonishment, dwindled to almost nothing. And now having been to the San Francisco office and learned all that could be learned from the management, including the late superintendent, and being thoroughly dissatisfied with the management or policy followed for at least two years past. I feel that I have not only a right to criticize, but also a duty to perform to the public and the shareholders to make my convictions known.
Wm. P. Blake

EDITOR: ENGINEERING AND MINING JOURNAL:

Sir: Prof. Wm. P. Blake in his reply to my communication in your issue of December 29th refers to his covert expressions concerning the Silver King Mine. In his letter in the issue of December 22nd, as being a statement that "was general and retrospective," and "a criticism upon the policy of making surface improvements while development work below was neglected". A reference to the issue of the Journal of December 22nd will show how far from being general or an honest, open criticism his observations were. Professor Blake may have had no intention to arraign me individually may be quite true, but his allusions certainly implied an arraignment of somebody, and as I was so largely a responsible factor in the management at the mine, and as his expressions were in a great measure reflections upon operations at the mine, there was little else for me to do than require him to "come from cover."

While Professor Blake did undertake to inform himself in San Francisco, in early November last, relative to the company's affairs. Such information as he acquired, and which was freely and correctly given to him, could not possibly place him in possession of all the details of condition and circumstances which has attended the operation of the property the past two years-the time to which he specially refers. He has visited the immediate vicinity of the property twice since early 1883, and on one of those occasions, nearly four years ago, I was pleased to have him accompany me in a trip through the mine. His frequent visits to the property for the purpose of making a description of it, and the publication of such description nearly six years ago, are matters of ancient history when considering recent operations at the mine. He could not then have acquired a correct conception of the conditions and circumstances that time and a vast amount of work have developed. In this to a certain extent, he made verbal acknowledgment to me recently. Without having visited the mine occasionally the past two years, Prof. Blake cannot have more than a cursory idea of the underground conditions. In the absence of any such personal inspection and investigation, and the apparent lack of information relative to the exigencies, which necessitated the provision of the more recent surface improvements, I question his ability to establish himself as one competent to pass judgment upon the policy exercised in the working of the mine. It is not at all likely that because Prof. Blake magnanimously undertakes to perform a duty to the public and the shareholders by expressing his convictions, the character of the management will be in any way qualified, for the correctness of his convictions may be questioned quite as much as the policy of the management.

In a reasonably long experience I have no recollection of the public presentation annually of such lengthy full and detailed reports of the year's operations as has been presented by the Silver King Mining Company. These reports, supplemented by the daily records of operations of every kind as the mine and mills for over five years past, which are on file in the Superintendent's office, offer tangible and public evidence of policies pursued.

Prof. Blake devotes considerable space to the new engine and change in fuel, and I trust has settled the questions of cost to his satisfaction. Of the forced necessity for these improvements, I have spoken in my last communication, and I now quote concerning the engine from the Superintendent's annual report of January 1st, 1889, as follows: " A test, under conditions of disadvantage to the new engine, showing a saving of 28-38 per cent in fuel over previous consumption."

Referring again to Professor Blake's communication and the paragraph beginning with the words, "I consider the management as open to severe criticism." etc. I will say first, that when it was decided to make the improvements above ground to satisfy the necessity, (such improvements being the engine and accessories)-the only improvements that can possibly be utilized by Professor Blake in his effort at criticism) there was an ore-body below "the 700-level bonanza" that still offered nearly two years extraction and which had been a potent element in the provision of dividends to wit, the 800 level. Second, that with the perspective possession of a large ore body above the 256 level that would greatly exceed the immediate requirements of the mill, with at least two or three years supply that it was considered could be safely counted upon, the requirements demanded by a "decisive and intelligent policy", seemed to be met. In the recent annual report, the superintendent says, in referring to the ground above the 256 level, "a year ago that section was the most promising in the mine. At that time the third floor 256 was showing its best! 103 feet above this and about half way to the surface, the 110 cross-cut had developed precisely the same character of ground with a satisfactory showing of mineral, while at the surface had occurred the valuable deposit extracted some years ago from the open pit. With these facts from which to judge and the continuance of the quartz porphyry column established without doubt it was but conservative reasoning to predict a large paving body and several years profitable work from the ground in question.

The 119 cross-cut was driven in the summer of 1887 and the conditions mentioned were matters of fact prior to undertaking any new improvements, even though they were a matter of necessity. Third, the stress laid upon the 700foot level bonanza by Professor Blake is entirely without point. The 700 level was practically exhausted nearly four years ago, at which time the 500, 600, and 800 levels stood in tact to say nothing of the upper levels. That the large block of ground, which seemed to give evidence of such profitable ore in the future should have since proved so disappointing is certainly not the fault of the management, or a circumstance to invite an unsubstantiated assertion of faulty policy and management.

Since the company paid its last dividends, as Professor Blake says, "over a hollow mine", about 1 million ounces of the silver have been produced. I now have cause to be pleased with visits which have been made to the property by several eastern stockholders, all within a year, who were given full access to everything and who made a close inspection in detail. I have not yet learned that they found any occasion to entertain Professor Blake's convictions.

(Signed)—Arthur Macy

Several Attempts to Reopen the Silver King

In the fall of 1894, Superintendent, W. S. Champion, resumed work in the new shaft. He reported finding a pocket of ore at a depth of 75 feet worth $40,000. However, the mine was again closed in 1895. From 1876 to 1887, the Silver King Company declared $1,950,000 in dividends. There were no dividends after 1887. A total of $300,000 in assessments was levied from 1888 to 1895; making the net dividends paid $1,650,000. The dividends were paid on 100,000 issued shares, and assessments were collected on 56,000 shares. The *Arizona Republic* reported on March 21, 1895:

"Superintendent Champion of the Silver King Mine left last night for San Francisco. The mine has been closed, and will not be re-opened *until silver takes its place* among the valuable metals of the world." [1]

Repeal of the Sherman Silver Purchase Act

The Sherman Silver Purchase Act, which was passed under President

Benjamin Harrison, a Republican, in 1883, had stabilized the national currency by establishing a Silver Standard. This was accomplished through purchases by the U. S. Treasury Department, and issuance of silver coins and silver standard notes. This act was repealed on October 30, 1893, and signaled the collapse of the silver boom. It was said that the repeal of this act was to impede the Western States from becoming too powerful, and perhaps superseding the power of the Eastern States. President Grover Cleveland, a Democrat, who superseded President Harrison in office, was afraid of a depression during his second term, and asked the Congress to repeal the Sherman Silver Act and replace it with a gold standard. This repeal brought about a collapse of many of the West's large silver operations, including the Silver King Mine, and effectively ended the silver boom in this country.

Major Events that caused the demise of Western Silver Mining

1849	The 16 to 1 ratio established-Silver and Gold minted at the ratio of 16 to 1.
1878	Bland-Allison Act-silver price rises-value of silver dollar drops-inflation follows
1890	Sherman Silver Purchase Act passed (Bland–Allison Recinded) silver rises briefly then declines sharply
1889	"Panic of 1889" begins with declining price of silver
1893	Silver Purchase Act repealed-silver prices decline rapidly-mines close-Panic of 1893 begins
1900	Gold Standard passed

Dr. Jones and the El Medico/Silver King claim

Among the Pinal County's historic files I located the attempt to reopen the Silver King Mine in 1905. The El Medico plat stated that Dr. Charles H. Jones, B. Matley and J. P. Welles paid for Mineral Survey No. 2052, conducted on March 25-27, 1905, by James C. Dobbins, U. S. Deputy Mineral Surveyor. The survey conducted on the claim known as the El Medico took place in the Pioneer Mining District of Pinal County, Arizona Territory. The claim was named after Dr. Charles Henry Jones of Tempe, hence the name El Medico. The survey was a very detailed drawing of the Silver King Mine area, describing all the mine buildings within the 600 by 1500 foot claim area. This document stated that the claimants made improvements consisting of, "A Discovery Shaft 4x6 by 10 feet deep, value $80 and a Tunnel 4x6 by 56 feet long, value $560." A patent application was signed by Frank Sangalle, U. S. Surveyor General for Arizona Territory. The name El Medico is still used today to describe the primary claim of the Silver King Mine. This information was confirmed by the *Arizona Republic* newspaper article titled, "Mining in Pinal County is Active," August 23, 1905.[2] However, I was unable to locate any information on just how much work was actually done at the Silver King at this time.

Dr. Jones and his partners had applied to patent the Silver King claim, but the final patent process of actually receiving the patent remains a mystery. All the steps in the patent process were completed. While pursuing the patent process, the above partners were sued by the Silver King Mining Company, who apparently still had legal claim to the mine. The plaintiffs in this case were the former President of the Silver King Mining Company, James Barney, and one of the original locators, Isaac Copeland. The suit was filed in the Territorial Court of Arizona, and a decision was rendered in favor of the Silver King Mining Company for a judgement of $50,000. However, Dr. Jones and his companions apparently settled the judgement for $10,000, and they paid for a lease option and possession of the mine. The court papers stated that they (Silver King Mining Company) *were entitled to a patent*, but fate has a strange way of interfering with things. As a well-known quip states, "The best known plans of men and mice sometimes go astray," as we shall see below.

Dr. Charles H. Jones 1890— Arizona Historical Foundation

A Curse on the Mine?

Things certainly went astray for the new legal owners of the Silver King Mine. First J. P. Welles died of a heart attack in 1906. Meanwhile, B. Mattingly apparently sold his share of the mine to George D. Christy, their attorney, who came down with typhoid fever, and for weeks his life hung in the balance. During the time Christy was ill, Dr. Jones died on March 7, 1907, of pneumonia.. The *Arizona Republic* on March 7, 1907, stated that their friends warned the men about the fatality of owners of the mine, but apparently this went unheeded. The widows of the above men and Christy's wife thought the mine had a *curse* on it and did not complete the successful reopening of the mine. They later sold it to a New York company, who also used the name Silver King of Arizona Mining Company, and the widows bade it good riddance.[3] The estate papers of

Dr. Jones mentions the $10,000 put forward by him for the mine, and that his widow Eleanor P. Jones filed a claim for the money from the estate.

Dr. Charles Henry Jones was born in 1866. He went to the Pulaski Academy and Adrian College in New York. He completed medical school at the University of Minnesota. Dr. Jones came to Tempe, Arizona Territory in 1892. He was very active in the community and a supporter of the Tempe Normal School, which later became Arizona State University. There was a large obituary in the *Tempe News* on March 15, 1907. It stated:

> "In his profession, he ranked among the ablest physicians and surgeons in Arizona, and was prominent in medical circles. The Jones Prize for Scholarship, which is awarded each year has been a healthy stimulus to many a student." It further said this about his compassion as a physician. "His patience and painstaking care knew no distinctions, and whether his patient was a wealthy citizen or a poor Indian from the reservation, he was treated with the same considerate care and kindly interest." Dr. Jones is listed in the journal, *Medicine in Territorial Arizona.* It states he was president of the Arizona Medical Association in 1897 and active in the "varied aspects of medicine." [4]

After Doctor Jones and his associates failed to successfully reopen the mine the *Arizona Silver Belt Newspaper* reported on December 23, 1908 that Scofield and John H. McCabe took a lease and bond on the mine. An article on December 24, 1908 "Good Assays From Old Silver King," stated, "assays on the shaft at sixty-five feet were as follows: Copper, $18.93; silver, $428.96; gold, $6.17 and iron ten cents."[5]

In 1913, Henry Lomker, better known as "the professor," lived at the Silver King Mine. He apparently lived there several years. The *Arizona Silver Belt* newspaper described Lomker as follows.

"He gave the best years of his life to the study of this proposition. Quaint and quixotic, he is nevertheless a deep student of many things—geology, mineralogy, German philosophy, and literature. He has always been confident that the mother lode, or fissure vein, would be found. It has become a ruling passion with him. It is true that the practical miners laugh at him, because in running a tunnel he swung it in a circle instead of along straight lines. But when it is remembered that his object was to explore the formations in the vicinity of the Silver King Mine, his methods may not have been as impractical at might appear at first blush." [6]

It is not known if he was successful at his endeavors, as I could find no further mention of him in subsequent stories.

The 1916 Reopening Attempt

In 1916, the mine was again reopened by the Silver King of Arizona Mining Company, purchased after much research of the legal owners, and incorporated under the laws of Arizona, with a capitalization of $1,500,000. The officers and directors were E. Couper-Thwaite, President, Alex. B. Downe Vice-President, Francis C. Masson, Secretary & Treasurer, Curtis, G. Rorebeck, Director and Walter Ainsworth, Director. Alex B. Downe, an Australian, was a mining engineer of worldwide experience and had originally secured an option on the mine for $10,000 from the Dr. Jones estate about 10 years prior, but his option ran out during the panic of the early 1900s. He examined the mine and was convinced *it was still a bonanza*. After a ten-year struggle with all the various claimants, he was able to acquire the Silver King with the aid of English capitalists who backed him.

ANALYTICAL REPORT OF 1916

An analytical report on the mine was prepared and advertised by H. D. Wells & Company, a stock and bonds brokerage of 25 Broad Street, New York City, during 1916. The stock was then listed on the New York Stock Exchange and according to the analytical report it had an active market.

SILVER KING OF ARIZONA MINING COMPANY

"Silver King of Arizona Mining Company—Once Arizona's richest and most famous mine with a production of over $10,000,000 of high-grade silver ore from operations in a comparatively small area of its upper levels—this was the SILVER KING. Destined in the opinion of management and many other competent authorities, to again become one of the state's largest producers of silver and copper, probably surpassing, as it did in the early days, its neighbor Magma, formerly known as the Silver Queen—this is the SILVER KING. Silver King is a mine and not a prospect. Its production record, its location and its management all warrant the expectation that it will in the very near future add an important chapter to mining history. Its stock, therefore, appears to offer an investment and speculative opportunity of unusual merit and possibilities." [7]

The mine did not however reopen during 1916 as it changed hands.

The Mine Changed Hands Again During 1917

During 1917, the Silver King was sold to A. W. Hillebrand, who became president, John Fowle, vice-president and W. F. Ainsworth, secretary-treasurer. E. J. Stern, E. S. Booth, G. D. Christy and W. W. Lawhon were appointed directors. Ainsworth had been affiliated with the previous owners, and Christy was the attorney and part owner of the mine during 1907, just after the death of Dr. Jones. It took about 2 years to clear up the claims of the 27 would-be claim jumpers and vendors.

INCORPORATED UNDER THE LAWS OF THE STATE OF ARIZONA

NUMBER B4413 — SHARES 100

Silver King of Arizona Mining Company

CAPITAL STOCK $ 2,000,000

PAR VALUE OF SHARES ONE DOLLAR EACH

This Certifies that --JOHN WADEMAN-- is the owner of -ONE HUNDRED- fully paid and non-assessable Shares of the Capital Stock of Silver King of Arizona Mining Company transferable on the books of the Corporation by the holder hereof in person or by duly authorized Attorney upon surrender of this Certificate properly endorsed.

In Witness Whereof the said Corporation has caused this Certificate to be signed by its duly authorized officers and to be sealed with the Seal of the Corporation this day of 19 JUN 26 1918

W. F. Ainsworth, SECRETARY — A. W. Hillebrand, PRESIDENT

Countersigned Security Transfer and Registrar Co. Transfer Agent and Registrar JUN 27 1918 Asst. Secretary

1918 Silver King Stock Certificate — Sam Michael Collection

In 1917, the old main shaft, 987 feet deep, was pumped out and repaired. Small high-grade ore deposits, overlooked by former operators, were mined on the 120 foot level. (This level was referred to as both the 114 and 120 foot level by various sources but I refer to it as 114 based on the information inn the original reports.) A small flotation mill was completed at the site in 1918 to treat the ore and low grade dump material. About 35 tons a day were treated, with a reported extraction of 90 percent. Robert Bowen, a former mine foreman and superintendant, revealed just before he died that there was a very rich vein on the 400 foot level, that was sealed up when the mine was closed in 1887. He stated that he never told anyone because he wanted to claim the mine for himself when it became legal to do so. Unfortunately Bowen died before he was able to claim the mine. Another *dream* dies with a miner/prospector. A small rich vein was located on the 400 foot level and yielded some rich ore that was shipped to the smelter. In January 1918, a 24 ton mill was installed at the Silver King Mine, and for several months the mill turned out about $100,000 net per year. Shipments of concentrates continued intermittently to July 1919. At that time, the management claimed to have developed 10,000 tons of ore, averaging over $20 per ton. Numerous newspaper articles in the *Arizona Republic* dated 1916, 1917, and 1918 made reference to the reopening of the Silver King. They included the following headlines: "Old Silver King May be Reopened By Eastern Party; Mining In Pinal County Is Active; Silver Ore Body Opened In Silver King Body; Old Silver King Comes Back As Producer; Silver King Soon Be Scene Of Big Mining Activity; and Bonanza Silver Ore Fourth Level Old Silver King."

Plan for the Increase of Capital Stock of the Silver King of Arizona Mining Company 1917

To the Stockholders of the Silver King of Arizona Mining Company:

The new directors and the new officers, who were elected on April 9, 1917,to manage the Silver King of Arizona Mining Company, examined the affairs of the Company and found that it was in a deplorable financial condition.

The Company did not at that time own the Silver King Mine or any other mining claim whatsoever. It had only an option or right to purchase the mines, which had been forfeited and cancelled. The Syndicate, which had been formed to finance the Company upon the election of new officers, agreed to furnish the funds necessary to secure the title to the mines and to operate the same provided that 400,000 shares of treasury stock were available for that purpose. According to the report made by the retiring Treasurer to the stockholders at the meeting in April, more than the required stock should have been in the treasury. The new directors thereupon entered into an agreement to sell the Syndicate 400,000 shares of the treasury stock of the Company, and the Syndicate agreed to purchase the amount deemed necessary to secure title to the mines and complete their development. The Syndicate thereupon advanced the funds necessary for the immediate operation of the Company. The new officers found only 155,710 shares in the treasury, although the report of the auditing committee appointed by the stockholders shows that the treasury should have contained upwards of 400,000 shares. Pursuant to the order of the Board of Directors legal proceedings were instituted to recover the missing treasury stock. The owner of certain mining claims comprising approximately sixty acres adjacent and contiguous to Silver King mine, being one of the principal stockholders of the

Company, offered to convey to the Company the said mining claims for 500,000 shares of capital stock of the Company, and to donate all of said stock to the Company to be used in providing for the options of the Syndicate, in lieu of the missing treasury stock, and any balance to be held subject to the order of the Board of Directors. The Syndicate agreed to perform its contract notwithstanding the missing treasury stock provided that it was given an option on additional 100,000 shares at One Dollar ($1.00) per share. On account of the present high price of silver, the Board of Directors accepted this loan in order that the Company might secure title to the mines, install machinery, and be in position to operate immediately upon the approval of the plan by the stockholders.

The Company now is the Owner of the Silver King Mines

The Syndicate thereupon advanced the necessary funds and the officers secured title to the Silver King and other mining claims upon which the Company had an option, and had the deeds recorded in the name of the Company. The Syndicate also advanced sufficient funds to purchase machinery and to operate the Company until after September 27, 1917.

Respectfully submitted,

SILVER KING OF ARIZONA MINING COMPANY,
A. W. HILLEBRAND, *President*
W.F. AINSWORTH, *Secretary*

Chest Containing Silver Found in Old House at Silver King

The *Superior Sun* reported this story on June 15, 1917. "Chest Containing Hundreds of Dollars Worth of Silver Found in Old House." While doing some renovating to an old stone house in the Old Silver King camp, Mrs. Hoskins, wife of Walter Hoskins the well known mining man, accidentally unsealed an improvised vault built into the stone wall of the house in which she found a box completely filled with high-grade silver ore. Mrs. Hoskins also found a package of course placer gold. Miners placed the value on the silver at $2000, and $50 on the gold. This treasure had been hidden away for about 40 years. The house used to belong to the oldest settler at Silver King, who arrived there about 42 years ago. "Ore from the old silver producer was so rich that it was lifted to the surface in great blocks and actually reduced by sawing into slabs for convenience in shipping." The article went on to say that the identity of the owner will probably be never known and that this treasure trove may belong to one of the early promoters of the Silver King, or could have been the cache of a high-grader of the early days. It also stated that Hoskins will have little to do but advertise her find, and if no one shows up to prove the property theirs, she will have legal possession.

The $500,000 Bond Issue of 1919

In 1919 a $500,000 bond issue financed the mine, and a new shaft was started 150 feet northwest of the ore chimney. The old shaft was kept unwatered and ore from the 114 and 400-foot levels was treated at the mill. In October, a crosscut from the old shaft on the 400-foot level connected with the 415 foot level of the new shaft. Operating account showed expenditures, $311,403; income and receipts, $326,006; cash balance, $27,704.

Information from the 1920 Report:

"Property: The Silver King Mine, 11 claims, over 170 acres, 3 miles from Superior, Pinal County, on the Stoneman Grade. It is an old silver mine, popularly credited with a production of $10,000,000 and known to have paid $1,950,000 in dividends up to July, 1887. Opened to a depth of 1,000 feet, the mine was originally worked for gold and silver, copper ore in the form of tetrahedrite coming in at a depth of about 310 feet and continuing to depth of 510 feet.

Geology: The conditions are described in Bulletin No. 540, U. S. Geol. Survey. Though currently spoken of as a vein, the ore body is a stockwork, in which the extraordinary rich ores, carrying various silver arsenical and antimonial suphides of silver, with argentite and stromeyerite, apparently gave out at the bottom levels. Below this level, lower grade but payable concentrating ore, amenable to flotation may be expected and lateral development is also expected to develop more ore. Old records show that copper ore was mined on the 88-foot level. Company was to sink a 3-compartment 1,000-foot shaft and upon its completion planned erecting a 500-ton mill. The shaft was down 636 feet at the end of June 1920.

Equipment: Includes gasoline hoist, compressor, pump, 5-ton auto truck and a 23-ton plant using a Marcy mill and flotation apparatus. The mill started operating in November 1918. Shipments during 1919 totaled $97,606, and in 1918 netted the company $46,983. In July 1920, John Fowle, Vice President and General Manager, was appointed receiver at an annual salary of $4,000 a year. Judgments against the company for $36,731, resulted in foreclosure sale, July 1921, to Walter F. Ainsworth, New York, for $250,000." [8]

On August 15, 1920, a ledge of high-grade silver ore had just been struck at about 200 feet down in the new three-compartment shaft. This was a new discovery, as the rich silver ore previously mined came from the 400-foot level in the old workings. The report of the State Mine Inspector for 1918, states that at the beginning of the year 22 men were employed at the surface and 18 underground. At the last inspection 17 men were on the surface and 14 underground. At the time of both inspections the property was producing about 500 tons per month.

Silver King of Arizona Mining Company

Stockholders Annual Report March 11, 1920 (Report Excerpts)

A resolution was adopted changing the authorized capital stock of the company from

2,500,000 shares of the par value of $1.00 each to 500,000 shares of the par value of $5.00 each.

John Fowle, the General Manager, made his annual report which showed that the new three compartment shaft, ground for which had been broken April 10, 1919, had reached a depth of 550 feet on December 31st, 1919. A report from the mine on this date shows that the shaft is down to the 635-foot level; the station at the 617-foot level completed, and the cross-cut to the old shaft begun.

Ore to the value of $94,455.83 was mined during the year 1919.

Total income and receipts for 1919 $326,006.29 (Includes sale of 10-year 7% $213,400.00 in Convertible Bonds and $15,000.00 in Treasury Stocks.

Hidden Treasure Uncovered by "Locating Needles" in Old House in Pinal

This was the story in the *Superior Sun* on February 25, 1921.

> "J. Renteria, who was investigating the ruins of an old house in the Ghost Town of Pinal, found an improvised casket in the wall. An old man who used 'locating needles' accompanied Renteria. These instruments are supposed to point to gold, according to the old gentleman." The article said that the "needles" were applied and indicated a certain spot. "In a few minutes Renteria found a small chest made of rough boards. The lid was removed and $4150 in gold pieces, some silver money and a three-pound silver bar and two sterling plates were found in the chest. The gold and silver was of mintage of the middle eighties and in perfect condition."

No one seemed to know who occupied the house when Pinal was the boom mill town of the Silver King. The treasure was found at the base of a side adobe wall. The house was located near Queen Creek and close to the site of one of the old mills. The house had originally been a one-room place; the kind that was usual occupied by the average prospector of that era, and was slightly removed from what was the residential part of the old town. The Silver King once more yielded treasure to a lucky seeker. This find started quite a little rush of treasure seekers to the long abandoned town, but no additional discoveries were reported.[9]

1921 Silver King Mine Status

The following correspondence about the old mine was located in the archives of the Arizona Bureau of Mines.

> "In reply to your letter dated May 11, I wish to say that the old Silver King of Arizona Mine was one of the bonanza producers in the State along in the 80s.

In its palmy day, the mine produced on an average 100 tons of high grade silver ore daily. The office of the new company is located at No.1 Broadway. Catlin, Fore & Boyd, 25 Broad St., New York, are the syndicate managers. The company was incorporated in 1921 in Delaware, with a capitalization of $2,000,000 shares at one-dollar par value. The property consists of 11 claims about three miles from Superior, in Pinal County, in this State. The Silver King of Arizona Mining Company became bankrupt in 1920. Henry C. Carr reported in 1924 that at least $2,000,000 worth of commercial ore remains in the mine, although it is of too low grade to stand shipment as mined. It is believed that deeper development may expose additional amounts of milling ore. Although apparently a new company, we have on file no clippings whatever relating to this property, dated after May 1, 1924." [10]

1935 Opening by Dr. F. W. Johnson and Bat Gays

Silver King Again Opens

SUPERIOR, Apr. 8.—(AP)—New life was coursing through the veins of the historic old Silver King mine near here today after a 15-year period of inactivity.

The mine, which in its half century of existence produced more than $10,000,000 in wealth, is being reopened by a group of Superior men headed by "Bat" Gays.

The first shaft to be sunk in 15 years is being driven.

Since it was discovered in 1873 the Silver King has had a career of ups and downs. The fluctuating price of silver has caused it to be opened on several occasions only to be closed.

The present high price of silver is responsible for its reopening

"Silver King Again Opens,"— Arizona Republican 1935

"Silver King Mine Will Open April 1," was the title of an article by the *Arizona Republican* on March 19, 1935. It stated that operations would begin about April 1, headed by Dr. F. Walter Johnson, a surgeon from Pomona, California. It further stated that Dr. Johnson was associated with New York mining men and will spend considerable money in operations. It also stated that the first car of ore was smelted at the Magma Smelter. An article in the *Arizona Gazette* on April 9, 1935 stated: " New life was coursing through the veins of the historic old Silver King Mine here today after 15 years of inactivity." The article goes on to report that a group of Superior men led by 59 year old "Bat" Gays made attempts to reopen the mine. It stated that a new shaft called the "Gays Shaft" was sunk 75 feet where water was encountered, and it said that the present high price of silver was responsible for its reopening. It stated that water-pumping equipment was being sent to the mine. "Bat" Gays was a well-liked Pinal County deputy-sheriff for several years, who was assigned to the Superior area. [11]

(Apparently the water caused major problems in the new shaft, and I could not find any articles that stated the reopening was successful in this 1935 venture.)

Another article in the *Arizona Republic* the previous day, April 8, 1935, stated that the "New Deal" of President Roosevelt raised that prices of silver, and prospectors swarmed over the mountains in search of new silver claims.

Th *Arizona Republic* on May 13, 1935 in stated that during the Silver Jublee celebration at Superior held May 1935, Bat Gays took visitors to the Silver King and gave them pieces of silver ore as souvenirs. The visitors also went to Old Pinal City where a few houses still stood. The article also said a group called the Whiskerino Club also attended with all the merchants and club members, about 150 strong, sporting about two weeks of whisker growth. The article said that fines would be assessed against the members who violated the agreement not to shave until after the celebration was over.[12]

On February 8, 1940 the *Casa Grande Dispatch* reported: "Mining Area At Superior Sees Action." It stated: "James Herron and associates have taken a lease on the Silver King Mine. The property is owned by Bat Gayes of Superior." Another article by the same paper on February 15, 1940, stated: "the dump at the Old Silver King Mine has been leased to G. H. Middleton. James Herron and associates have taken a bond and lease on the mine, and have started un-watering with a crew of five men employed. Bat Gayes and associates of Superior own the silver property."[13] James Herron was a four-term sheriff of Pinal County, and his descendents still run the Herron Cattle Ranch near Picket Post Mountain. I was unable to ascertain how much work was done on the Silver King, but one of his sons stated that they ran into water problems and had to shut down. Things were quiet at the Silver King during the war years and shortly thereafter. The next information about the mine that I was able to locate was the following report in 1959.

**"DEPARTMENT OF MINERAL RESOURCES
FIELD ENGINEERS REPORT
OCTOBER 22, 1959**

Leasee:	Bob Olson, Box 1054, Superior, Arizona (1940-1942)		
Owners:	Foxworth-Galbraith	(26%)	
	Middleton (Grace)	(10%)	now Grace Middleton
	Gayer (D.W.)	(76%)	now Dick Lobb"[14]

Apparently the Foxworth-Galbraith interest was acquired by Mrs. Grace Middleton, Box 496, Superior, and the Bat Gayes interest by Dick Lobb of Superior, in 1940-1942. Olsen stated that he shipped 15 tons of ore averaging $100.00 per ton in silver and lead from a rich veinlet on the 280 foot level of the new shaft. He cleaned out the new shaft and retimbered it to the 300 foot level. He stated that two "old timers" who had driven a drift on the 800 foot level, reported 8 to 10 inch vein of high grade was present. The seam varied somewhat in width. They had sorted the ore. Olsen also stated that between the 200 and 300 ft. levels some 40 sets of good ore was left when the old shaft caved in about 1910. Olsen sunk an old 100 foot winze from the 200 foot level to 285 foot level on the "vein." He found 8 feet of ore averaging between 8 and 10 ounces of silver. The ore was mainly along the hanging wall. The ore zone dipped 80 degrees to SW and the strike was about North-South vein east of the Grace Middleton house. Olsen said the old shaft was 1000 feet deep and the new shaft about 600 feet. The new shaft was sunk about 1913 and the old shaft about 1905 to 1909.

After the company headed by Hillebrand and others worked the Silver King Mine, the Lobb family of Superior apparently gained the mining claim rights to the El Medico claim about 1940, which was the main shaft of the Silver King. They remained the main owners of the mining claim, until the claim on the mine expired in the 1990s, after the death of Chester Lobb. The Deen family of Kearny-Riverside, Arizona then acquired the claim in 1996-1997.

**"STATE DEPARTMENT OF MINERAL RESOURCES
OF ARIZONA
FIELD ENGINEERS REPORT
FEBRUARY, 25, 1959"**

"Mine: Silver King
District: Mineral Hills (Pioneer) Pinal Co.
Engineer: Lewis A. Smith
Subject: Mine Visit
Claims: 14 (unpatented)
Owners: Dick Lobb and Grace Middleton, Superior, Arizona. Mrs. Middleton, Box 496, Superior, AZ
Status: Idle
Work: 1 deep shaft (not accessible), 1 shaft 75 feet deep, 1 shaft 20 ft. deep, 1 tunnel 80-100's numerous cuts and shallow workings.
Minerals: Gold, silver, copper, some lead. (Chalcopyrite, bornite, stromeyerite, galena and equivalent oxides) Native silver is reported to have been prevalent at one time."

"The physical condition of the mine is not very good because no work has been done there for a long time. One steelhead is still present as well as one stone building." [15]

Grace Middleton—The "Queen" of Silver King

Grace Middleton, along with her husband, Gordon, came to the Silver King and leased 14 mining claims, including the Silver King. Gordon died in 1957, but Grace held onto the dream that the Silver King would yet produce millions more in silver. She lived in the "old mansion," which was actually the old mine office and visitors house during boom times, until Gordon died. Grace then moved out of the mansion and built a small cabin nearby with wood taken from the abandoned houses at the old town. Her cabin had only the conveniences of 100 years ago. It had no electricity, gas or phone. She heated the cabin with a wood burning stove and chopped her own kindling. Twice a month Grace went to Superior to take hot showers, eat restaurant food, and do her shopping. She would then hire a delivery truck to take her and her goods and a barrel of drinking water to the Silver King. Grace was an avid reader and would read at night by the light of her kerosene lamp. She kept about five dogs and several cats for company. She was a former cosmetologist and psychology instructor, and people said that she talked liked an educated person, although she lived like a pauper.

Grace prospected and mined all over the Silver King area. She was said to be equally adept with a bulldozer or blasting with dynamite. During her later years at the Silver King she used to give dollar tours of the mine area. She also sold ore samples from the Silver King mine dumps. Sometime during her stay, after Gordon's death, she teamed up with Richard Lobb, son of George Lobb the founder of Superior, to try and make the Silver King operational again. This did not work, however, and after her death it is believed that Richard Lobb took over her claims. Richard Lobb incidentally, had been born at Silver King in 1885, during the boom years. Grace Middleton died in the 1970s still carrying with her the dream that the Silver King would come back to life and produce it's millions again.[16]

The Guzman Era at the Silver King

About 1977, Mike Guzman Sr. and his son Mike Guzman, Jr. took over the claims at the Silver King. They owned the Guzman Construction Company, which still does business in the Superior-Miami-Globe area. The Guzman's are a major company doing business in excavation, concrete, asphalt and rental equipment. They moved a lot of the ore dumps at the Silver King site and used the cyanide leaching process to try and extract the silver from the ore. An Arizona Bureau of Mines report dated January 5,1981, stated "Guzman Construction is heap leaching dump materials from the Silver King Mine near Superior. Leaching is done with cyanide. Recovery of values from the pregnant solution (primarily silver) is accomplished with a modified Merrill-Crowe system (Escapule plant)." It also states that 3-7 people were employed there. In 1986, the Guzmans gave up their lease option on the claims at Silver King, and the claims reverted back to Chester Lobb, whose family held the leases since the 1940s.[17]

Bob Schoose and Ernie Saviano Attempt to Open the Silver King

Bob Schoose, in March of 1998, stated to me that he obtained the rights to the Old Silver King Mine from Chester Lobb, about 1990, on a lease basis with partner Ernie Saviano. They were to work the silver ore left at the 100-300 foot levels, based on the Old Silver King reports, where silver was taken out at these levels during the 1920s. Schoose decided to attempt to reopen the mine; however, Ernie Saviano wanted to take the silver out of the water that had been leached through the natural process over the last 70 years. Based on cost estimates and environmental concerns, they were unable to extract the silver from the water. Bob Schoose is part owner of Goldfield Ghost Town on the Apache Trail (State Route 88) outside of Apache Junction. Ernie Saviano then went onto other interests, such as trying to reopen the old Bull Dog Mine next to Goldfield Ghost Town, which was a producing gold mine in the 1890s, and is next to the A & A Gravel operation off Route 88 prior to Goldfield. The Bulldog Mine is thought to be the Lost Dutchman Mine by some people, as it had an 18 inch gold vein. This was one of the clues that the Dutchman left. But, this is a very controversial issue and many others doubt that it was the Dutchman's mine.

Bob Schoose 2005 — Author's photo

Silver King Mine Production Records 1875—1928
by J. B. Tenney

Year	Price	Ounces of Silver	Gross Value
1875	1.24	40,323	50,000
1877-79	1.16	706,157	819,142
1880	1.15	439,689	505,642
1881	1.13	574,049	648,675
1882	1.14	714,912	815,000
1883	1.11	533,787	592,503
1884	1.11	429,559	476,811
1885	1.07	764,832	818,370
1886	.99	656,566	650,000
1887	.98	709,134	694,951
1888	.94	319,149	300,000
1889	1.00	55,000	55,000
Total 1875 - 1889		5,943,157	$ 6,526.094
1918	1.00	37,000	37,000
1919	1.12	126,892	142,119
1920	1.09	65,872	71,800
1928	.58	3,000	1,755
Total 1918 - 1928		232,764	$ 252,674
Grand Total:		6,175,921	$ 6,778,768

Note: Other figures differ somewhat from Tenney's totals, and these figures do not count the money made on gold and other metals.

Chapter Fifteen

In Search of History—Camp Picket Post to Camp Pinal

While I was gathering information for this book I decided to try and follow the old Stoneman Trail and Grade from Picket Post Mountain up over King's Crown Mountain and on to old Camp Pinal. This route I thought would give me a better perspetive of the history of the area surrounding Silver King Town and the ghost town of Pinal City. The Stoneman Grade still goes up past the Silver King Mine, but it is hardly discernible from that point, and difficult to access without permission. A locked gate at the entrance to the Silver King Mine prohibits entry at this point. However, at times you can access it via the top, from Route 60, just past Superior and the Queen Creek Tunnel on the old Forest Service road that winds back past the open pit marble mine. There you can hike a short distance to the top of the grade, which affords you a spectacular view of the valley below. The Stoneman Grade winds downward from there, but it is washed out in many places and is not a recommended hike for the casual hiker. However, this area is sometimes closed due to construction at the Omya Marble Mine.

Camp Supply on Top of Stoneman Grade

In December of 1997, I located the Pump Station spring, and what I believe are the remnants of the stone ovens of Stoneman's 1870-71 Camp Supply. It is located on top of Stoneman's Grade, next to the spring. I also found some very old stone foundations close to the spring. While looking for old ruins there I found some old burro shoes, similar to the kind used by the Army during the 1870s, and other remnants that indicate that it was the old Camp Supply. Historian Clara T. Woody described Camp Supply as follows: "Established at the same time as Camp Pinal were two sub-camps, one on Queen Creek was called Picket Post, below the Butte of the same name. The other one was at the top of the western slope of the Pinals, about four miles northwest of Camp Pinal, and was called Camp Supply. Between these two camps a pack trail was built called the Stoneman Grade. The 'Grade' was only about five miles long but it established the route. In all three of these camps, stone ovens were built lined with bricks of baked clay, found in the vicinity. Ruins of some of these ovens are still visible." [1] The Stoneman Grade was a frequently traveled route by Silver King and Pinal miners, as well as the freighters and other travelers, during the glory days of the Silver King. It was said that the tough, hard working mining man often walked up this trail, many miles to Globe, to drink and carouse with the ladies on his off day, returning back to the mine or mill to toil once again.

Dan Thrapp writes of the Stoneman Grade in *Al Sieber- Chief of Scouts*: "On our way up the west side of the mountains to the old abandoned Camp Pinal, we came upon an old oven which had been built by the troops when digging the trail. In it were entombed the bodies of two prospectors, who had probably been jumped by Indians while they were asleep. The end of the box that had held tomatoes closed the opening and was inscribed with their names." [2]

The Stoneman Trail and Pinal Ranch

In December of 1997, Jack Carlson and I were trying to locate the continuation of the old Stoneman Trail, from the top of King's Crown Mountain onto Pinal Ranch. I was directed to speak with Tom Clary, the present owner of Pinal Ranch. Mr. Clary, a geologist, bought part of the ranch property in 1968, and another parcel, including the old ranch house in 1985. He gave me a tour of the original house, made of wood, built about 1878, and the second house built a year or so later made of adobe. We toured the houses and he described their history showing us the old gunports. The exterior and interior walls of the Pinal Ranch House are made of adobe and are 14 inches thick. He took us to the Old Stoneman Camp, just west of his ranch, where there were still some old foundations present, and the gravesite of a soldier who had died there. During General Stoneman's brief headquarters stay at Old Camp Pinal, there were about 400 soldiers stationed there, including Camp Picket Post and Camp Supply. Tom said that his father, Alvin F. Clary had been a government hunter, and trapped

Alvin F. Clary and mountain lion cir. 1940s
— Tom Clary Collection

10th largest lion and normal size lion, 1999— Author's photo

and shot predators including many mountain lions in the area from 1930 until he retired. Tom also trapped some predators at the Pinal Ranch and showed me the stuffed head and fur of the tenth largest mountain lion on record that was trapped in Arizona. He said that he received a $1600 reward from a local rancher for trapping that lion. He had to prove that it was a particular lion with a missing claw, and sure enough he showed me the missing claw. This particular male lion had been killing cattle in the

Tom Clary and mountain lion cir. 1970s — Tom Clary Collection

area, and was a very large cat with a reddish brown coat. I took photos to compare it with an average size lion that he had trapped. Tom also told of the days when the Pinal Ranch was an apple producer and sold apples commercially. He showed me the old apple orchard labels from the Craig and Irion Ranch. Tom also showed me the old huge cork oak tree in the yard, described in Helen B. Craig's book, *Within Adobe Walls 1877-1973.*

The Continued Search for the Stoneman Trail

Tom then referred us to Monty Lockhart, who managed the old J. I. Ranch just down the road from him. Monty, who knows the range land very well, showed us two locations where he thought the old Stoneman Trail came down from the mountains into Pinal Ranch. On December 29th and 30th of 1997 I hiked these old trails. On the 30th, I found what must have been a rest stop on one of the trails. While there my hiking companion Jack Carlson and I found several items indicative of the 1870s and 1880s. These relics included broken burro and mule shoes and pieces of bottles with the thick glass, identical in style to those that I found at the Old Silver King townsite. On March 21, 1998, while trying to locate the other parts of the Stoneman Trail, Jack Carlson and I located some other similar items on the eastside of Devils Canyon. This was east of the creek going towards State Route 60. It was at a small mesa near the top of Devil's Canyon that looked like a resting-place or stopover. These items, along with other historical information, now confirmed this part of the trail to us. On April 16, 1998, Jack Carlson, Clyde Tressler and I located the rest of the Stoneman Trail, from the Pump Station to the road leading to the Power Plant, North of Route 60.

This completed the location of the Stoneman Trail from the Silver King Mine to the old Pinal Ranch (or so I thought), as the route from the Silver King to the Pump Station had been located in 1996. The Pump Station at the summit of the Stoneman Grade was used to provide water for the Silver King hoist and pumps located at the mine during the late 1880s. It is my opinion that the Pump Station was the site of Stoneman's Camp Supply, and this location fits the written description found by the author in various historical documents. The spring described by Stoneman's men and the location of the Pump Station are identical.

The Stoneman Trail Mystery Photograph

During December of 1997, I discovered an old photo of some horse back riders going up the "Stoneman Grade" to Silver King. Jack Carlson and I then began our search for the site of this photo, since it was the only photo that I could locate of people on the Stoneman Grade. Little did I know that this search would take almost two years! Jack Carlson and I walked up and down the known Stoneman Grade and Trail in search of the exact site that this photo was taken. We also rode up and down Route 60, in search of the unique rock formations and small hill that was in the photo. All our efforts seemed to be in vain. We showed the photo around to some of the local historians and old timers from Superior to Globe, but no one could identify the location.

Stoneman Trial mystery photo cir. 1880s — *Arizona State Archives*

After many months of searching and hiking on the known Stoneman Grade, Jack Carlson made a discovery at the State Archives. He spoke to Laurie Devine, the Archivist of the Photograph Collection, who located another copy of this photo. This time it had a partial description of the location on the back. We had been looking in all the wrong places. It seems that there was another earlier branch off the Stoneman Trail. This one went down from the top of the hills to Devils Canyon below and came out near old Camp Pinal, General Stoneman's Headquarters.

On March 24, 1999, Jack Carlson, Clyde Tressler and I hiked up the streambed of Devil's Canyon searching for the old photo site. We were about to call it quits when Jack Carlson located the much illusive place where the photo was taken. The master hiker had found the right location! [3] All the rock formations matched perfectly! Two pine trees had grown up near this location which partially obscured a perfect duplication of the scene, but we were able to recreate a pretty good facsimile of the original photo.

Searching the long abandoned trail I located some 1880s type items. I found an old mule shoe, and part of an old burro shoe, as well as pieces of glass from bottles made in the 1880s, while Jack and Clyde located the rest of the trail. This in my mind confirmed the fact that this trail must have been used also. It led south and downwards from the top of Devil's Canyon, where the other part of the trail that we had identified earlier went farther east. This part of the missing trail and the mystery photo had at

Author at mystery photo site, 1999 — Jack Carlson photo

long last been found! A few weeks later we searched this area again. We located more broken horse and mule shoes, and a whiskey bottle from the 1880s.

Clyde Tressler at Stoneman Trail 1999 — Author's photo

The History of the Pinal Ranch

The original Camp Pinal buildings were abandoned by the military in 1872. Sometime prior to 1878, Andrew Hawkins and Tom Buchanan took up residence as squatters at old Camp Pinal. Robert Irion bought out Hawkins' half interest and added an adobe addition to the existing log building. Shortly thereafter, a large adobe structure with many rooms and 14 inch thick walls was built next to the original log-adobe building. Both structures are still standing and in use. Robert Irion and Tom Buchanan

1870s Pinal Ranch wood house 1996— Author's photo

became partners in the "stop-over" between the rich Silver King Mine and the Globe mining areas. Not long afterward, Irion bought out Buchanan. In 1895 the Irions moved to Tempe, and Dudley L. Craig, Robert Irion's

1870s Pinal Ranch adobe house 1996 — Author's photo

stepson, took over the ranch. He constructed a narrow one-lane wagon road from Pinal Ranch to Globe about 1906 on the old horse and mule trail. Although it began as a cattle ranch, in 1906 the ranch started selling apples commercially from their orchard of about 1000 apple trees, using the "Pinal Ranch" apple crate label, with Irion and Craig as owners. Then in 1922, the apple crate label was changed to show D. L. Craig as the owner. Pinal Ranch was also used as a weather observation place from 1893 to 1953. The records reflected that the Pinal Ranch, at 4520 feet, received about 25 inches of rain per year. In 1973, the Craig family sold the ranch.

Tales of "Packing In" to Pinal Ranch

Robert Irion, his wife Elizabeth Craig Irion, her son Dudley Craig and Robert's sister Miss Anna Irion moved to the Ranch in 1878. (See Elizabeth Irion's story in the Chapter on Pioneer Women.) In her book, *Within Adobe Walls 1877-1973,* Helen Baldock Craig, who married Dudley Craig's son Gerald, wrote of "Packing In" to the ranch. In those days there was no wagon road to the ranch, and everything had to be packed in, either by horse, mule or burro. Pinal Ranch was located about mid-way between Silver King Town and Saxes Station, near present day Miami, and the trail went right past the ranch. It became a natural stopping station for travelers and freighters to stop and rest, change animals and get a meal. The Irion family provided many a meal and earned a nice income during the Silver King years and beyond. During the Silver King days, it was said that 200 pack animals passed through the ranch daily.

Old Pinal Ranch Apple Label— Tom Clary Collection

Eight miles west of Globe, at Saxes Station, there was a hotel for guests. From there, travelers went back and forth by stage to Globe, for their journey over the mountains by pack animal to Silver King. From Silver King it was a stagecoach ride into Florence and Phoenix. The saddle trains had about five to ten horses and carried passengers along with their luggage. One round trip was made daily between the Silver King and Saxes Station. Each train had an Indian boy as a helper and a packer. The freight pack trains consisted of 20 to 30 burros or mules. These pack trains went past the ranch until about 1919. Robert Stead owned one of the earlier trains. On one of the trips, Stead's brother, who was working for him on the train, drowned in a pool in what is now Devil's Canyon and is buried at Pinal Ranch. Mrs. Irion, in her notes, stated that it was 28 miles from Globe City to the Silver King by trail, *but 120 miles by wagon.* This states the importance of the Stoneman Trail. It was said that the old Stoneman Trail, to Pinal Ranch and beyond to Globe, was still used by burro and mule freighters until the 1920s and 1930s.

The Old Estey Organ goes Over the Trail

The mules and burros carried heavy loads. According to Helen B. Craig, they carried about 300 to 350 pounds for a mule, and 200 to 250 pounds for a burro. She stated that the heaviest load to go over the trail was two barrels of whiskey that weighed 510 pounds. The bulkiest load was the old Estey Organ, which was delivered to the ranch by pack mule, with two men walking beside it to steady it, one of whom was Robert Irion. This old organ stood in the living room at the Pinal Ranch for 93 years. It was a beautiful organ in a handsome walnut case. The organ was loaded on a ship in Boston, taken around the Horn to San Francisco, then by railroad train to Casa Grande and then by freight team to Silver King. From there it went over the trail, which was very narrow and steep at times, and arrived in first rate playing condition at the Pinal Ranch.[4]

Traveling the Old Globe to Superior Trail

A recollection of a Mr. McPherson, an early settler of Superior, relates the travel from Globe to Superior in 1912. " We left Globe early one morning in June on the road to Superior. We rode a surrey to the Irion Ranch, and from there we rode saddle horses through Irion's Canyon, or through Devil's Canyon, and up over Oak Flats to the Bellamy Trail, which was clear on top of Apache Leap those days." (There was an old trail that ran from the south side of where Route 60 is now and Oak Flats, down the mountain to Superior) "Oh Boy! T'was some trail in places, as the horses had to leap or slide off a rock two feet or more down to the next one. It was some relief when we got to what is known as the park, and you know that we were all tired. It sure was music to the ears to hear the old L. S. & A. (Lake Superior & Arizona Mine), later known as the famous Magma Copper Mine, now the property of Broken Hill Properties (BHP). As we passed the plant we went across the creek and onto the stables, which stood where the Piggly Wiggley (General Store) is now. After dismounting and turning the horses over to the corral boss, who was Phil Ruiz, and then on to see Ernest Kellner, who ran the Piggly Wiggly store then." [5] This old trail was pointed out to me by Monty Lockhart, and it would be difficult to hike down those dangerous precipices, even today.

The other name for this trail was the Swift Trail named after Mr. T. T. Swift, the Supervisor of the Tonto National Forest Service at that time. He hired about 30 Indians who helped build and finish the trail, down the mountainside from Oak Flats to Superior. The *Globe Record* published a story about the trail around 1920 that stated "the trail through Devil's Canyon to Superior is being improved and said that Mr. T. T. Swift was laying out the grade so that it can be improved for wagons and automobiles."[6]

However, it would not be until 1922 before a road was built between Superior and Globe called "The Million Dollar Highway." For the first time in 44 years, it was no longer necessary to go over the trails by horseback from Superior, and the old trails fell into disuse and became overgrown and faint. Part of the Stoneman Trail is now only rarely used by cattlemen during a round-up or by very adventurous hikers. The remains of the Swift Trail can still be seen on the south side of Route 60. .

Boyce-Thompson Southwestern Arboretum

Picket Post Mountain lies just south of Queen Creek. It is 4,375 feet above sea level and is 2,011 feet in elevation from the base. The flat top and prominence of Picket Post are visible for several miles. At the junction of Queen Creek and Silver King Wash, is the Boyce-Thompson Southwestern Arboretum, named for Col. William Boyce Thompson, original owner of the Magma Mine at Superior. Col. Thompson, in addition to being a mine owner, was a conservationist. He wanted to restore the area around Picket Post Mountain back to what it was, before the area was overgrazed and shredded of its trees by the miners and settlers of the ghost town of Pinal City who used all available wood for it's stoves, furnaces and steam boilers. Col. Thompson bought the land from the Forest Service and dedicated the Arboretum in 1929.[7] The Arboretum is still used for botanical experiments. Many examples of the world's vegetation are being grown there today, and the grounds contain many fine specimens of trees, including what I believe to be the world's largest Eucalyptus tree. The arboretum continues to flourish today under the capable directorship of Mr. William Feldman, and continues to draw many worldwide visitors annually. I have personally visited the Arboretum many times and enjoy the splendor of the plants under the shadow of the magnificent Picket Post Mountain.

The Stoneman Grade Tale of the "Lost/Found/Lost Again Penny"

On July 10, 2000 my grandsons Tony and Michael San Felice ages 11 and 10 respectively, accompanied me in a search for the Old Stoneman Grade near the Silver King Mine site. We were trying to verify the lower part of the grade near the mine site. We located some objects using a metal detector such as an old cartridge and part of an old mule shoe, such as the kinds used in the 1880s. As the metal detector was acting up we gave up the search.

On the way back home Tony stated that he had found a 1845 penny on the Stoneman Grade but thought that it was so old that it was worthless, so he tossed it off the trail near some cactus plants. Well, you can imagine the expression on my face when he said that. I told him that what he found was probably worth a lot of money, not to say the historical significance. I decided that I would get another metal detector, and we would try to search for the missing penny.

On July 12 we started out searching the area near where Tony said he had tossed the penny. We located the area and it was near two very large saw-tooth yucca plants. My friend Jack Carlson was with us. He had Tony toss another penny near where he said that he had tossed the old coin. Well, as luck would have it this area was full of iron and copper ore, and the metal detector was sounding off every minute with a "new find."

This was exasperating to all of us. We did not find the "penny" or any other thing of value in this area so we gave up after about two hours in the Arizona summer heat. We even searched all around the two yucca plants visually. We could not even locate the penny Tony had just thrown. We decided that the old penny will have to wait for another day when I could rent a detector that discriminates ferrous metals from coins. We decided to wait for cooler weather, as we thought the summer monsoons would bring some rain and cooler weather to King's Crown Mountain, where the Stoneman Grade is located. However, it never cooled off that summer and we never did locate that old penny. It still lies among the cacti on the Stoneman Grade waiting for some future adventurer to find. Perhaps my grandson Tony will find it again someday!

Chapter Sixteen
1996 Silver King Reopening by the Deens

The Deen family, Ron Sr., sons Ron Jr., Joe and Gary, come from old mining families of the Riverside and Kearney, Arizona areas. They are all involved in the mining industry in one way or another. They had been aware of the Silver King properties as a mining possibility for several years starting the early 1980s. They started to review the property for acquisition, but the property was "all tied up," according to Ron Sr. In 1994 they started seriously looking at the Silver King, but it wasn't until 1996 that they started the process and contacted (the late owner) Chester Lobb's attorney to obtain legal rights to the properties. According to Ron Sr., they soon found out the property was no longer legally part of the Lobb estate. It seems that the estate had failed to file renewal papers on the Silver King properties including the El Medico claim, and had in fact let the mining claims expire for some years, and did not renew them within the proper allotted time.[1]

L-R: Ron Jr., Ron Sr., & Joe Deen 1999— Author's photo

The Deens then filed the proper papers making them the legal claimants to the Silver King properties and the El Medico claim, according to Ron Deen Sr.[2] I was out at the mine when I saw the survey activity, and asked Ron Sr. what they were doing. He said they were re-staking the property and had a certified survey conducted to re-certify the corners of the properties

Background on the Deen family and Partners

The Deen family came to Arizona by way of Texas. Ron Deen Sr. and his father Mack Valentine Deen came to Arizona from Texas in 1936, during the depression years. Mack tried mining and worked in the mines at

Globe for a while. He even tried his luck at being a cowboy. Mack Deen owned the old Riverside stage station. This is the same Riverside stage station that was the scene of several stagecoach holdups in the 1880s. The property is still in the family, however the flood of 1993 washed away the 100-year-old adobe building. The well for the stage station is still there. In fact, the little community of nearby Riverside still uses the well as their source of drinking water.

Mack Deen cir. 1950 — Deen Family Collection

Ron Deen Sr.

Ronald Redell Deen Sr. was born April 19, 1936 at Ft. Stockton, Texas. Ron Sr. graduated from the Ray High School in 1949. In 1954 Ron Sr. completed a year of college at the University of Arizona, on a scholarship from the Kennecott Mining Company. He then went to work at the Ray Mine Complex in the assaying, engineering and sampling departments for 30 years. Ron Sr. has also worked as a quality control tecnician. When the Ray Mine was temporarily shut down in 1982, Ron Sr. went to work

L-R: Joe and Ron Deen Sr. 1997— Author's photo

with his sons in their construction business. He went back to work at Ray in 1983 as a quality control technician for ASARCO, after they bought the Ray Mining Complex. He worked in the engineering department worked as a warehouse technician. He retired in 2002. Ron Sr. married Marlene Ruby Putnum, who was born in 1938. They lived at Ray, Arizona until the family moved to Riverside in 1964. Ron and Marlene had three sons, Freddie Joseph Deen, born July 3, 1954, Ronald Bedell Deen Jr., born April 26, 1958 and Gary Todd Deen, born May 17, 1962. They were all born at the old mining town of Ray, Arizona. The old mining towns of Ray and Sonora are now just a memory. Where there were once little thriving mining towns, there is now the large open pit mine of Ray, owned by the Asarco Mining Company. A Mexican company called El Groupo recently bought out Asarco. The three sons went to elementary school in Ray and Junior and Senior High School at Kearney.

Joe Deen—A Profile in Courage

Joe Deen (manager of the Silver King until 2000), graduated from school in 1972, and went to work at the Kennecott Mining Company at the Ray Mine complex. In 1975 Joe became a deputy sheriff for the Pinal County Sheriff's Department for about 18 months. Joe said that the deputy work only paid about $6000 per year, and that he went back to mining as it paid better. For a five year period he owned a construction business that basically did road building. Joe's company rebuilt the unpaved road from Young, Arizona to State Route 260, which is the paved road to Payson. Joe also said that he did a lot of work for the U.S. Forest Service at this time. Joe then went to work for from 1990 to 1999 for Asarco at the Ray Mine Complex at Kearny, Arizona. He worked as a shift foreman prior to being laid off in 1999.

Joe Deen with Silver King bullion 1997
— Author's photo

During October, 2000 Joe Deen had to find work, since the Silver King was shut down, and went to work at the Barrick Gold Mine in Elko, Nevada. This was high in the Nevada mountains, and during the winter Joe called back and said the snow was so deep it covered his car. Another member of the group, George Sites, went to Honduras to work in a gold mine also. Joe's twin sons went to Elko with him, but Logan, one of the twins, joined the army and Trevor stayed with Joe. To illustrate the dangers of mining, Joe stated that one of the mine bosses at Elko, where he worked, was trying to assist another miner when his safety line broke and he fell several hundred feet down a shaft to his death.

While at Elko, Joe was injured at the mine when a transmission fell on him from a mucking machine, then suffered from a mysterious illness which affected his body's muscular structure. This illness that left him weak all the time, and he had to return to Arizona. This illness was finally diagnosed as ALS or Lou Gehrig's Disease. This crippling disease affects all the muscles and body tissues. Joe lost muscle tissue, however he took an experimental treatment for liver diseases. Joe did not give in to this terrible debilitating disease. Joe moved with his wife, the former Debbie Harmon, to Kearny along with one of his twin sons Trevor and other son Adam. The other twin, Logan, served in the Army in Iraq. (Logan went to Iraq during the war with the 101st Airborne Division) Joe also has a married daughter, Lisa, and grandchildren.

Tony San Felice (Author's son) and Joe Deen 2003
— Author's photo

Shortly after Joe came back home to seek treatment he had to be medivaced by helicopter to a Tucson hospital due to severe breathing problems. Joe had a tracheotomy performed and was put on continuous oxygen. Joe was told that he did not have long to live, but did not accept this, and sought experimental treatment from a man who had ALS for 35 years, who had developed a serum that was able to keep him alive.

Joe took the same serum and surprised everyone, especially the Tucson doctors, who thought for sure that he would not survive very long. Joe has tried many different treatments, and initially made some progress, then was confined to a bed, where he remains, as of this writing in 2005. He has tried special diets and treatments for liver disease, which the Deens' think is the cause of Joe's ailment. One of the things that has helped Joe is the use of colloidal or molecular silver, which Joe took internally, and also used to sterilize his breathing apparatus.

Though he was getting weaker all the time, Joe came to the mine on occasion to observe the de-watering, even in a wheelchair. He gave advice on developing the Plan of Operations. As I was writing this book, Joe recalled many details of the underground operation years after he had been down in the mine. He maintained a positive attitude that assisted all of us associated with the mine. Joe's special courage gave us confidence during the struggle to complete the various phases of the long drawn-out Plan of Operations. He never gave up hope of one day seeing the reopening of the mine. He had wonderful recall of details regarding mining and milling operations and assisted with the smelting process of converting raw ore concentrates into silver bullion. Joe never gave up hope and is an inspiration to all of us who know him. We respect and admire his courage and willingness to live, despite difficult odds. Joe can talk with some effort, but his mind is still active, and he is as sharp mentally as ever.

Joe continues to barely survive, and some of the treatment protocols made him gain some weight back. As of March 2005, Joe is still with us, but is getting weaker and had to be put on a feeding tube. However, he still can talk and always requests updates on the bond issue and the process of locating and meeting with potential investors. He remained an intricate part of the process and his input was appreciated and sought.

Ron Deen Jr.

Ronald Redell Deen Jr. graduated high school in 1977 and went to Central Arizona College at Casa Grande. He then became a rodeo rider on the professional rodeo circuit for 4 years. After that he worked in the family construction business for about 10 years. Ron Jr. worked in heavy construction and built stock tanks for the United States Forest Service. He also worked for the Bureau of Reclamation at a stone quarry near Blyth, California called Palo Verde. There were dangereous situations at this quarry, as previous workers left live explosives in the quarry, but Ron

Ron Deen Jr. at Silver King 150 ft. level with 1800s ore cart— Author's photo

Jr. survived the explosive situation. He then worked drilling water wells. Ron Jr. worked for the Magma Mine Company as an underground mechanic and as a mill mechanic. Ron Deen Jr. currently works for the El Groupo Company at Hayden, Arizona. Ron is married and lives with his wife, the former Linda Cole, on his ranch at Kearny, Arizona. Ron also has a son by another marriage and Linda has a daughter by another marriage, as well as a grandson, Clay.

Gary Deen

Gary Deen is the third son and also a partner in the mine. He grew up and graduated from the Ray High School in 1980. Gary received a Boilermaker Apprenticeship from the Nine Western States Boilermaker Apprenticeship Committee. From 1980 to 1986 he traveled through Arizona and New Mexico, building and repairing power plants and smelters. In 1987 he worked at Magma Copper Company at San Manuel, Arizona, as a boilermaker in the underground mine and in the open-pit mine. In 1988, Gary married Billie Stambaugh, daughter of Bill Stambaugh, and the great-granddaughter of Theodore Swift, who was a forest ranger and supervisor of the Tonto National Forest in 1907. Theodore Swift was the founder of the Swift Trail on Mount Graham, Arizona. The Old Swift Trail from Superior, Arizona to Oak Flats was also named for him. This was a steep mule trail that ran along the south side of Queen Canyon up to Oak Flats, and was the only route towards

Gary Deen & Bill Stambaugh at Silver King 1997 — Author's photo

Globe other than the old Stoneman Trail, which led upwards from the Old Silver King Town and mine. Gary left Magma Copper Company in 1990 and became a cotton farmer in Marana, Arizona. [3] Gary's father in law, Bill Stambaugh, was an asset to the Deens and greatly supported the re-opening of the Silver King.

Dick Bynum

Richard (Dick) Mark Bynum, one of the partners, was born January 1, 1945 in Salem, Oregon where he worked in the logging industry for several years. Dick received an Associate Degree from the Eugene, Oregon Community College in Flight Technology and went to work as a pilot delivering the U. S. Mail. He met his future wife Janet Thurston while delivering mail in Ely, Nevada, and they were married in 1971. They have two sons, David and Brian. Dick came to Arizona in 1985, and worked at the Ray Mine Complex in various capacities, including solvent extraction. Dick has had job titles that included Metallurgy Technician, Furnace Supervisor and Flotation Tank Technician while at the Ray Complex. He also worked as a locomotive loader and ore processor. He met the Deens at the Ray Mine Complex. They became lifelong friends, and eventually partners in the Silver King Mine. Dick learned many mechanical skills while working in the logging and mining industry, and is capable of repairing most equipment used around the mine. Bynum retired in 2002.[4]

Dick Bynum with Silver King bullion 1997 — Author's photo

George Sites

George Sites is also one of the partners. He was born January 30, 1937 in Hervitia, West Virginia. George came to Arizona with his family in 1945, and they settled in Buckeye. George went to Florida where he worked in a furniture factory, then took up the trade as a welder-machinist. While in Florida he also worked as a steel fabricator. He returned to Arizona where he went to work for the mining industry. While working for the mines George developed a copper-thermite welding process. In 1967, he diamond-drilled or core-drilled all over the State of Arizona. He also had his own water drilling company at one time. During 1968, George went to work for the Inspiration Mine & Consolidated Copper Company in Miami, Arizona as a machinist. While at the Inspiration Mine he maintained the large railroad locomotives. He retired from there as Field Maintenance Foreman in 1992. George has attended various schools in metallurgy and machinery at Florida State University and Arizona State University. In 1986, he went to Alaska for the Inspiration Company, where he operated a huge gold mining dredge in the ocean. Then in 1994, he went to Honduras to build a complete processing mill for gold. [5] George lost his wife Candy to an illness during 2004.

George Sites on Silver King hoist 2000
— Author's photo

Also, one of Joe's cousins, who helped work on the mill, was killed in a plane accident in Alaska. Another person who worked at the mine was killed in an auto accident.

Claim Jumpers at the Silver King

As with many of the mining claims of the 1880s, when there is a lot of money to be potentially made, someone will try to take it away by crossfiling a claim, or as they said in the old days "claim jump you". This did in fact occur, when a claim dispute arose out of just such a case and someone tried to "file on top of our claim" stated Ron Deen Sr.. The Deen family as of this writing does hold possession of the properties, and in their mind they are the only legal claimants to the property. To quote Ron Sr., " We have no hard feelings, and only want to do the work honestly and get the ore processed, so we can all begin to make this thing pay off."

Finding the Deens Underground at the Silver King

During October 1996, I went up to the Old Silver King Mine and discovered that someone was working in the old mine. It turned out to be the Deen family. Thus began my association with the Deen's. I was able to locate several of the old mining reports of the 1880s, along with several old maps, that described in detail the workings of the underground mining operations, and made copies for them. In return, they began to explain the process of underground mining, processing the mine's dumps and tailings, and the complicated process of milling the ore. A friendship was formed, and I was able to learn about the old Silver King Mine, literally "from the ground up," as the saying goes, or in this case from the ground down.

Joe Deen and Author at Silver King 1996
— Author's photo

When I first met them, they were in the process of draining the water level down and had it drained whereby the old 256 foot level was now open. I wanted to see for myself the underground workings and climbed down the mineshaft. The way down was by a series of ladders fastened directly to the vertical walls of the shaft. These ladders went down 16 feet to a small platform, where another ladder went down another 16 feet, etc. I climbed down into the old mineshaft where almost no one had been since the old miners of the 1880s. As I reached the 256 foot level, Ron Deen Jr., Dick Bynum and a miner named Roy Wheeler were shoring up the old timbers and doing some exploratory work. They were examining the original quartz pipe. The Silver King Mine is a quartz pipe mine, where the body of ore runs vertically within a quartz pipe or chute. They were taking ore samples up to the surface to be assayed. At the 256 foot level the tunnels were filled with old timbers and the work was slow going. Going down was not too difficult, but the going up part was very exhausting for me. These series of ladders were called the old man-way of the original 1880s shaft, or a way to get out, if the hoist was not working. The Deens had replaced the old ladders with new wooden ones and they were fastened securely to the wall of the shaft.

On subsequent months that followed, after taking a mine safety course, I went into the mine on several occasions to photograph the old tunnels and their operations. I also went into the 114 foot level to photograph it. It was a much safer tunnel to examine than the 256 foot level as it was a hard rock tunnel which did not require timbers to shore it up. I personally obtained several specimens of excellent silver ore from these levels.

Roy Wheeler at 256ft level of Silver King mine 1997 — Author's photo

Roy Wheeler was somewhat of a superstitious person, as many of the old miners were, and I always had a ghost story to tell him about the miners who had died in "the hole." I used to enjoy a good laugh with the Deens when I told these stories. On occasions, I accompanied Roy to photograph the old tunnels, drifts and winzes. He was very knowledgeable of the workings of an underground mine. Roy was always concerned with our safety, as well as with the underground noises and strange occurrences.

A Conversation With an Old Miner—Antonio Cisneros

On March 20,1997, while at the Silver King Mine, I was told that an old miner was working his claim, near the base of the King's Crown Mountain, not far from the Silver King. I went over and spoke with him. He said that his name was Antonio Cisneros, but that I should call him Tony. He said he was born in 1914, was 82 years old, and that he was born in Sonora, Mexico, but moved to Bisbee, Arizona when he was one and a half years old. Tony stated that he went into the Army in 1945 and then went to work in the mines at Bisbee. Tony said that he came to Superior in 1948 and became friends with Chester Lobb, who gave him 5 claims near the Silver King Mine, when he (Lobb) decided to move to Oklahoma. Tony said he obtained 4 more claims near the Silver King, including the one where I met him, the Big Dog #1 claim. He said that he had done the necessary assessment work and paid a $100 fee on the claims since 1948. He had filled two burlap sacks, and was digging near an old stone wall and gathering small pieces of what looked like black ore. He said it was black silver. Ron Deen Jr. was there and said that it looked like silver ore. Tony told me he had been coming to this area for several years, and first came to this area with his father, also named Antonio. He said that his father had been familiar with the area near the Silver King and Superior since the late 1800s. He had been a miner at Bisbee and many other mines in Arizona, including the mines at Jerome. He pointed out the area he was working. It was an old mine dump area, where in the past a small smelter had been set up, and where the ore from nearby shafts and hillside open dig areas was brought to be smelted into bars.

Antonio Cisneros at Silver King 1997
— Author's photo

Tony spoke of the old days when the Silver King area had more population than Superior. He showed me the areas on the surrounding hillsides, where the parameters were for silver ore. Tony spoke of the past when much silver and gold was mined in this area, and stated that there still was much ore left, not only in the tailings but also in the hills. He said he had killed many large rattlesnakes in this area over the years. Tony says he gives most of his ore away to family and friends, and remembers once giving a large piece of ore to an assayer who melted it down to a piece of pure silver. In the 1980s, some mining people from the Copper Queen Mine came to his claims and said he had some very good ore. Tony said that through the years he has searched the areas of the Tortilla and Eagle and Buffalo Mountains south of Superior for the elusive mother lode. He also has searched Iron Mountain in the Superstition range and indicated that he had found several abandoned mines on that mountain. He says that the Silverado and Fortuna mines, west of the Silver King, were once very rich mines. Tony recalls prospectors in the past taking ore from the various nearby hills and mountains with their burros. He said that when he first came to Silver King many of the original building were still there, and that people from all over came to see the "big house." (The original two story mine office).[6]

Ghost Voices at the Silver King Mine

The mill at Silver King consists of several pieces of valuable used equipment. By the winter of 1998, the mill was in operation. Joe Deen, the owner, thought it best to get someone to provide security for the mill, equipment and concentrates. He brought a man to the mine area simply known as Richard who had been living in Superior. Richard had a 40 pound mixed breed dog he called Vassie who followed him everywhere. Joe provided Richard with a place to stay in the camper that was at the King, and also brought a porta-potty for him.

Richard soon became known to everyone who came to the Silver King as "Two-Gun," since he wore two guns all the time, even when Joe and the other workers were there during daylight hours. At night "Two-Gun" and Vassie were the only occupants at Silver King, or so we all thought. However, it seems that late at night when "Two-Gun" and Vassie kept watch, "Two-Gun" began to hear strange noises. He thought at first it was the wind, because the wind can howl at times, up through the canyon where the Silver King Mine is located, and on up King's Crown Mountain.

Two Gun Richard at Silver King 1999
— Author's photo

Eventually "Two-Gun" began to believe that the noise was not the wind, but voices. In late March, one night, he heard noises outside the camper where he and Vassie had gone to go to sleep for the night. Vassie was whining and "Two-Gun" listened to what he thought were voices, just outside of the camper. At first he thought someone had come up to Silver King, either Joe or someone else. Surely it must be one of the Deens or Dick Bynum, since only they had keys to the gate, one-half mile away. He went out to check and the voices stopped, but no one was there. On each successive night the voices came and went. He was able to say that they sounded just like women's voices. "Two-Gun" was beginning to get spooked. On his final night at Silver King, the women's voices were loud and clear just outside of the camper. Richard would not repeat what the voices said, but he was very spooked. The next morning he told Joe that he was quitting at the Silver King. The voices were just too much for "Two-Gun," and he packed up and left. He said he could handle any human intruders, but the voices were just too much for his mind, and he and Vassie just could not get any sleep.

Joe Deen also stated that he had heard strange voices around the old mine. In 1997, while sleeping in a camper next to the old main shaft, Joe heard someone striking rocks with a hammer. He stated that the sounds were coming from down in the mine, at the 114 foot level. Joe said that on another occasion, after working at the mine, the workers were having a cookout when he heard voices in the area of the old shaft. The next morning Joe asked his brother Ron if he had heard the voices last night. Ron said yes, but that he though Joe was talking to a visitor. Another time, Joe sent Logan, his son, to check and see if the main gate to the mine road was locked at dusk. In a few minutes, Logan came flying up the road on a three-wheeler and said that he heard voices coming from the old Silver King Cemetery. Joe stated another time when he was at the mine showing a salesman around, the salesman walked over to the Glory Hole, and when he came back he asked Joe who the old Mexicans were in the area of the camper. When Joe went to check, no one was there. Dick Bynum related this story about voices. One time when he went down to the old outhouse, by the ruins of the old town, he heard voices coming from the back of the outhouse. When he went to check, no one was there. This made a believer out of Dick.

At one time or another, I thought I heard voices near the old Silver King Cemetery. To me they sounded like small children playing. On one occaision at dusk, I took my wife Winnie to see the headstone and iron fences at the old cemetery. She told me that she thought she heard something, and said that the place was too spooky for her. I couldn't get out of there fast enough for her. Were the voices real, or was the solitude at a lonely old mining camp too much for the imagination of "Two-Gun" Richard and the others? Other people have said that they have heard voices at the Silver King Cemetery, which is nearby to the old townsite. The ghosts of the past have become very real for some of us!

Testing the Silver King Ore Values

For about a year I gathered and cobbed what I thought was pretty good silver ore from the dumps, which I stored next to my house. If anyone looked at the ore, it looked pretty much like some yard rock, unless you looked closely. Therefore, I was never worried about it. After I gathered about a ton of ore, Joe Deen and I took the ore back to the Silver King Mill and crushed it. Joe ran it through the mill process where it assayed out at about 900 ounces of silver per ton. Joe gave the concentrates to a man who said he could smelt it into silver bars. This man, however, absconded with the concentrates, never to be seen again—at least as of this writing. In the old days a man caught with another mans silver would be "hung high until dead." All that work gone for naught, as the saying goes.

L-R: Rusty San Felice, Ron Deen Sr. & Tony San Felice at Siver King mine 1999 — Author's photo

While the work at the mine went on I had the opportunity to search the mine for artifacts from the 1880s, of which I did find some relics of days gone by. All in all, I was gathering first hand information as to how the mine originally operated, from the research, as well as the bits and pieces of equipment left at the mine, by discussing exactly how today's mine operation worked.

The Deen's also explored the other shafts near the old main shaft to check them for high-grade ore, but no high grade was located. They worked the ore from the glory hole and located a very nice vein of azurite in the light gray porphyry, but no high-grade ore. My research revealed two more shafts near the glory hole. These were opened after 1900 and during the 1930s. In January of 2001, one of the old shafts was relocated when the rocks on top of it collapsed, revealing its location.

The New Head-Frame is Erected

As time went on, the Deens refined the milling of ore, and obtained some pretty good concentrates from the dumps, as the assay reports showed.

But, the Deens thought they could get a better grade of ore from within the mine, so they decided to build a fabricated steel head-frame. They obtained all the various machinery to develop the main shaft, with an automatic hoist and cage. Then, they could more efficiently and safely send down the miners and supplies and bring up the ore. Building the new steel head-frame resembled a giant version of the old Erector Set toys of my youth. It was very interesting to watch the progress of the bolting and welding of the various pieces until it finally began to take shape.

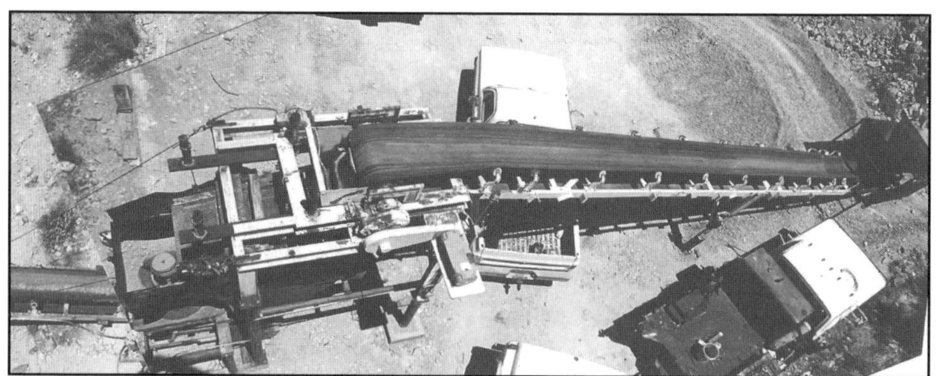

Silver King ore crusher 1998 — Author's photo

On October 25, 1999, the new head frame was erected at the Silver King. George Sites had worked for about four months welding the steel I-beam frame together, along with Dick Bynum and the Deens. The Deens hired a large crane with a 100 foot boom to erect the head frame. It stands 66 feet high from the baseplate, and 14 and 1/2 feet down to the mineshaft opening. It was erected to bring up the ore from the shaft and to dump the ore into a large ore bin that is next to the head frame. On November 22, 1999 the supporting frame was attached on the eastside. It was made out of 10 and ½ inch steel pipe, 3/8 inch in thickness. It is 61 feet long and has a 22 foot spread at the foundation. It is angled to give maximum support to the head frame, when the skipjack brings up the ore. Work at the mine commenced on the mineshaft, down past the 114 foot level. New support framing was added to keep the shaft from collapsing inward. Plywood was added on one side of the mineshaft walls to keep debris from entering the skip and man cage.

Hoisting the ore

The main hoist house is 24 feet by 24 feet. The hoist is a Vulcan/Denver power down/power up; safety checked hoist, with 600 feet of ¾ inch steel cable with a

26-ton pull capacity. The engine shaft (SK main shaft) is 10 foot by 5 foot with two compartments 5 foot by 5 foot each. A water pump is installed at the 400 foot level of the mine to draw water from the mine in order to keep the mine dry, as there is an underground water supply that dumps water into the mine at about 60 feet below the surface. The man compartment of the man cage/skip jack lift is 4 feet by 4 feet secured by a steel closing device and surrounded by steel walls. The skipjack is installed with a safety device that releases safety catches into the 4"X 4" wood safety rails installed vertically along both sides of the main shaft. This safety feature meets industry standards. Additionally there are wooden ladders in the man compartment of the mine which extend down 300-feet. The headframe is made of 10-inch steel H beams, 63 feet high, reinforced and braced by 10-inch schedule 80 steel pipe, 70 feet long installed at an angle to reinforce the main head frame per industry standards. The head frame sets on a reinforced two-foot concrete slab, 20 feet by 20 feet.

The Ore Crushing Process

a. The ore is brought out of the mine by a skip jack ore bucket 4 feet by 4 feet with a 4 ton capacity, which then dumps the ore via a drop lever into the 50 ton steel ore bin.

b. The ore then slides down a metal chute onto a 30 inch by 70 foot conveyer belt, which takes it to a 25 ton ore hopper and then to the Pioneer Jaw Crusher with a 10 inch by 24 inch crushing steel jaws, where it is sprayed by water, then goes to a 3 foot by 6 foot by 4 foot surge bin.

c. The ore then goes to a 30 inch by 60 foot conveyer belt, which feeds a 30 inch by 60 foot conveyer belt into a Allis Chalmers double 6 foot long by 3 foot wide steel screen, where it is again sprayed by water.

d. The ore then goes to a 30 inch by 30-foot conveyer belt, which takes it to an ore stockpile.

e. The ore is then transferred from the ore stockpile by a front-end loader to an 8 foot by 12 foot 25 ton hopper. From the hopper it goes via an 18 inch by 30 foot conveyer belt to the variable speed 6 foot by 2 foot internal Hardy Ball Mill, where it is pulverized and water is added.

f. The ore is then treated with the proper additives and the metal is extracted via the flotation process . *At this point the ore is called concentrates or cons.*

g. The solution is then always wet until it reaches the 10 foot by 20 foot steel dryer holding tank. After the cons dries sufficiently into a solid material they are stored in 55 gallon drums. The cons are stored into secure "C" containers waiting for shipment to the smelter.

The Process of Milling Ore

After the completion of a flotation mill at the Silver King Mine, just below the old main shaft, the process of milling ore at the Silver King Mine was explained to me by the Deens in March of 1998, as follows:

Step one. The ore dump material is first run through the Denver Jaw Crusher and next through a second ore crusher called an Allis Chalmer Ore Crusher (reduction crusher), which reduces the ore from 10 inch to 1/4 inch minus grade or mill feed.

Step two. The 1/4 inch minus ore then goes to a 1000 ton ore stockpile. This ore is moved by a front-end loader to an ore bin, which feeds 195 lb. a minute. From this 1/4 minus moves ore up a conveyor belt to the Harding Ball Mill.

Step three. In the Harding Ball Mill (6 foot by 2 foot), water is added to the ore and the ore is ground and reduced into a pulp. Next, it is sent to a cyclone feed sump (ball mill discharge sump), where it is pumped back to two cyclone sumps for classification. The pulp is then sent back to the mill overflow (which is 150 minus grade), then goes to a conditioning tank. The ball mill holds 7 tons of steel balls (3-6 inchs in diameter) which grinds the ore to 150 minus mesh grade. This liberates all the values in the ore to be extracted in the Denver Float Cells.

Step four. The ground ore goes to the conditioning tank (150 mesh grade), where the prep chemical reagents (chemicals used to float the minerals are called a prep) are added. This makes a froth, which allows the metal particles to float up. The other chemicals are collectors, which collect the metals to be extracted in the float process. These metals are silver, gold, platinum, lead, zinc and copper. The conditioning tank gives time for the values in the pulp to react prior to being extracted in the Denver Concentrator Float Cells. These cells hold 2000 gallons of pulp. As the ore changes, the process of what type of chemicals and the amount will change also.

Step five. This material then goes to the Denver Float Cells where the metals are extracted in the froth, which then is pumped to a thickener tank. In the thickener tank the metals settle out (this material is called concentration or "cons"). The "cons" are pumped from the thickener tank to a Denver Belt Filter Plant (1 by 3 foot) where they are discharged into 55 gallon drums. Next, the wet concentrates are dumped into a dryer tank where the rest of the water is evaporated off. The "cons" are ready for shipment to a smelter, where they will be refined and the precious metals separated. The Denver Sub A-Floatation Cells handle 14 tons of 100-mesh grade pulp and water per hour. The thickener tank stores the "cons" to be processed to the filter plant. It can hold 2600 gallons of thickener in the 6 by 8 foot tank. The 55 gallon drums hold about one ton of "cons" which is expected to yield X ounces of silver, X ounces of gold, X lb. of lead and zinc, and X lb. of copper, based on their most recent assay reports. The "cons" then will be smelted into bars called "Doré Bars," consisting of all the precious metals. The Doré Bars will be sold to a refinery, and further refined into gold, silver, copper, lead, zinc and copper bars.

Step six. The tailings from the Denver Float Cells go to a Dyster Concentrating Table, where the larger pieces of metal are gravity separated. The bearing tailings are then pumped to a tailing pond, where the water is then reclaimed back to the mill. The silica will then be stripped out of the tailings, to be used in the smelting process.

Logan & Trevor Deen at Silver King 1999 — Author's photo

The mill has 20th Century technology, allowing for the extraction of most of the values in the ore, which the processes of the past could not do. Therefore, this makes the mill not only more efficient, but able to pay a profit. The make-up of chemicals in the process has a life span of 36 hours and is made up daily. It takes about 60 lbs. of chemicals to make the process work, at a cost of about $2.40 per lb. These chemicals are environmentally friendly, based on results of various tests made by the geologist.

The concentrates are checked periodically by using a Stereomaster II Microscope, made by Fisher Scientific, Model SPT-ITH, with a magnification of 15X in each view lens. Looking through this device you are able to see the various metals and other minerals in the concentrates. The cons are checked both wet and dry, which shows a different consistency. The mill is inspected by the State Bureau of Mines and a Federal Mine inspector, once the mine becomes fully operational. The entire process for the bench testing of the Old Silver King ore, and extracting the values of the precious metals in order to determine if the ore was worth it, cost several thousand dollars.

Exploratory Mining Down Under

In order to make the mine pay off, the Deens decided to conduct exploratory mining at the 114 foot level. Ore was gathered from the various drifts/tunnels to obtain samples for assays. At the same time, the 114 foot level was rehabilitated in order to make work in the tunnels more safe. New beams were added at the entrance to the 114 and shored up any place that looked like a problem area. They went about the task of cleaning up the tunnels, including widening some, to create room for working. The old ore cart tracks were repaired and new tracks were added

Johnny Chavez at Silver King 114 ft. level 2000 — Author's photo

where appropriate. A machine called a mucker was built and installed at the 114 foot level, in one of the old tunnels. The mucker is designed to collect up all the loose ore in the tunnel. The mucker drags the ore through the tunnel by means of a drag (scoop-like piece of sheet metal), and a dragline made out of steel cables connected to a winch. The winch pulls the drag full of ore up a ramp, and deposits the ore in the ore cart. The ore cart is pushed manually to the shaft entrance where it is dumped into the skipjack. The skipjack full of ore is hoisted up the shaft. The skipjack has a tripping device on it, which automatically flips the skipjack open when it reaches a predetermined level at the ore bin. The ore is deposited into a large ore bin above ground next to the main shaft. Several cartloads of ore were taken up to the surface in this manner in order to determine where the actual mining was to take place. The different loads were then tested for percentage of silver content.

Joe Deen at 114 ft. high-grade silver vein 2000 — Author's photo

In January 2000, while checking out the various tunnels at the 114 foot level, a rich vein of silver ore was located. Could this be the same rich vein that the miners of the 1880s followed down several hundred feet? Assays indicated that it was very rich Argentite, consisting of 70 percent silver, and was contained in the old quartz pipe. The Deens continued to rehab the mine while at the same time taking out rich ore pockets at the 114 foot level during the spring of 2000. They reinforced the tunnels (or drifts) and stopes, laid tracks for the ore cart, and set up slushers to remove the ore along with the waste rock. This process required the expertise of underground miners. There were usually six men working during this clean-out of the 114 foot level.

During August of 2000, the Deens decided to see if they could have some of this rich ore smelted. They ran this ore through the mill process and took the concentrates to a private smelter plant near the very small town of Wickieup, Arizona, owned by Bob Gardner. They were successful with the first smelt, removing the copper, but had difficulty getting all of the lead out. The large cupels made for removing the lead broke during the process. They took 2.5 tons of concentrates to the smelter, which reduced down to 1 ton of rough copper, and ingots amounting to 630 pounds of silver, gold, and lead. The second smelting in the cupel furnace reduced the ingots to 344 pounds of silver, gold and lead. These resulted in Dore´ bars weighing about 25 pounds apiece. Every smelting process burns off some of the precious metals, according to Joe Deen. The next step was to get the rest of the lead separated from the gold and silver without losing too much of the precious metal. Some of the small high-grade silver bars were taken to Lois Wallace at Sunburst Jewelry, Inc. Ms. Wallace made some superb rings and jewelry pieces from the silver.

Louisa Barber & Dennis Foreman at Silver King 2000 — Author's photo

Bud Hansen in Silver King shaft 1999 — Author's photo

Everything seemed to be falling into place. The Deens thought they were getting things going and were on the verge of producing silver and gold. They thought they could now pay back the debts for the start-up of the mine and mill, and proceed with the underground exploration and expansion. But the Deens were in for a huge setback. An unexpected event happened in the development of the mine.

The Cease and Desist Order and the Shutdown of the Mine

During August 2000, Forest Service officials came to the Silver King Mine with an Assistant U.S. Attorney from the U.S. Attorney's Office of the District of Arizona. A Cease and Desist Letter was handed to the Deens, which ordered the Silver King Mine to cease and desist operations, or a court injunction would be obtained, and they were subject to a huge fine daily of thousands of dollars. The Deens stated that under the 1999 U.S. Federal Court decision Shumway VS United States, Ninth Circuit Court of Appeals decision, that they had a right to operate. This attorney stated that Shumway was being sent to a lower court, and therefore was not applicable. Basically the Shumway Case was detailed as follows: Ray and Molly Shumway owned seven mill sites in the Tonto National Forest in Arizona. The case involved issues regarding the extent to which an owner on an unpatented mining claim or mill site has surface rights, despite the absence or approval of a Forest Service operating plan and bond. It also included the constraints on disapproval. The court found for the Shumways, and issued a 19 page landmark decision on the mining rights of a mine and mill site claimant on federal lands. The courts issued a summary judgement, which allowed the Shumways to maintain their mill site rights. This case is a landmark decision for small miners' rights. Therefore, I have included the critical excerpts from it.

Joe Deen at Silver King 2000
— Bob Stambach photo

The sections of the Shumway decision below clarify the Deens' position on mining claim rights.

The Shumway Decision section 14939, page 17, paragraph two states: "Mine and mill site operators must give the Forest Service a notice of intent to operate, and based on this notice if the District Ranger determines that such operations will likely cause significant disturbance of surface resources, the operator shall submit a plan of operations, unless the mine operation falls within a small group of exceptions." A plan of operations describes the type of operation proposed and the manner conducted. The plan of operations must be approved by the District Ranger who must "analyze the proposal, considering the economics of the operation, along with the other factors in determining the reasonableness of the requirements for surface resource protection." The filing of the detailed plan of operation with the Forest Service was the key disagreement between the Deens and the Forest Service.

The Shumway decision affirmed the Mining Law of 1872, and stated in section 14923, page 6, paragraph one: " Despite much contemporary hostility to the Mining Law of 1872 and high level political pressure by influential individuals and organizations for its repeal, all repeal efforts have failed, and it remains the law. The miners' custom that the finder of valuable minerals on government land is entitled to exclusive possession of the land for purposes of mining and to all the minerals he extracts, has been a powerful engine driving exploration and extraction of valuable minerals, and has been the law of the United States since 1866. Mill site claims, which are ancillary to mining because grinding the ore is often necessary to extracting the valuable minerals, follow substantially the same rules as mining claims."

Shumway states on page 9 of the Court's decision, in section 14927, paragraph two: "Mining claims located after the effective date of the 1955 Act are subject, prior to issuance a patent, to a right of the United States to manage surface resources, and for the government and whomever it permits to do so, to use the surface, so long as they do not endanger or *materially interfere* with prospecting, mining, or processing." Furthermore, Shumway on page 14 of the decision, section 14935, paragraph three, states: "The owner of a mining or mill site claim does not need a patent, or a vested right to issuance of a patent, to possess and use the property for legitimate mining or milling purposes. A mining or milling claim is 'property' in the fullest sense of the word. Despite the absence of a patent, the government cannot take the valid mining claim for public use without paying compensation. The owner of a perfected mining or mill site claim is not required to secure patent from the United States; but so long as he complies with all provisions of the mining laws, his possessory right, for all practical purposes of ownership, is as good as though secured by patent."

In concluding the Shumway decision, the three judge panel of Judge Joseph T. Sneed and Andrew J. Kleinfeld, Circuit Judges and Judge Evan J. Wallach, Judge of the United States Court of International Trade, sitting by designation, rendered its concluding remarks in an opinion by Judge Kleinfeld, and concurrence by Judge Wallach, as stated in section; 14941, page 19 paragraph one. " Failure to file an approved operating plan cannot, ipso facto, cause a forfeiture of the bona fide claim owner's equitable title and possessory right. The Shumways are not guests at their mill site, but property owners."

Judge Wallach concurring in the opinion in favor of the Shumways stated on page 19, last paragraph. " I concur in this opinion. The Forest Service would do well to remember that miners explored and built much of the western United States using long-established techniques, some of which are dated from the Middle Ages. Many of those methods are still perfectly workable. Mining equipment, especially that used in grading, crushing, and grinding the ore may well be old, battered and rusty and yet still be entirely serviceable, particularly for small operators. It is hardly junk." [7]

The Deens believed that the Shumway decision gave them the right to operate, without interference from the Forest Service, but did not have the necessary funds to take their case to court. Therefore, they felt they had to comply with all of the rigid requirements of filing a plan of operation, even though they felt that they were right under the mining laws. During this process, as you shall see below, the Deens tried to prove that a patent existed, which would exempt them from the Forest Service requirements. Also, they tried to prove that the mine had a perpetual claim for over 100 years, therefore exempting them from the FS requirements. Another issue discussed with the FS, was the fact that the work site was less than 5 acres, therefore, could be exempted from filing a plan of operation. As you will see below, none of these options proved successful. The Deens felt that they had no options left, unless they went to federal court, but they did not have the requisite funds, so they decided to file a plan of operation and comply with the cumbersome regulations of the Forest Service.

The Plan of Operations

The Forest Service officials stated that the Deens could not operate the SilverKing without an approved Plan of Mining Operation from the Forest Service. The Deens said that they Silver King Mining Corporation (SKMC) had submitted four plans of operation, but the Forest Service (FS) would not approve their plan, even though they had met all the Forest Service requirements. Prior to this, the Deens thought they could operate without the plan of operations from the Forest Service, since the Silver King was a mining claim prior to the Forest Service regulations being enacted, the Deens stated that the 1872 mining act would prevail and they could operate. However, since the previous claim owner's attorney allowed the claim to lapse, the Silver King lost continuity of ownership, and this gave the Forest Service the right to enforce regulations on the claims. The Deens stated that the Forest Service added new conditions each time they submitted a plan, but by their own regulations the FS is supposed to approve a Plan of Mining Operations within 30 days of submission, but of course the Plan took over four years. A lot of these problems were caused by miscommunication on both parties, as well as the Deens receiving inaccurate information from parties that they believed. Also, the law regarding FS Plans of Operation for mining on FS land is not very clear and the regulations are not specifically spelled out. The FS regulations contain certain requirements, documents, permits and studies to be performed, if the District Ranger deems it necessary. In other words, certain requirements could have been waived, but the SKMC was required to comply with the regulations, which are intended for huge operations, but could also apply to small miners, based on the way they are written and interpreted.

These new conditions stated by the FS officials, after rejecting the Deens initial plans of operation included many environmental and numerous other issues, along with a bond requirement. The Deens stated that the FS imposed strict regulations on them, while other nearby mining operations did not have to meet the same rigid requirements. The Deens said that they only wanted to comply with reasonable regulations as set forth by the Shumway decision of the U.S.Court of Appeals for the Ninth Circuit and the Mining Law of 1872.

The Deens told the FS officials that they only wanted to get the mine working so they could pay off their debts for the mine and make a little money, while at the same time hiring some out of work miners from the "hard-times" nearby area of Superior. All the small miners like the Deens want is the opportunity to seek and find the metals that helped make Arizona one of the greatest mining states, while at the same time have the opportunity to do what they have done all their lives. That is, mine the earth and follow the mining laws that were set forth in 1872, which allows them to make a living at mining. All small miners follow the American dream of maybe striking it rich one day.

These current mining regulations severely restrict small mining operations, and need to be changed and the FS regulations need to differentiate between small mining operations, exploration, and large operations for which many requirements are necessary to protect the environment. Research showed that modern day small mining operations does indeed fall under too many regulations and regulatory agencies. So much that it makes the small mining operation adhere to the very strict regulations that were enacted for the large mining operations.

Recent decisions by the courts have amended the FS regulations in favor of the small miner. A 2004 case involving a Northern California small dredging operation found in favor of the small miner named Terry McClure. The McClures invoked the Mining Law of 1872, which was upheld by Judge Lawrence Karlton who overturned a lower court magistrate stating, "the magistrate erred in concluding they were not authorized to do so by the Mining Law of 1872." The judge stated that the appellants did not have to obtain a special use authorization for mining as this (mining) activity is specifically excluded from special use permits. This precludes the small miner from filing a Notice of Intent letter or form (requirement) to the FS and an approved Plan of Operation.

If I have learned one thing by studing this process, it is that Congress needs to address these issues, and the small miners need an *advocate* to champion their cause. Otherwise, mining in the United States will continue to benefit big business and foreign investors, and the opportunity for the average person to search for riches in the ground will be a thing of the past. This country was built upon individual speculation and achievement, and it would bode well for our legislators to remember this legacy.

The U. S. Attorney's Mandate of August 2000

On August 28, 2000, a meeting was held in the office of Assistant U. S. Attorney for Arizona, Ashley D. Adams. At this meeting, several representatives of the FS were present. The Deens were represented by an attorney, and had several interested people, including the author, present to assist with the discussion of this meeting. A letter was given to the Deens by the U. S. Attorney, which included 22 complex issues that must be met by the Deens in order for them to reopen the mine.

Development of the Plan Of Operations

During October of 2000 the Deens obtained the services of a researcher skilled in developing planning documents. This person proceeded to develop a working plan in accordance with the Forest Service requirements and the 22 issues outlined in the cease and desist letter of August 28, 2000. This plan took many weeks to write, as additional issues were brought forth by the Forest Service, including a requirement that the Silver King enter into a Consent Order with the Arizona Department of

Environmental Quality (ADEQ). This consent order mandated that the Silver King comply with 32 environmental issues during a period of 360 days, whereby the ADEQ would allow the Silver King to operate, prior to filing an application for a Aquifer Protection Permit, and paying a fee of several thousand dollars.

After much research and writing, the (Supplemental) Plan of Operations was submitted to the FS in December of 2000. The FS requested many more documents or additional information. What started as a four page outline from the Forest Service, as the basics of a plan of operations, had turned into a sophisticated detailed document and a delay in the reopening of the mine. The plan during December 2000 now contained over 600 pages, with photographs, maps and supplemental documents. The FS also required the Silver King to apply for an NPDES Permit, which mandated a detailed Storm Water Prevention Plan. This was completed and a permit was issued by the U. S. EPA. Also required was an Air Quality Permit, an explosives permit from the BATF, and an archaeological survey by an FS approved consulting firm. An archeological firm (SWCA, Inc.) was hired by the Deens to conduct a historical survey, which it completed and submitted to the Forest Service. The Forest Service accepted the report, and the State Historical review group, known as SHPO approved it. The explosives permit was obtained, as was the Air Quality Permit.

All of the 22 issues brought forth by the FS, other than the Environmental Assessment (EA), were responded to by the SKMC and submitted by March of 2001, and were accepted by the District Ranger. The Deens hired an independent environmental firm (Bender Environmental Consulting, Inc.) to complete this complex Environmental Assessment (EA) study, which took 18 months. By November of 2001, a draft EA with attachments was submitted to the Forest Service. After a revision another draft of the EA was submitted to the FS. All of the issues required by the FS were completed and accepted, and the FS accepted the (EA). The plan documents and correspondence now increased to over 1000 pages.

Finalizing the Plan of Operation

After a completed EA was submitted to the Forest Service, all of the issues required were completed and accepted. The FS placed a legal notice in the East Valley Tribune on May 27, 2003. The notice began a 30-day comment period. The EA was sent to those who had responded to the initial public scoping letter and to all Native American Indian Tribes affected. No negative comments were received.

The FS then completed a review of the EA process and concluded that the EA did not require an Environment Impact Study. On January 27, 2004,

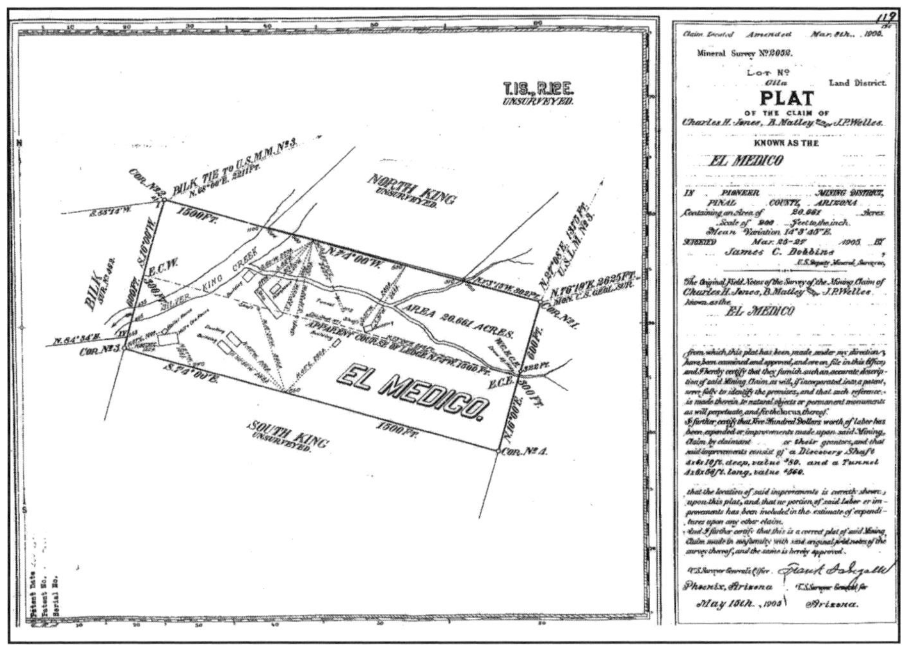

Silver King Survey Plat 1905 — Author's collection

Karl P. Siderits, Forest Supervisor of the Tonto National Forest issued a Finding of No Significant Impact Letter (FONSI). This letter signifies that no significant impacts will be made on the environment by operating the Silver King Mine. The FONSI letter listed several issues that were included in the EA and other guidelines that the SKMC are to comply with, including an 18 month operation period for the mine to operate consistent with the ADEQ Consent Order (previously signed by both the ADEQ and the SKMC), with the stipulation that the SKMC can apply for an extension pending compliance with the ADEQ Consent Order. The FS and SKMC recognized that the Plan of Operations is a working document, and by law can be amended from time to time, as activities or circumstances change at the mine, and this supplement document, along with the completed EA amends the Plan Of Operations (POO) to reflect the recent FONSI letter of January 27, 2004.

The Plan of Operations was synopsized from over 1000 pages of documents and working papers to 16 pages of text and 16 charts, maps and tables. The only other requirement left was establishment of reclamation bond by the FS, which the SKMC must comply with, and the mine can legally reopen. The reclamation bond requirements were developed and approved, which covered all phases of the mine shutdown. However, of the 200 bonding agencies that are approved by the FS, not one would issue a bond to the SKMC in the amount of $138,800 as was now required. As of 2005 the SKMC was now actively searching for private funding for the bond and cost of reopening the mine

XII.

That all of the allegations of plaintiff's complaint herein are true, and that none of the allegations of defendants' answer and cross-complaint are true.

CONCLUSIONS OF LAW.

As conclusions of law from the foregoing facts, the Court finds:

I.

That the plaintiff is, and the defendants are not, entitled to the area in conflict in this action.

II.

That the plaintiff is entitled to a patent to the said Silver King mining claim, as described in these findings.

III.

That the plaintiff is the owner, and entitled to the possession of said Silver King mining claim and premises, and every part thereof, by virtue of a compliance by the plaintiff and its grantors and predecessors in interest, with the mining laws of the United States of America and of the Territory of Arizona, and that the defendants, and every of them, be barred and forever estopped from having or claiming any right, title, or interest whatever in or to said premises, or any part thereof adverse to this plaintiff.

Let judgment for the plaintiff herein be entered accordingly.

Edward Kent

Judge

Dated this 22nd day of May, 1906.

1906 El Medico Court Decision — Author's collection

The Search for a Patent for the Silver King

During the development of the plan of operations, members of the Silver King group thought that if they could prove that a patent for the mine was indeed issued, that the SKMC would not fall under the jurisdiction of the FS. They stated they would be out from under what they referred to as "oppressive and unnecessary regulations." Additionally, all the mining claims around the El Medico Claim were patented. Therefore, George Sites, Ron Deen Sr. and Dennis Smith embarked on a "holy grail" search for the missing patent. Their search led them to many places, and they obtained many documents relating to the patent issue. But, they did not actually find that a patent was issued. All necessary documents were filed during 1905 and 1906 when Arizona was still a territory. The records were shifted from place to place, so it is possible that all the requirements for a patent were met. A patent *may have even been issued,* or pending issue awaiting only a fee payment from the original Silver King Corporation, or Dr. Jones and his associates, who met their untimely demise before the patent issue could be resolved, as described in the preceding chapter. Below is a copy of the 1905 El Medico Plat Survey by the US Deputy Surveyor and a copy of the 1906 court decision, Finding of Fact and Conclusions of Law, from the Third Judicial District Court of the Territory of Arizona. As of 2005 a patent was not located.

Funding Required to Reopen the Silver King

The mine owners have invested one and one half million dollars in the mine and could have reopened the mine but exigent circumstances have exhausted their cash reserves. The owners know the mine can make money and millions of dollars in ore remain in the mine. Only ten percent of the ore body was taken from the mine. Assay reports from underground exploration and dump material, as well as historical assays and financial reports on the mine confirm the analysis of the potential wealth still in the mine. The mine is not a prospect but a working mine with all of the equipment in place ready to operate with a minimal start up process.

As of this writing several potential investors have been contacted by the mine's owners but the offers for financing have all come with serious detriments regarding ownership of the mine. Some of these potential investors and their offers were tantamount to outright seizing of the mine with a bare minimal cash investment. After a review of these offers they were not taken seriously. An extensive business plan and financial prospectus has been developed. With the price of silver and gold rising during the 21st century, the wise speculator could possibly amass a fortune by investing in the Silver King. The owners are waiting for just the right serious and trustworthy investor. Who knows, maybe the Silver King will rise again to take its rightful place among the King of Mines.

Silver King high grade silver from 114 ft. level, 2000 — Author's photo

Silver King Hoist House & Head Frame, 1999 — Author's photo

Footnotes

Chapter One—Discovery of The Silver King Mine

1. "Notes of Travel through the Territory of Arizona, An Account of the Trip made by General George Stoneman and others in the Autumn of 1870," *Arizona Miner* newspaper, Prescott, Arizona Territory, 1870, and *The Smoke Signal*, magazine, Tucson Corral of the Westerners, Tucson, Arizona, No. 10, Fall 1965, page 14.
2. *Early Military Posts in Arizona*, unpublished manuscript, Clara T. Woody File, Arizona Historical Society, Tucson, Arizona, p. 4.
3. *Pioneer Women*, unpublished manuscript, Clara T. Woody File, Arizona Historical Society, Tucson, Arizona, p.
4. *Pinal Drill,* February 26, 1881, Pinal, Arizona Territory.
5. *Arizona Municipalities* magazine*,* James M. Barney, August 1940, P.10.
6. *Garrisons of the Regular U.S. Army, Arizona 1851-1899*, S. C. Agnew, Council on Abandoned Military Posts,
P. O. Box 171, Arlington, VA, 1974.
7. *The Smoke Signal*, pub. By The Tucson Corral of the Westerners, Tucson, Arizona No. 10, Fall 1965 page 14.
8. *Dictionary of the American West,* Win Blevins, Sasquatch Books, Seattle, Washington, 2001, Pages 275-76.
9. *"History, Geology, and Vegetation of Picket Post Mountain"***,** Frank S. Crosswhite, article in *Desert Plants***,** a Booklet Published by the University of Arizona for the Boyce Thompson Southwestern Arboretum, Vol. 2, 1984, p. 73-77.
10. *Arizona Historical Review, Vol. VI,* No. 4 P.12.
11. *Globe, Arizona,* Clara T. Woody & Milton L. Schwartz, Arizona Historical Society, Tucson, Arizona, 1977, p.
12. Adapted in part from "Apache survivors leaped to deaths, and the mountain wept," Lowell Parker*, Arizona Republic Newspaper,* August, 5, 1975, Phoenix, Arizona; *Anecdotes of the Early Days,* December 20, 1918, source unknown; and notes of John D. Walker files, Greg Davis collection, Tempe, Arizona.
13. *Pioneering in Arizona,* Emerson O. Stratton & Edith Stratton Kitt, Arizona Pioneers Historical Society, Tucson Arizona, 1964, P.36-37.
14. *Tales of the Superstitions.* Robert Blair, Arizona Historical Foundation, Tempe, Arizona, 1975, P. 75, and *Superstition Mountain Journal Volume 18*, "The Weiser-Walker Map", by Gregory E. Davis, Superstition Mountain Historical Society, Apache Junction, Arizona, 2000, P. 7.
15. *Globe, Arizona—The Life And Times of A Western Mining Town*, Robert Bigando, Mountain Spirit Press, Globe, Arizona, 1990, Pgs. 2-3.
16. *Globe, Arizona—The Life And Times of A Western Mining Town*, Robert Bigando, Mountain Spirit Press, Globe, Arizona, 1990, Pgs. 10-12.
17. *Arizona Municipalities* magazine*,* James M. Barney, August 1940, P.10.
18. Clara T. Woody's notes from her files at the Arizona Historical Society at Tucson.
19. *Arizona Gazette,* Phoenix, Arizona, September 17, 1875.
20. *Arizona Daily Citizen***,** April 3, 1875, Tucson, Arizona.
21. *Reminiscences of Emerson O. Stratton***,** as told to his daughter, Edith Stratton Kitt, Summer of 1925, Clara T. Woody files, Arizona Historical Society, Tucson, Arizona.
22. *Reminiscences of Emerson O. Stratton***,** as told to his daughter, Edith Stratton Kitt, Summer of 1925, Clara T. Woody files, Arizona Historical Society, Tucson, Arizona.

23. "One Hundred Years Of Yesterdays-The Silver King Mine: Its Life And Its Death," Carlotta Silvas-Martin, *Superior Centennial Report*, Superior, Arizona, 1982, P.2.
24. Clara T. Woody files, Arizona Historical Society, Tucson, Arizona.
25. *Statistics of Mines and Mining in the States and Territories West of the Rocky Mountains;* Eighth Annual Report of Rossiter W. Raymond, United States Commissioner of Mine Statistics, published in Washington, DC, 1877, pages 345-346.
26. Professor William P. Blake's 1883 account of "How the Silver King Mine Was Discovered," unpublished manuscript, New Haven, Conn., Arizona Bureau of Mines files, Phoenix, Arizona.
P. 10.
27. Professor William P. Blake's 1883 account of "How the Silver King Mine Was Discovered," unpublished manuscript, New Haven, Conn., Arizona Bureau of Mines files, Phoenix, Arizona. P. 10.
28. Hilzinger, J. George, *Treasure Lane: Arizona Advancement Co.,* Tucson, Arizona, 1897.
29. Professor William P. Blake's 1883 account of "How the Silver King Mine Was Discovered," unpublished manuscript, New Haven, Conn., Arizona Bureau of Mines files, Phoenix, Arizona. P. 10-11
30. *Globe, Arizona,* Clara T. Woody & Milton L. Schwartz, Arizona Historical Society, Tucson, Arizona, 1977, p.
31. *Tales of the Superstitions,* Robert Blair, Arizona Historical Foundation, Tempe, Arizona, 1977, P.150.
32. *The Arizona Sentinel*, Yuma, A. T. August 3, 1878.
33. *Good Man, Bad Men, Lawmen and a few Rowdy Ladies*, John A, Swearengin, Florence, Arizona, 1991, P.22-23.
34. "The Silver King Mine, An Arizona Bonanza," *Phoenix Points West*, January 1964, Phoenix, Arizona.
35. *Wells Fargo in Arizona Territory*, John & Lillian Theobald, Arizona Historical Foundation, Tempe, Arizona, 1978, P. 124.
36. *Destination Tombstone, Adventures of a Prospector,* Edward Schiefflin-Founder of Tombstone, as compiled by Marilyn F. Butler, Royal Spectrum Publishing Mesa, Arizona, 1996, P.77-82.

Chapter Two—When Silver Was King—The 1880s Silver King Mine

1. *Pinal Drill,* September 15, 1883, Pinal, Arizona Territory.
2. *Pinal Drill*, September 23, 1883, Pinal, Arizona Territory.
3. *Pinal Drill,* May 14, 1881, Pinal, Arizona Territory.
4. *Western Mining,* Otis. E. Young, Jr., University of Oklahoma Press, Norman, 1970, P. 194-196.
5. Professor William P. Blake's 1883 account of "How the Silver King Mine Was Discovered," unpublished manuscript, New Haven, Conn., Arizona Bureau of Mines files, Phoenix, Arizona. P. 14-16.
6. Professor William P. Blake's 1883 account of "How the Silver King Mine Was Discovered," unpublished manuscript, New Haven, Conn., Arizona Bureau of Mines files, Phoenix, Arizona P. 19-20.
7. Professor William P. Blake's 1883 account of "How the Silver King Mine Was Discovered," unpublished manuscript, New Haven, Conn., Arizona Bureau of Mines files, Phoenix, Arizona P.25.

8, "The Silver King," from a *Pinal Drill* reprint, September 23, 1882, from an article originally printed in the *Denver News,* 1882.

Chapter Three—Development of the Silver King Mine

1. *Blown To Bits In The Mine,* Eric Twitty, Western Reflections Pub. Co., Ouray, Colorado, 2001, p.35-36.
2. Robert Bowen family notes on the Silver King Mine. This may have been the man killed by Apaches when the Silver King was located years earlier.
3. *Mining Library- Volume V-Details of Practical Mining*, Compiled from the *Engineering and Mining Journal*, McGraw Hill Book Co., New York, NY, 1916. P. 513-514.
4. *San Diego Union,* January 19, 1877.
5. The *San Francisco Stock Report* of 1877.
6. *Arizona Citizen,* June 30, 1877.
7. *Pinal Drill Newspaper*, December 17, 1881. The language of the day was maintained to give the reader the exact flavor of the English language used in the 1880s by many editors.
8. *Pinal Drill Newspaper,* September 22, 1883, Pinal City, Arizona Territory. Report of the Silver King Mine of activities and a tour given to visiting dignitaries.and the above newspaper.
9. *Pinal County Record Newspaper,* Pinal City, Arizona Territory.
10. *Pinal County Record Newspaper* published the annual report of the Silver King Mine for 1886.

Chapter Four—Processing Ore at The Silver King Mills

1. Silver King Mine, *U.S.G.S Bulletin 540* by F. L. Ransome, and unpublished manuscript by J. B. Tenny, circa 1920s.
2. *Mining & Scientific Press,* July 20,1878, San Francisco, California, P.34.
3. *The Miners,* Time Life Books Inc., Chicago, Ill. p. 120-121.
4. *Pinal Drill,* December 17, 1881, Pinal, Arizona Territory.
5. *Mining & Scientific Press,* San Francisco, California April 28, 1883.
6. *Mining & Scientific Press,* San Francisco, California January 19, 1884.

Chapter Five—The Mining Men of the Silver King Mine

1. *Globe, Arizona—The Life And Times of A Western Mining Town*, Robert Bigando, Mountain Spirit Press, Globe, Arizona, 1990, Pgs. 16-17.
2. *Globe, Arizona—The Life And Times of A Western Mining Town*, Robert Bigando, Mountain Spirit Press, Globe, Arizona, 1990, Pgs.16-20.
3. *Western Mining,* Otis. E. Young, Jr., University of Oklahoma Press, Norman, 1970, Pg. 178.
4. "Pinal And Silver King", *Pinal County Recorder,* September 11, 1885, Pinal, Arizona Territory.
5. *Arizona's Eighties*, 1880s Anonymous File, Unpublished Manuscript, Arizona Historical Society, Tucson, Arizona.
6. *Gold & Silver Mining In Arizona, 1848-1945,* AZ State Historic Preservation Office, Melissa Keane & A. E. Rogge, Phoenix, Arizona, 1992, Pg. 112
7. *No Place for Angels, Roscoe G. Wilson's Stories of Old Arizona Days,* Arizona Republic, Phoenix, AZ and Arizona Silhouettes, Tucson, Arizona, from the files of the

Arizona Historical Society, Central Division, Tempe, Arizona, Pgs. 243-247.
8. *Blown to Bits in the Mine,* Eric Twitty, Western Reflections Publishing Company, 2001, P. 35.
9. *Western Mining,* Otis. E. Young, Jr., University of Oklahoma Press, Norman, 1970, Pgs. 183-185.
10. "The Miners", *Time Life Books Inc*., Chicago, Ill. 1976, Pgs.104-109.
11. *Territorial History of the Globe Mining District,* Unpublished Thesis by Alexander Ziede, University of Southern California, August 1939, Pgs. 87-88, Arizona Historical Society Files, Central Division, Tempe, Arizona.
12. This story comes from a descendant of John Meehan, Mike Benefield of Prinefield, Oregon. His grandmother lived at Silver King during the 1880s and was married to a Meehan. I verified this by census records.
13. *Arizona Eighties*, 1880s Unpublished Manuscript, Anonymous File, Arizona Historical Society, Tucson, Arizona.
14. *Reminiscences of an Arizona Pioneer-Personal Experiences of Albert Simpson Neighbors*, as Interviewed by Charles M. Wood, Unpublished Manuscript, April 20, 1926, 47-48, Clara M. Woody Files, Arizona Historical Society, Tucson, Arizona.
15. Jack Fraser File, Greg Davis Collection, Tempe, Arizona.
16. *Blown To Bits In The Mine,* Eric Twitty, Western Reflections Publishing Co., Ouray, Colorado, 2001, p.35.
17. For a more complete explanation of setting the fuses and blasting, see "Chapter VII, The Western Miner & His Methods," in *Western Mining* by Otis E. Young Jr and *Blown To Bits In The Mine* by Eric Twitty, "Chapter One."
18. *Arizona Eighties*, Unpublished Manuscript 1880s, Anonymous File, Arizona Historical Society, Tucson, Arizona.
19. *Arizona Silver Belt,* June 14,1884, Globe, Arizona.
20. *Rails to Carry Copper,* Gordon Chappell, Pruett Publishing Company, Boulder, Colorado, 1973, Pg. 4.
21. *Arizona Municipalities* magazine, James M. Barney, August 1940, Pg. 26.

Chapter Six—Mule Skinners, Freighters and Cowboys

1. *Pinal Drill,* Pinal, Arizona Territory 1883 Pg.9-15.
2. *Arizona Graphic, Dec. 23, 1899, Vol. 1, No. 15*
3. *An Arizona Wagon Freight Train of 1877,* Unpublished Manuscript, Geraldine Craig File, Arizona Historical Society, Tucson, Arizona.
4. *Pinal Drill,* 1883.
5. Marshall Trimble, *Arizona-A Calvacade of History,* Treasure Chest Publications, Tucson, Arizona, Pgs. 95-96
6. *Forgotten Towns of Arizona,* James M. Barney, Arizona Municipalities, Phoenix, Arizona, 1940.
7. Nell Murbarger, *Ghosts of the Adobe Walls*, Westernlore Press, Los Angeles, California, 1964 pg. 174.
8. Pinal Drill, July 23, 1881, Pinal, Arizona Territory.
9. *A Pack Train,* Unpublished Manuscript, Geraldine Craig File, Arizona Historical Society, Tucson, Arizona.
10. "Life At Old Silver King Was Primitive, But Zest Filled," Georgia Moore, Superior 1882-1982 Centennial Brochure, Pg. 14, Gladys Walker Files, Superior, Arizona.
11. "Globe's Colorful Pioneer Days", *Arizona Record* March 29, 1956, Globe, Arizona,

Clara T. Woody, Pg. 6.
12. *Pre-Historic and Historic Gila County,* Dan Rose, Republic and Gazette Printery, Phoenix, Arizona, 1935. Pg.
13. *Unpublished Reminiscences of Perry Wildman,* a merchant at Silver King and Pinal City in the 1880s, as told at Pioneer's meeting, Dec. 29, 1926, Arizona Historical Society, Tucson, Arizona.
14. *Pinal* Drill, September 30, 1882, Pinal. Arizona Territory.
15. "Pinals Boom And Bust," *Arizona Republic Sunday Magazine*, July 18, 1965, Phoenix, Arizona.
16. Jack Carlson provided assistance with this complicated process, of determining ownership of the old ranch, during 2003.
17. *Matt Cavaness Memoirs*, Superstition Mountain Historical Society, Apache Junction, Arizona.
18. *Arizona Daily Star*, "Matt Caveness Tells of Balmy Days of *Pineal*, September 9, 1910.
19. *PinalCounty Record,* Pinal Arizona Territory, February 2, 1886.
20. *Arizona Municipalities* magazine, James M. Barney, August 1940, *Pg. 26.*
21. *The Brand Books of the 1880s of the Arizona Livestock Sanitary Board*, Phoenix, Arizona & the State Archives Library of Arizona, Phoenix.
22. *Superstition Mountain – A Ride Through Time*, James A. Swanson and Tom Kollenborn, Arrowhead Press, Phoenix, Arizona, 1981, Pg. 107.
23. *Arizona Cattle Log*, January 1967, Phoenix, Arizona.
24. *Emigrant Knight of Cornwall*, Ann Knight Rose, in Collaboration with Gladys Peadon, self-published 1991, Pgs. 33-65, Monte Vista, Colorado.
25. *Arizona Cowboys*, Dane Coolidge, U. of Arizona Press, Tucson, Arizona.
26. *Two G. W. Coles-Pioneers in the West*, Dick and Judy Chamberlain, Flournay, California, 1990, Pg, 19-21, and an unpublished manuscript by George Drakovich, circa 1924, Pg. 5, from the Gladys Walker Collection, Superior, Arizona.

Chapter Seven—Life at Silver King Town and Pinal City
1. *Pinal Drill.* July 7, 1883, Pinal, Arizona Territory.
2. *Pinal Drill*, August 8, 1883, Pinal, Arizona Territory.
3. "Life At Old Silver King Was Primitive, But Zest Filled," *Georgia Moore, Superior 1882-1982 Centennial Brochure, P.14,* Gladys Walker Files, Superior, Arizona.
4. "History of Grazing on the Tonto," a lecture by Senior Forest Ranger Fred W. Croxen, at the Tonto Grazing Conference in Phoenix, Arizona, November 4-5, 1926.
5. *One Hundred Yesterdays,* W. Earl Merrill, self-published, Mesa, Arizona 1972, Pages 31-32.
6. *Pinal County Record,* Dec. 25, 1885, Pinal City, Arizona Territory.
7. *Pinal Drill,* December 25, 1880, and Jan. 1, 1880, Pinal, Arizona Territory.
8 *Pinal Drill*, August 13, 1881, Pinal City, Arizona Territory.
9. *Arizona Weekly Enterprise,* September 5, 1885, Florence, Arizona Territory.
10. *Reminiscences of Perry Wildman*, Arizona Historical Society Files, Tucson, Arizona.
11. *Pinal Drill*, Sept. 9, 1882, Pinal, Arizona Territory.
12. *Pinal Drill,* February 26, 1881, Pinal, Arizona Territory.
13. *Emigrant Knight Of Cornwall,* Ann Knight Rose, published in 1991, Monte Vista, Colorado, p.12-13.
14. *Tales of the Silver King Mine*, unpublished manuscript by Charles P. Mason, nephew

of one of the original owners, circa 1900, Bureau of Mines and Mining Museum, Silver King File, Phoenix, Arizona.
15. *Unpublished manuscript by Perry Wildman*, Gladys Walker collection, Superior, Arizona.
16. Reminiscences of Perry Wildman, as told to the Arizona Pioneer Historical Society on Dec. 29, 1926, P.1-6, *Arizona Historical Society Files, Tucson, Arizona.*
17. *Reminiscences of Perry Wildman*, as told to the Arizona Pioneer Historical Society on Dec. 29, 1926, P.1-6, Files of the Arizona Historical Society, Tucson, Arizona.
18. *Pinal Drill*, August 18, 1883, Pinal, Arizona Territory.
19. *Pinal Drill,* August 25, 1883, Pinal, Arizona Territory.
20. *Reminiscences of Perry Wildman*, as told to the Pioneer's Meeting, Dec. 2., 1926, Files of the Arizona Historical Society, Tucson, Arizona, Pg. 1.
21. Recollections of a Mr. McPherson from an undated Term Paper by George Drakovich, from the Gladys Walker Private Collection, Superior, Arizona.
22. *Pinal County Record*, February 19, 1886, Pinal, Arizona Territory.
23. William L. Wardwell Letters, Arizona Historical Society, Tucson, Arizona, letter dated January 25, 1886.
24. William L. Wardwell Letters, Arizona Historical Society, Tucson, Arizona, letter dated January 25, 1886.
25. William L. Wardwell Letters, Arizona Historical Society, Tucson, Arizona, letter dated February 1, 1886.
26. William L. Wardwell Letters, Arizona Historical Society, Tucson, Arizona, letter dated February 1, 1886.
27. William L. Wardwell Letters, Arizona Historical Society, Tucson, Arizona, letter dated February 1, 1886.
28. William L. Wardwell Letters, Arizona Historical Society File, Tucson, Arizona February 15, 1886.
29. William L. Wardwell Letters, Arizona Historical Society File, Tucson, Arizona, February 15, 1886.
30. *Arizona Weekly Enterprise,* April 16, 1887, Florence, Arizona Territory.
31. *Pinal Drill,* May 28, 1881, Pinal, Arizona Territory.
32. "One Hundred Years Of Yesterdays- The Silver King Mine: Its Life And Its Death," Carlotta Silvas Martin, *Superior Centennial repport, Superior, Arizona 1982*, P.2.
33. *Arizona Business Directory and Gazetteer,* W. C. Turnell Pub., Bacon & Co. Printers, San Francisco, California, 1881, Pg. 156.
34. *Pinal Drill,* July 15, 1882, Pinal, Arizona Territory.
35. *An Arizona Wagon Train of 1877*, Unpublished Manuscript, Geraldine Craig File, Arizona Historical Society, Tucson, Arizona.
36. *Tucson Daily Citizen,* May 28, 1930 " One Time Busy Little City of Pinal Fades Away as Silver Mine Gives Out."
and *Ghosts of Adobe Walls,* Nell Murbarger, Westernlore Press, Los Angles, 1964, Pgs. 171-172.
37. Nell Murbarger, *Ghosts of the Adobe Walls*, Westernlore Press, Los Angeles, California, 1964 pg. 171-2. Also the *Pinal Drill* 1883.
38. *Arizona Cowboys*, Dane Coolidge, University of Arizona Press, and Tucson, Arizona, 1984, p18-19.
39. Unpublished Reminiscences of Perry Wildman, a merchant at Silver King and Pinal City in the 1880s, as told at Pioneer's meeting, Dec. 29, 1926, Arizona Historical Society,

Tucson, Arizona.
40. *Pinal Drill,* September 15, 1883 Pinal, Arizona Territory.
41. *Pinal Drill,* September 15, 1883 Pinal, Arizona Territory.
42. *Pinal Drill,* June 16, 1883,Pinal, Arizona Territory.
43. "One Hundred Years Of Yesterdays- The Silver King Mine: Its Life And Its Death," Carlotta Silvas Martin, *Superior Centennial Report, Superior, Arizona, 1982*, P.2.
44. *Pinal Drill,* August 6, 1881, Pinal, A. T.
45. *Pinal Drill,* August 27, 1881, Pinal, Arizona Territory.
46. "The Town History Forgot," Lee Lucas, *Desert Magazine*, February 1968.
47. *Sheriff Thompson's Day,* Jess G. Hayes, University of Arizona Press, Tucson, Arizona, 1968, P. 172-178. This book was based on the career of John Henry Thompson, a frontier lawman from Gila County.
48. Recollections of a Mr. McPherson, from an undated Term Paper by George Drakovich, from the Gladys Walker Private Collection, Superior, Arizona.
49. *When All Roads Led to Tombstone, A Memoir by John Pleasant Gray,* Edited and Annotated by W. Lane Rogers, Tamarack Books, inc., Boise, Idaho, 1998, p.85-86.
50. *Pinal Drill,* February 26, 1881, Pinal, Arizona Territory.
51. *Pinal Drill,* February 12, 1881, Pinal, Arizona Territory.
52. *Pinal Drill*, March 12, 1881, Pinal, Arizona Territory.
53. *History of Mining in Arizona,* from *Chapter 8, The History of Mining at Superior* by Gladys Walker and T. G. Chilton, published by the Mining Club of the Southwest Foundation and American Institute of Mining Engineers, Tucson, Arizona, 1991, Pg. 231.
54. *Pinal Drill,* April 9, 1881, Pinal, Arizona Territory.
55. *Ghost Towns of Arizona,* James E. & Barbara H. Sherman, University of Oklahoma Press, Norman, Oklahoma, 1974, Pg. 126.
56. *Arizona Place Names,* Will C. Barnes, University of Arizona Bulletin No. 2., University of Arizona, Tucson, Arizona, 1935, Pgs. 200 & 352.
57. *Rails to Carry Copper*, Gorden Chapel, Pruett Publishing Company, Boulder, Colorado, 1973, Pg. 5.
58. "One Hundred Years of Yesterday-The Silver King Mine: Its Life and Its Death," Carlotta Silvas-Martin, *Superior, Arizona Centennial Report, 1982*, P.1.

Chapter Eight—Pioneer Women of Silver King and Pinal

1. *PreHistoric and Historic Gila County Arizona,* Dan Rose, Republic and Gazette Printery, Phoenix, Arizona, and Trail to Yesterday Books, Tucson, Arizona, 1975, Pg. 22.
2. *Pinal Drill,* October 16,1880, Pinal, Arizona Territory.
3. *Pinal Drill,* October 7, 1882, Pinal, Arizona Territory.
4. *Reminiscences of Mary E. Bailey*, unpublished manuscript by Abby Overly, 1937, Arizona Historical Society, Tucson, Arizona.
5. *Pinal Drill* of September 8, 1883 Pinal, Arizona Territory.
6. Information from the family of Harry Thompson, whose mother lived at Silver King.
7. Information from the family of Harry Thompson, whose mother lived at Silver King.
8 *Mrs. Robert A. Irion,* Unpublished Manuscript, Geraldine Craig File, Arizona Historical Society, Tucson, Arizona.
9. *Engineering and Mining Journal—Letters From the West,* February 18, 1882, Vol. 33, Pgs. 91-92.
10. *The Earp Brothers of Tombstone*, Frank Waters, C.N. Potter, NY, NY, 1960 p. 224.
11. *I Married Wyatt Earp, The Recollections of Josephine Sarah Marcus Earp,* Edited by

Glen Boyer, University of Arizona Press, Tucson, Arizona, 1993, Pg. 43-44; *The Earp Brothers of Tombstone*, Frank Waters, University of Nebraska Press, Lincoln, Nebraska. 1976, Pg. 209-213; *The Illustrated Life and Times of Wyatt Earp,* Bob Boze Bell, Boze Books, Cave Creek, Arizona, 1993, Pg. 97; *Superior, Arizona Centennial Report of 1982*, " One Hundred Years of Yesterdays, The Silver King Mine: Its Life and Death," Carlotta, Silvas-Maria, p. 3; *The Earp Brothers Of Tombstone,* Frank Waters, C. N. Potter, NY, NY, 1960; and *Wyatt Earp-Frontier Marshall,* Stuart Lake, Houghton Mifflin, NY, NY, 1931.

Chapter Nine—A Pioneer Family—The Bowens of Silver King

1. Personal Recollections of the Bowen family history as compiled by Velma Bowen Tucker of Tempe, Arizona, circa 1990.
2. *Pinal Drill,* September 29, 1883, Pinal, Arizona Territory.
3. Personal Recollections of Harry Brown included in the Bowen family history as compiled by Velma Bowen Tucker of Tempe, Arizona, circa 1990.

Chapter Ten—Stagecoaches, Bullion & Bullets

1. *Anonymous File-Arizona Eighties,* Arizona Historical Society Tucson, Arizona.
2. *Wells Fargo in Arizona Territory,* John & Lillian Theobald, The Arizona Foundation, Tempe, Arizona, 1978.
3.*Pinal Drill* and *Pinal County Record*, 1881-1886, Sam Michael Collection, Mesa, Arizona.
4. *Wells Fargo in Arizona Territory,* Arizona Historical Foundation, Tempe, Arizona, 1978, Pg. 62.
5. *Pinal Drill*, March 3-10, 1883, Pinal, Arizona Territory.
6. *Arizona Silver Belt*, February 28, 1885, Globe, Arizona Territory.
7. *Pinal County Record*, December 25, 1885, Pinal, Arizona Territory.
8. *Pinal County Record*, November 20, 1885, Pinal, Arizona Territory.
9. *Pinal County Record*, November 27, 1885, Pinal, Arizona Territory.
10. *Pinal County Record,* December 25, 1885, Pinal, Arizona Territory.
11. *Pinal County Record*, January 22, 1886, Pinal, Arizona Territory.
12. *Pinal County Record*, July 9, 1886, Pinal, Arizona Territory.
13. *Arizona Eighties*, Anonymous Files, Arizona Historical Society, Tucson, Arizona.
14. *Globe, Arizona—The Life And Times of A Western Mining Town*, Robert Bigando, Mountain Spirit Press, Globe, Arizona, 1990, Pgs. 37-38.
15. *Pinal Drill,* December 17, 1881, Pinal, Arizona Territory.
16. *Pinal Drill*, July 7, 1881, Pinal, Arizona Territory.
17. *Wells Fargo in Arizona Territory*, Arizona Historical Foundation, Tempe, Arizona, 1978, Pg. 51.
18. *Wells Fargo in Arizona Territory*, Arizona Historical Foundation, Tempe, Arizona, 1978, Pgs.53-54.
19. *Frontier Times Magazine*, Western Publications, Austin, Texas, Summer 1960, Pg. 52.

Chapter Eleven—Frontier Justice and the Outlaws

1. *Prehistoric and Historic Gila County Arizona,* Dan Rose, Republic and Gazette Printery, Phoenix, Arizona, 1935, Pg. 19.
2. *Pinal Drill*, July 16,1881, Pinal, Arizona Territory.
3. *Pinal Drill,* July 16, 1881, Pinal, Arizona Territory.
4. *The Lost Peralta-Dutchman Mine,* Walter Gassler, 1983, published by the Superstition

Mountain Historical Society as a previous unpublished historical document, Goldfield, Arizona, Pg. 46-47.
5. *Pinal Drill,* January 8, 1881, Pinal, Arizona Territory.
6. *Pinal Drill,* March 4, 1882, Pinal, Arizona Territory.
7. *Unpublished Manuscript of Charles P. Mason,* as told to Mrs. George F. Kitt, 1924, Arizona Historical Society, Tucson, Arizona, Pg. 1.
8. *Pinal Drill,* December 26, 1882, Pinal, Arizona Territory.
9. *Wells Fargo In Arizona Territory,* John & Lillian Theobald, Arizona Historical Foundation, Tempe, Arizona, 1978, Pg. 85-86.
10. *Tales of the Silver King Mine,* unpublished manuscript, Charles P. Mason, Circa 1924, Arizona Historical Foundation, Hayden Library, ASU, Tempe, Arizona.
11. *Forgotten Towns of Arizona,* James M. Barney, Arizona Municipalities, Phoenix, Arizona, August 1940, Pg. 26; and *Ghosts of the Adobe Walls,* Nell Murbarger, Westernlore Press, Los Angeles, California, 1964, Pg. 173; and *Reminiscences of Perry Wildman,* as told at Pioneers' Meeting, December 29, 1926, Arizona Historical Society, Tucson, Arizona.
12. Dr. James Larr, in *Popular Science Monthly* August 1885, as published in the *Pinal County Record,* September 11, 1885, Pinal, Arizona Territory.
13. *Pinal County Record,* September 17, 1886, Pinal, Arizona Territory.
14. *Globe, Arizona-The Life and Times Of A Western Town,* by Robert Biando, Mountain Spirit Press, Globe, Arizona, 1990, Pg.37.
15. *Sheriff Thompson's Day,* Jess G.Hayes, University of Arizona Press, Tucson, Arizona, 1968, P. 40.
16. *Arizona-Prehistoric, Aboriginal, Pioneer, Modern,* Vol. II, by James H. Mclintock, Clarke Publishing Co., Chicago, Illinois, 1916, Pgs. 460-461; *August Murder,* Unpublished Manuscript, Clara T. Woody Collection, Arizona Historical Society, Tucson, Arizona, Box 5; Wells *Fargo in Arizona Territory,* John and Lillian Theobald, Arizona Historical Foundation, Tempe, Arizona, 1978, Pg. 63; *Arizona Gazette,* August 31, 1882, Phoenix, Arizona Territory; *Pinal Drill,* October 21, 1882, and *Pinal Drill,* October 27, 1883, Pinal, Arizona Territory.
17. *The Territorial History of the Globe Mining District,* Unpublished Thesis, Alexander Zeide, U. of Southern California, August 1939, Arizona Historical Society Library, Central Location, Tempe, Arizona.
18. *The Territorial History of the Globe Mining District,* Unpublished Thesis, Alexander Zeide, U. of Southern California, August 1939, Arizona Historical Society Library, Central Location, Tempe, Arizona.
19. Joe Deen's grandfather Mack Deen later owned the Riverside Stage station. Information from interview with Joe Deen.
20. *Tales of the Silver King Mine,* Unpublished manuscript by Charles P. Mason, Bureau of Mines Files, Phoenix, Arizona.
21. *Arizona Legends and Lore*, Dorothy Daniels Anderson, Golden West Publishers, Phoenix, Arizona, 1993, Pg. 140.
22. *Arizona Legends and Lore*, Dorothy Daniels Anderson, Golden West Publishers, Phoenix, Arizona, 1993, Pg. 140.
23. *Pinal Drill,* August 25, 1883, Pinal, Arizona Territory.
24. *Pinal Drill*, September 8, 1883, Pinal, Arizona Territory.
25. *Tales of the Silver King Mine,* unpublished manuscript, Charles P. Mason, Circa 1924, Arizona Historical Foundation, Hayden Library, ASU, Tempe, Arizona P. 137.

26. *Pinal Drill,* November 24, 1883, Pinal, Territory; *Arizona Weekly Enterprise,* August 10 and 18, and September 8, 1883, Florence, Arizona Territory; *Tales of Arizona Territory,* Charles D. Laurer, Golden West Publishing Company, Phoenix, Arizona, 1990; and *Wells Fargo in Arizona Territory,* John and Lillian Theobald, Arizona Historical Foundation, Tempe, Arizona, 1978. "Red Jack" Elmer's last name was also spelled Almer. He was also known as "Red Jacket" and used other names.
27. *Arizona Republic*, June 13 1935 Phoenix, Arizona, 10.
28. *Arizona Enterprise*, September 25 1885 Prescott, A. T.
29. *Arizona Enterprise,* August 15, 1885, Prescott, Arizona Territory.
30. *Arizona Enterprise,* November 20, 1886, Prescott, Arizona Territory.
31. *Tales of Arizona Territory,* Charles D. Laurer, Golden West Publishers, Phoenix, Arizona 1995, Pg. 91.
32. *Arizona Enterprise* June 2, and June 9, 1888, Florence, Arizona Territory.
33. *Florence Tribune,* August 6, 1898, Florence, Arizona Territory.
34. *Arizona Gazette*, June 12, 1936, Phoenix, Arizona, p.8.
35. *Arizona: Pre-historic-Aboriginal-Pioneer-Modern, Vol. II,* James H. McClintock, S. J. Clarke Publishing Co., Chicago, Ilinois, 1916, Pg. 469; and *Reminiscences of Mike Rice,* Unpublished manuscript, Arizona Historical Society, Tucson, Arizona.
36. *Wells Fargo in Arizona Territory,* John and Lillian Theobald, Arizona Historical Foundation, Hayden Library, Tempe, Arizona, Pgs. 115-117.
37. *The Peralta-Reavis Land Grant*, from the Robert Blair files of the Arizona Historical Foundation, Tempe, Arizona; *The Baron of Arizona*, E.H. Cookridge, John Day Co. NY 1967; *The Peralta Grant, James Addison Reavis and the Barony of Arizona*, Donald M. Powell, University of Oklahoma, 1960.
38. *Sheriff Thompson's Day*, Jess G.Hayes, University of Arizona Press, Tucson, Arizona, 1968, P. 36-37.
39. *The Clantons of Tombstone,* Ben T. Traywick, Red Maries Book Store, Tombstone, Arizona, 1996, Pgs.185-218.
40. *Phineas T. Booglebrand*, a poem by Dan Wright, Superior, Arizona, 1996.

Chapter Twelve—Tragic Tale of the Lost Soldiers Mine

1. The Lost Soldiers story was compiled as a result of ten years research. Various sources include: Unpublished Manuscript— *Story of the Lost Dutchman,* by George "Brownie" Holmes, Phoenix Arizona, 1944, P. 18; *Jim Barks Unpublished Notes; Tales of the Superstitions* by Robert Blair; *The Lost Dutchman Mine* by Sims Ely; *Curse of the Dutchman's Gold* by Helen Corbin; and Tom Kollenborn's News Articles in the *Apache Sentinel.* I also found numerous other sources and unpublished manuscripts of old timers of Silver King and Pinal City of the 1880s that relate the story of the two soldiers.
2. "Dutch John Pipps and the Lost Dutchman Gold Mine," *Superstition Mountain Historical Society Journal, Volume 13, 1995,* by Gregory E. Davis; "John Pipps and the Two Soldiers Lost Mine," *Apache Signal*, August 28, 1985, by Tom Kollenborn, Apache Junction, Arizona, P. 3; *Phoenix Daily Herald,* September 15, 1886, Phoenix, Arizona Territory, P. 3. and *Arizona Silver Belt*, October 9, 1886, Globe, Arizona Territory.
3. *The Curse of the Dutchman's Gold,* Helen Corbin, Foxwest Publishing, Phoenix, 1990, p. 144; *The Pinal County Record*, October 2, 1885, Pinal, Arizona Territory; I also have a copy of the coroner's inquest, filed October 23, 1885, in the Justice Court of Pinal County, Arizona.
4. John Chuning is listed in the Pinal County Register as being a resident of Pinal City

during the 1880s; Helen Corbin published a copy of Chuning's death certificate in her book, *The Curse of the Dutchman's Gold*, on page 166.
5. Unpublished Manuscript- *Story of the Lost Dutchman*, by George "Brownie" Holmes, Phoenix Arizona, 1944.
6. *The Sterling Legend*, Estee Conatser, Gem Guides Book Company, Pico Rivera, California, 1972, P. 37..
7. I personally examined the Robert Bowen notes. They contained the same basic story as is contained in this chapter. Also Velma Bowen Tucker recalled the story told by her father, Thomas Bowen, as it was related to him by his father Robert Bowen, who was at the Silver King Mine when the soldiers came to the mine.

Chapter Thirteen—Silver King Mine's Impact on Nearby Mining
1. *Pinal Drill* August 8, 1883, Pinal, Arizona Territory.
2. *Geology And Ore Deposits of The Superior Mining Area, Arizona,* by M. N. Short, F. W. Galbraith, E. N. Harshman, T. H. Kuhn, and Eldred D. Wilson, *Arizona Bureau of Mines, Geological Series, No. 16, Bulletin No. 151,* Published by University of Arizona, Tucson, Arizona, October, 1943, p. 142.
3. *Pinal Drill,* February 4, 1882, Pinal, Arizona Territory.
4. *Pinal Drill,* September 11, 1880, Pinal, Arizona Territory.
5. *Pinal Drill*, August 1, 1883, Pinal, Arizona Territory.
6. *Pinal Drill, August 22, 1883, Pinal, Arizona Territory.*
7. *Pinal Drill*, August 1, 1883, Pinal, Arizona Territory.
8. Pinal Drill, *August 1, 1883,* Pinal, Arizona Territory.
9. *Mining and Scientific Press*, July 16, 1887.
10. *Pinal Drill*, March 11, 1882, Pinal, Arizona Territory.
11. *Pinal Drill*, March 11, 1882, Pinal, Arizona Territory.
12. *Pinal Drill,* February 4, 1882, Pinal, Arizona Territory.
13. *Mining and Scientific Press*, July 16, 1887.
14. *Pinal Drill*, September 11, 1880, Pinal, Arizona Territory.
15. *Pinal Drill,* September 8, 1883, Pinal, Arizona Territory.
16. *Pinal Drill,* March 10, 1883, Pinal, Arizona Territory.
17. *Rock to Riches*, Charles H. Dunning, and Southwest Publishing Company Phoenix, Arizona 1959, Pages 379-380.
18. *Florence Blade Tribune,* August 3, 1951, Florence, Arizona, and January 1997 *Arizona Highways* magazine, on pages 38-43.
19. *Shaft Furnaces Beehive Ovens,* Lynn R. Bailey, Westernlore Press, Tucson, Arizona 2002, P. 199-205; *Pinal Drill*, July 9, 1881; and the *Mining and Scientific Press*, June 25, 1881.
20. *The History of Mining at Superior,* Gladys Walker and T. G. Chilton, 1991, p. 231.
21. *Geology And Ore Deposits Of The Superior Mining District,* 1943, Arizona Bureau of Mines and Mining Museum files.
22. *Globe Arizona,* Clara T. Woody & Milton Schwartz, Arizona Historical Society, Tucson, Arizona, 1977, p. 17-
23. Notes of Gladys Walker, Superior, Arizona Historian, April 2002.
24. *Railroads of Arizona, Vol. 11*, David F. Myrick, Howell-North Books, San Diego, California, Pgs. 687-694.
25. *FlorenceTownsite A. T.,* Industrial Development, Authority of the Town of Florence, Inc., 1976 p. 10-11.

26. *FlorenceTownsite A. T*—Industrial Development Authority of the Town of Florence, Inc., 1976 Pgs. 13-15.
27. *Railroads of Arizona, Vol. 11,* David F. Myrick, Howell-North Books, San Diego, California, Pgs. 635-636.
28. *Pinal Drill,* November 11, 1881, Pinal, Arizona Territory.
29. *Gila Centennials Historical Celebration and Pageant Booklet*, edited by Nick Ragus, Frances Gerhardt, John Woody, Jimmie Fulton, Ester Preston and Clara T. Woody, Globe, Arizona, 1976, P. 17-18.
30. *History of Mining in Arizona, "Chapter 2, Arizona's Silver Belt"* by Wilbur A. Haak, Mining Club of the Southwest Foundation, Tucson, Arizona 1991, P. 33.
31 *History of Mining in Arizona, "Chapter 9, History of Smelting in Arizona*" by Forrest R. Ricard, Mining Club of the Southwest Foundation, Tucson, Arizona, 1987, P. 196-7.
32. *Gila Centennials Historical Celebration and Pageant Booklet*, edited by Nick Ragus, Frances Gerhardt, John Woody, Jimmie Fulton, Ester Preston and Clara T. Woody, Globe, Arizona, 1976, P. 21.
33. *Al Sieber Chief Of Scouts,* Dan Thrapp, University of Oklahoma Press, Norman, OK, 1995, P. 384-389

Chapter Fourteen—The Death and Re-births of the Silver King Mine

1. *Arizona Republic,* March 21, 1895, Phoenix, A. T.
2. *Arizona Republic* August 23, 1905, Phoenix, A. T.
3. *Arizona Republic* on March 7, 1907, Phoenix, A. T.
4. *Medicine in Territorial Arizona,* Frances E. Quebbeman, Arizona Historical Foundation, Phoenix, Arizona, 1966, P. 221-225.
5. *Arizona Silver Belt,* Globe, Arizona Territory, December 23, 1908, Page 6, and December 24, 1908, Page 1.
6. *Arizona Silver Belt,* November 4, 1913, Globe, Arizona Territory, p.2.
7. *Analytical Report, Silver King of Arizona Mining Company,* H. D. Wells & Company, New York City, NY 1916.
8. *Silver King Mining Company Report of 1920*, Arizona Bureau of Mines and Mining Museum files, Phoenix, Arizona.
9. *Superior Sun*, Superior Arizona, February25, 1921, P.2.
10. *Arizona Republican,* March 19 1935, and *Arizona Gazette* April 8, 1935, Phoenix, Arizona, p. 2.
11. *Official Letter the Silver King Company of 1924*, files of the Bureau of Mines and Mining Museum, Phoenix, Arizona.
12. *Arizona Republic*, May 13. 1935, Phoenix, Arizona, P.3.
13. *Casa Grande Dispatch,* February 8, 1940, Casa Grande, Arizona.
14. Department of Mineral Resources Field Engineers Report October 22, 1959, Bureau of Mines and Mining Museum, Phoenix, Arizona.
15. Mineral Resources of Arizona Field Engineers Report February, 25, 1959 Bureau of Mines and Mineral Museum files, Phoenix, Arizona.
16."Historic Silver King Mine May Yet Be Rich In Ore," *Superior Sun,* November 22, 1962, Superior, Arizona, and "Silver King's Queen", *Tucson Daily Citizen,* November 6, 1972, Tucson, Arizona.
17. "Guzman's Feat is the All-American success story," *Arizona Silver Belt,* April 16, 1981, Globe, Arizona, p.7, and a letter from Wade Church's Law Office, dated November 19, 1986, Phoenix, Arizona.

Chapter Fifteen— In Search of History Camp Picket Post to Camp Pinal

1. *Early Military Posts in Arizona,* unpublished manuscript, Clara T. Woody File, Arizona Historical Society, Tucson, Arizona, P. 2.
2. *Al Sieber- Chief of Scouts,* Dan Thrapp, University of Oklahoma Press, Norman, Oklahoma, 1995, p. 143.
3. Jack Carlson is an avid hiker and one of the best trail locators that I have met. Jack, along with Elizabeth Stewart co-authored the book, *Hiker's Guide to the Superstition Wilderness,* and I knew that if anyone could locate this mystery photo location, Jack could.
4. W*ithin Adobe Walls, 1877-1973***,** Helen B. Craig, Phoenix, Arizona, 1975, P. 56-60.
5. Recollections of a Mr. McPherson from an undated Term Paper by George Drakovich, from the Gladys Walker collection, Superior, Arizona.
6. Adapted from a story called *The Swift Trail* by Mary Bagwell, Superior, Arizona, date unknown.
7. *"History, Geology, and Vegetation of Picket Post Mountain"***,** Frank S. Crosswhite, article in *Desert Plants***,** a booklet published by the University of Arizona for the Boyce Thompson Southwestern Arboretum, Vol. 2, 1984, P. 73-77.

Chapter Sixteen—1996 Silver King Reopening by the Deens

1. Interview with Joe Deen 1997.
2. Interview with Ron Deen Sr. at the mine 1997.
3. Interview with Joe Deen and Ron Deen Sr. 1998.
4. Interview with Dick Bynum 2000.
5. Interview with George Sites 2000.
6. Interview with Antonio Cisneros March 21, 1997. Cisneros has since passed away but his claims are still active and belong to the family.
7. U.S. Court of Appeals for the Ninth Circuit, USA V. SHUMWAY, Case Number 96-16480, date filed 12/28/99. (This landmark case can be found on the Internet site http;www.ce9.uscourts.gov)

Bibliography

Published Books

Chapter One—Discovery of The Silver King Mine

Agnew, S. C, *Garrisons of the Regular U.S. Army, Arizona 1851-1899*, Council on Abandoned Military Posts, P. O. Box 171, Arlington, Virginia, 1974.

Anderson, Dorothy Daniels, *Arizona Legends and Lore*, Golden West Publishers, Phoenix, Arizona, 1993.

Brand Books of the 1880s of the Arizona Livestock Sanitary Board, Phoenix, Arizona, and the State Archives Library of Arizona, Phoenix.

Bailey, Lynn R., *Shaft Furnaces Beehive Kilns*, Westernlore Press, Tucson, Arizona 2002.

Barnes, Will C., *Arizona Place Names*, University of Arizona Bulletin No. 2., University of Arizona, Tucson, Arizona, 1935.

Bell, Bob Boze, *The Illustrated Life and Times of Wyatt Earp*, Boze Books, Cave Creek, Arizona, 1993

Bigando, Robert, *Globe, Arizona—The Life And Times of A Western Mining Town*, , Mountain Spirit Press, Globe, Arizona, 1990.

Bird, Allan G., *Silverton Gold*, self-published, Silverton, CO, 1999.

Blair, Robert,*Tales of the Superstitions.*, Arizona Historical Foundation, Tempe, Arizona, 1975.

Blevin, Win s *Dictionary of the American West,* Sasquatch Books, Seattle, Washington, 2001

Boyer, Glen, *I Married Wyatt Earp, The Recollections of Josephine Sarah Marcus Earp,* University of Arizona Press, Tucson, Arizona, 1993.

Butler, Marilyn F., compiled *Destination Tombstone, Adventures of a Prospector,* Edward Schiefflin-Founder of Tombstone, , Royal Spectrum Publishing, Mesa, Arizona, 1996.

Carlson, Jack and Elizabeth Stewart, *Hiker's Guide to the Superstition Wilderness,* Clear Creek Publishing, Tempe, Arizona, 1995.

Chamberlain, Dick and Judy, *Two G. W. Coles-Pioneers in the West*, Flournay, California, 1990.

Chappell, Gordon, *Rails to Carry Copper,* Pruett Publishing Company, Boulder, Colorado, 1973.

Cookridge, E.H., *The Baron of Arizona*, John Day Co. New York, NY, 1967.

Coolidge, Dane, *Arizona Cowboys*, University, of Arizona Press, Tucson, Arizona.

Conatser, Estee, *The Sterling Legend*, Gem Guides Books, Pico Rivera, California, 1972.

Corbin, Helen, *The Curse of the Dutchman's Gold,* Foxwest Publishing, Phoenix, 1990.

Craig, Helen B., W*ithin Adobe Walls, 1877-1973,*self-published, Phoenix, Arizona, 1975.

Denver Equipment Company, *Mineral Processing Flowsheets*, Denver, Colorado, 1962.

Dunning, Charles H., *Rock to Riches*, Southwest Publishing Company Phoenix, Arizona 1959.

Ely, Sims, *The Lost Dutchman Mine,* William Morrow Company, New York, NY. 1953.

Fetchenhier, Scott, *Ghosts and Gold - The History of the Old Hundred Mine*, self-published, Durango, CO, 1999.

Haak, W. A., *Copper Bottom Tales,* Gila County Historical Society, Globe, Arizona, 1991.

Haak, Wilbur A., *History of Mining in Arizona*, "Chapter 2, Arizona's Silver Belt," Mining Club of the Southwest Foundation, Tucson, Arizona 1991.

Hayes, Jess G., *Sheriff Thompson's Day,* University of Arizona Press, Tucson, Arizona, 1968.
Hilzinger, J. George, *Treasure Lane,* Arizona Advancement Co., Tucson, Arizona, 1897.
Industrial Development, Authority of the Town of Florence, Inc., *FlorenceTownsite A. T.,* 1976.
Keane, Melissa & A. E. Rogge, *Gold & Silver Mining In Arizona, 1848-1945,* Arizona State Historic Preservation Office Phoenix, Arizona, 1992.
Lake, Stuart, *Wyatt Earp-Frontier Marshall,* Houghton Mifflin, New York, NY, 1931.
Laurer, Charles D., *Tales of Arizona Territory,* Golden West Publishing Company, Phoenix, Arizona, 1990.
McGraw Hill Book Co., *Mining Library- Volume V-Details of Practical Mining,* compiled from the *Engineering and Mining Journal*, New York, NY, 1916.
Mclintock, by James H., *Arizona-Prehistoric, Aboriginal, Pioneer, Modern,* Vol. II, , Clarke Publishing Co., Chicago, Illinois, 1916.
Merrill, W. Earl, *One Hundred Yesterdays,* self-published, Mesa, Arizona 1972.
Meyerriecks, Will, *Drills And Mills, Precious Metal Mining Methods of the Frontier West*, self-published, Tampa, FL, 2003.
Myrick, David F., *Railroads of Arizona, Vol. 11*, Howell-North Books, San Diego, California.
Murbarger, Nell*, Ghosts of the Adobe Walls,* Westernlore Press, Los Angeles, California, 1964.
Powell, Donald M., *The Peralta Grant, James Addison Reavis and the Barony of Arizona,* University of Oklahoma, 1960.
Quebbeman, Frances E., *Medicine in Territorial Arizona,* Arizona Historical Foundation, Phoenix, Arizona, 1966.
Ricard, Forrest R., *History of Mining in Arizona,* "Chapter 9, History of Smelting in Arizona," Mining Club of the Southwest Foundation, Tucson, Arizona, 1987.
Richards, Robert B., Charles E. Locke, Reinhardt Schuhmann, *Textbook of Ore Dressing,* McGraw-Hill Book Company, New York, NY, 1940.
Rogers, W. Lane, *When All Roads Led to Tombstone, A Memoir by John Pleasant Gray,* Tamarack Books, Inc., Boise, Idaho, 1998.
Rose, Ann Knight*, Emigrant Knight of Cornwall,* in Collaboration with Gladys Peadon, self-published 1991.
Rose, Dan, *Pre-Historic and Historic Gila County*, Republic and Gazette Printery, Phoenix, Arizona, 1935.
Sherman, James E. & Barbara H., *Ghost Towns of Arizona,* , University of Oklahoma Press, Norman, Oklahoma, 1974.
Statistics of Mines and Mining in the States and Territories West of the Rocky Mountains; Eighth Annual Report of Rossiter W. Raymond, United States Commissioner of Mine Statistics, published in Washington, DC, 1877.
Stratton, Emerson O., & Edith Stratton Kitt, *Pioneering in Arizona,* Arizona Pioneers Historical Society, Tucson Arizona, 1964.
Swanson, James A., and Tom Kollenborn, *Superstition Mountain–A Ride Through Time*, Arrowhead Press, Phoenix, Arizona, 1981.
Swearengin, John A, *Good Men, Bad Men, Lawmen and a few Rowdy Ladies*, Florence, Arizona, 1991
Theobald, John & Lillian, *Wells Fargo in Arizona Territory*, Arizona Historical Foundation, Tempe, Arizona, 1978.

Thrapp, Dan, *Al Sieber Chief Of Scouts,* University of Oklahoma Press, Norman, OK, 1995.

Time Life Books Inc., *The Miners,* Chicago, Ill., 1967.

Traywick, Ben T., *The Clantons of Tombstone*, Red Maries Book Store, Tombstone, Arizona, 1996.

Trimble, Marshall, *Arizona-A Calvacade of History*, Treasure Chest Publications, Tucson, Arizona, 1989.

Twitty, Eric, *Blown To Bits In The Mine*, Western Reflections Pub. Co., Ouray, Colorado, 2001.

Walker, Gladys and T. G. Chilton, *History of Mining in Arizona,* from "Chapter 8, The History of Mining at Superior," the Mining Club of the Southwest Foundation and American Institute of Mining Engineers, Tucson, Arizona, 1991.

Waters, Frank, *The Earp Brothers of Tombstone*, C.N. Potter, New York, 1960.

Waters, Frank *The Earp Brothers of Tombstone*, University of Nebraska Press, Lincoln, Nebraska. 1976.

Woody, Clara T. & Milton L. Schwartz, *Globe, Arizona,* Arizona Historical Society, Tucson, Arizona, 1977.

Young, Otis. E. Jr, *Western Mining,* University of Oklahoma Press, Norman, 1970.

Manuscripts and Thesis

1. Anonymous File, *Arizona's Eighties*, 1880s, manuscript, Arizona Historical Society, Tucson, Arizona.

2. Bark, James, *Jim Barks Notes;* unpublished manuscript, circa 1930, Greg Davis Files.

3. Blake, William P., Professor, 1883 account of "How the Silver King Mine Was Discovered," unpublished manuscript, New Haven, Conn., Arizona Bureau of Mines files, Phoenix, Arizona.

4. Brown, Harry, "Personal Recollections of Silver King," included in the Bowen family history as compiled by Velma Bowen Tucker of Tempe, Arizona, circa 1990.

5. Cavaness, Matt, *Matt Cavaness Memoirs,* Superstition Mountain Historical Society, Greg Davis Files.

6. Craig, Geraldine, *Mrs. Robert A. Irion*, unpublished manuscript, Arizona Historical Society, Tucson, Arizona.

7. Craig, Geraldine, *A Pack Train,* unpublished manuscript, Arizona Historical Society, Tucson, Arizona.

8. Craig, Geraldine, *An Arizona Wagon Freight Train of 1877*, unpublished manuscript, Arizona Historical Society, Tucson, Arizona.

9. Drakovich, George, "Recollections of a Mr. McPherson," undated term paper, from the Gladys Walker Collection, Superior, Arizona.

10. Gassler, Walter, *The Lost Peralta-Dutchman Mine*, 1983, published by the Superstition Mountain Historical Society as a previous unpublished historical document, Goldfield, Arizona

11. Holmes, George "Brownie," unpublished manuscript, *Story of the Lost Dutchman,* Phoenix Arizona, 1944.

12. Johnson, Bruce , "A Brief History of the Pioneer Mining District and the Magma Mine," Magma Mine Engineer, May 11, 1974, from the files of BHP, Tucson, Arizona.

13. Kitt, Edith Stratton, *Reminiscences of Emerson O. Stratton,* as told to his daughter the Summer of 1925, Clara T. Woody files, Arizona Historical Society, Tucson, Arizona.

14. Mason, Charles P., *Tales of the Silver King Mine*, unpublished manuscript, nephew of one of the original owners, circa 1900, Bureau of Mines and Mining Museum, Phoenix, Arizona.

15. Overly, Abby, *Reminiscences of Mary E. Bailey*, unpublished manuscript, 1937, Arizona Historical Society, Tucson, Arizona.
16. Ransome, F. L., "Silver King Mine," *U.S.G.S Bulletin 540.*
17. Rice, Mike *Reminiscences of Mike Rice,* unpublished manuscript, Arizona Historical Society, Tucson, Arizona.
18. Tucker, Velma Bowen, "Personal Recollections of the Bowen Family History on Silver King," Tempe, Arizona, circa 1990.
19.Wildman, Perry, *Perry Wildman's Unpublished Reminiscences* , a merchant at Silver King and Pinal City in the 1880s, as told at Pioneer's meeting, Dec. 29, 1926, Arizona Historical Society, Tucson, Arizona, Arizona Historical Foundation, Phoenix.
20. Wood, Charles M. *Reminiscences of an Arizona Pioneer-Personal Experiences of Albert Simpson Neighbors*, unpublished manuscript, April 20, 1926, Arizona Historical Society, Tucson, Arizona.
21. Woody, Clara T., *Early Military Posts in Arizona,* unpublished manuscript, , Arizona Historical Society, Tucson, Arizona.
22. Woody, Clara T., *Pioneer Women*, unpublished manuscript, Arizona Historical Society, Tucson, Arizona.
23. Woody, Clara T., *August Murder,* the Mack Morris Mine robbery and murder, unpublished manuscript, Arizona Historical Society, Tucson, Arizona.
24. Ziede, Alexander, *Territorial History of the Globe Mining District,* unpublished thesis, University of Southern California, August 1939, Arizona Historical Society Files, Central Division, Tempe, Arizona.

Notes, Letters, and Interviews

1. Anecdotes of the Early Days of Silver King, December 20, 1918, source unknown, Arizona Bureau of Mines Files.
2. John D. Walker files, Greg Davis collection, Tempe, Arizona.
3. Clara T. Woody's notes on Silver King and Pinal, from her files at the Arizona Historical Society at Tucson.
4. Robert Bowen family notes on the Silver King Mine. The unknown prospector found in the old stone oven may have been the man killed by Apaches when the Silver King was located years earlier.
5. This story of the prizefighter at Silver King comes from a descendant of John Meehan, Mike Benefield of Prinefield, Oregon. His grandmother lived at Silver King during the 1880s and was married to a Meehan. I verified this by census records.
6. Jack Carlson provided assistance with this complicated process, of determining ownership of the old Cavaness-Quarter Circle U Ranch, during 2003 conducting a thorough research of old land title records.
7. William L. Wardwell Letters, Arizona Historical Society, Tucson, Arizona, letter dated January 25, 1886. Dr. Wardwell was a part-time physician at Silver King and Pinal during 1886 and wrte letters home to his family in new Yory almost weekly.
8. William L. Wardwell Letters, Arizona Historical Society, Tucson, Arizona, letters dated February 1, 1886,and February 15, 1886.
9. Information on Silver King from a letter received from the family of Harry Thompson, whose mother lived at Silver King.
10. Joe Deen's grandfather Mack Deen owned the Riverside Stage Station, from interview with Joe Deen during 2000.
11. Robert Blair files on the Peralta-Reavis Land Grant from the Arizona Historical Foundation, Tempe, Arizona.
12. Joe Deering's coroner's inquest on October 23, 1885, from the files of the Justice Court of Pinal County, Florence, Arizona.
13. Lost Soldiers information from an interview with Velma Bowen Tucker during

2001, who recalled this story told to her by her father Thomas Bowen, who received the information from his father Robert Bowen.
14. Information the Magma Railroad from the notes of Gladys Walker of Superior.
15. Interview with Joe Deen 1997 regarding his biographical information.
16. Interview with Ron Deen Sr. at the mine 1997 regarding his biographical information.
17. Interview with Ron Deen Sr. 2000 at Riverside, regarding Gary Deen's biographical information.
18. Interview with Dick Bynum 2000 at the mine regarding his biographical information.
19. Interview with George Sites 2000 at the mine regarding his biographical information.
20. Interview with Antonio Cisneros at the mine March 21, 1997. Cisneros has since passed away but his claims near the Silver Mine are still active and belong to the family.

Periodicals and Magazines

1. *Arizona Eighties* magazine, Anonymous Files, Arizona Historical Society, Tucson, Arizona.
2. Arizona Graphic *magazine, Dec. 23, 1899, Vol. 1, No. 15.*
3. Barney, James M., *Arizona Municipalities* magazine, August 1940, P.10.
4. Barney, James M., Forgotten Towns of Arizona, , *Arizona Municipalities,* Phoenix, Arizona, 1940.
5. Crosswhite, Frank S., "History, Geology, and Vegetation of Picket Post Mountain", article in *Desert Plants,* a Booklet Published by the University of Arizona for the Boyce Thompson Southwestern Arboretum, Vol. 2, 1984, p. 73-77.
6. Davis, Gregory E., *Superstition Mountain Journal Volume 18*, "The Weiser-Walker Map", Superstition Mountain Historical Society, Apache Junction, Arizona, 2000, P. 7.
7. Davis, Gregory E., "Dutch John Pipps and the Lost Dutchman Gold Mine," *Superstition Mountain Historical Society Journal,* Volume 13, 1995.
8. *Engineering and Mining Journal*—Letters From the West, February 18, 1882, Vol. 33, Pgs. 91-92.
9. Fraser, Jack, file notes, Greg Davis Collection, Tempe, Arizona.
10. Larr, Dr. James, *Popular Science Monthly* August 1885, published in the *Pinal County Record,* September 11, 1885, Pinal, Arizona Territory.
11. Lucas, Lee, "The Town History Forgot,", *Desert Magazine*, February 1968.
12. . Moore, Georgia, "Life At Old Silver King Was Primitive, But Zest Filled," *Superior 1882-1982 Centennial Brochure,* Pg. 14, Gladys Walker Files, Superior, Arizona.
13. Ragus, Nick, *Gila Centennials Historical Celebration and Pageant Booklet*, edited by Frances Gerhardt, John Woody, Jimmie Fulton, Ester Preston and Clara T. Woody, Globe, Arizona, 1976, P. 17-18.
14. Short, M. N., F. W. Galbraith, E. N. Harshman, T. H. Kuhn, and Eldred D. Wilson "Geology And Ore Deposits of The Superior Mining Area," Arizona Bureau of Mines, Geological Series, No. 16, Bulletin No. 151, Published by University of Arizona, Tucson, Arizona, October, 1943, p. 142.
15. Silvas-Martin, Carlotta "One Hundred Years Of Yesterdays-The Silver King Mine: Its Life And Its Death,", *Superior Centennial Report*, Superior, Arizona, 1982, P.2. Arizona Historical Society, Central Division, Tempe, Arizona, Pgs. 243-247.
16. Tucson Corral of the Westerners, *The Smoke Signal* booklet, published by, Tucson, Arizona No. 10, Fall 1965 page 14.
17. Wilson, Roscoe G., "No Place for Angels," *Stories of Old Arizona Days,* Arizona Republic, Phoenix, Arizona, and Arizona Silhouettes, Tucson, Arizona, from the files of the Arizona Historical Foundation.

Newspapers

1. *Arizona Citizen,* Tucson, June 30, 1877,
2. *Arizona Daily Citizen,* Tucson ,April 3, 1875,
3. *Arizona Daily Star*, Tucson, September 9, 1910,
4. *Arizona Enterprise*, Prescott, August 15, 1885, September 25, 1885, November 20, 1886, June 2, & 9, 1888
5. *Arizona Gazette*, Phoenix, September 17, 1875, August 31, 1882, April 8 1935, June 12, 1936
6. *Arizona Miner*, Prescott, 1870, semi-monthly and weekly,
7. *Arizona Record*, Globe, March 29, 1956
8. *Arizona Republic*, Phoenix, March 21, 1895, August 23, 1905, March 7, 1907, May 13, 1935, June 13, 1935, July 18, 1965, August 5, 1975,
9. *Arizona Republican*, Phoenix, March 19, 1935,
10. *Arizona Sentinel,* Yuma, August 3, 1878,
11. *Arizona Silver Belt,* Globe, June 14, 1884, February 28, 1885, December 23, 1918, November 4, 1913, April 16, 1981
12. *Arizona Weekly Enterprise*, Florence, August 10, 18 1883, September 8, 1883, September 5, 1885, April 16, 1887,
13. *Casa Grande Dispatch*, Casa Grande, February 8, 1940
14. *Florence Blade Tribune* Florence, August 3, 1951
15. *Florence Tribune,* Florence, August 6, 1898
16. *Mining and Scientific Press*, San Francisco, California, July 20, 1878, June 25, 1881, April 28, 1883, July 16, 1887
17. *Phoenix Daily Herald*, Phoenix, September 15, 1886
18. *Pinal County Record*, Pinal A. T., September 11, 1885, November 20, 27, 1885, December 25, 1885, January 22, 1886, February 2, 19, 1886, July 9, 1886, September 17, 1886
19. *Pinal Drill*, Pinal, A. T., January 1, September 11, October 16, December 25, 1880, January 8, February 4, 12, 26, March 12, April 9, May 14, 28, July 7, 9, 16, 23, August 6, 27, November 11, December 17, 1881, March 4, 11, July 15, September 8, 9, 23, 30, October 7, December 26, 1882, March 3, July 7, 16, August 1, 8, 18, 22, 25, September 8, 11, 15, 23, 29, October 27, November 24, 1883
20. *San Diego Union,* San Diego, California, January 19, 1877
21. *Superior Sun*, Superior, February 25, 1921, November 22, 1962
22. *Tucson Daily Citizen*, Tucson, May 28, 1930

Professional Reports

1. *Ore Deposits Support Hypothesis of a Central Arizona Batholith*, Technical Publication No. 63, American Institute of Mining and Metallurgical Engineers, New York, NY,1928.
2. *Silver King Mine and Copper Deposits near Superior Arizona*, F. L. Ransome, U. S. G. S. Bulletin 540.
3. *Gold & Silver Mining in Arizona 1848-1945*, Arizona State Historic Preservation Office, Arizona State Parks Board, Phoenix, Arizona, 1992.
4. *Geology and Ore Deposits of the Superior Mining Area*, Arizona, Arizona Bureau of Mines Bulletin No. 151 by M. N. Short, F. W. Galbraith, E. N. Harshman, T. H. Kuhn, and E. D. Wilson, Published by University of Arizona, Tucson, 1943.
5. Arizona Bureau of Mines and Mining Museum Official Files, Phoenix, Arizona.
6. *The United States Military In Arizona 1846-1945,* Arizona State Historic Preservation Office, Arizona State Parks Board, Phoenix, Arizona, 1993.
7. The *Mineral Industries of Arizona*, by J. B. Tenney, Arizona Bureau of Mines

bublished by University of Arizona, Tucson, February 15, 1928.
8. *Magma Copper Company Annual Reports* of 1915, Files of BHP Corporation, Tucson, Arizona.
9. *Analytical Report, Silver King of Arizona Mining Company*, H. D. Wells & Company, New York City, NY 1916.
10. *Silver King Mining Company Report of 1920*, Arizona Bureau of Mines and Mining Museum files, Phoenix, Arizona.
11. *Official Letter the Silver King Company of 1924*, files of the Bureau of Mines and Mining Museum, Phoenix, Arizona.
12. Department of Mines and Mineral Resources *Field Engineers Report October 22,* 1959, Bureau of Mines and Mining Museum, Phoenix, Arizona.
13. Mineral Resources of Arizona *Field Engineers Report February, 25, 1959* Bureau of Mines and Mineral Museum files, Phoenix, Arizona.
14. *History of Mining in Arizona,* James Brand Tenney, Assistant Geologist, Arizona Bureau of Mines, Tucson, 1927-29.
15. U.S. Court of Appeals for the Ninth Circuit, *USA V. SHUMWAY*, Case Number 96-16480, date filed 12/28/99. (This landmark case can be found on the Internet site http;www.ce9.uscourts.gov)

Business Directories and Census Reports

1. Arizona Business Directories and Gazeetters, 1880-1892, W. C. Disturnell, San Francisco, California.
2. Arizona Census, 1876 & 1882.
3. Great Register of Pinal County, Arizona Terrritory 1886, 1888, 1890, 1892, 1894, 1896,1898, & 1911.
4. Pacific Coast Directory, 1880-1881, L. M. McKenny & Co. San Francisco, California.
5. Pacific Coast Directory, 1888-1889, L. M. McKenny & Co. San Francisco, California.
6. San Francisco Classified Business Directory, 1880, San Francisco, California.

Glossary of Mining Terms

Adit- A Tunnel with one surface opening
Amalgamation- Process of extracting silver/gold from ores sampled.
Assay- Process used to determine value of ores sampled. God and silver values are determined by fire assay while sulfide values are determined by chemical titration.
Batholith- Large intrusion of molten rock beneath the surface of the earth.
Bonanza- Rich ores.
Breccia- Highly fractured rock or ore fragments cemented by mineral such as quartz.
Bullion- Silver and Gold in the form of bars or buttons. Also bars of lead containing gold and silver.
Bull Quartz- Milky white quartz considered to be waste rock.
Cap- Timber set across and on top of posts for support of bad ground.
Cage- The elevator that people and equipment ride up and down raises or shafts.
Caldera- Sunken block of ground related to the extrusion of molten rock from volcanism.
Chalcopyrite- Copper/Iron Sulfide.
Chute- Wooden or metallic box at end of ore pass designed to hold and transfer ore into waiting ore cars.
Claim- Legal land status obtained by placing a discovery post or monument with four corner stakes and two sideline stakes. 600 feet wide by 1500 feet long.
Cobbings- The breaking off of high-grade ore from lower grade rock.
Concentrate- The metals remaining after the milling process separates ores from undesirable minerals such as quartz..
Cribbing- Timber put on tops or outsides of posts and caps to hold bad or broken ground. Also can be timbers laid in squares, one on top of the other to make manways or ore passes.
Crosscut- Tunnel driven perpendicular to vein. A nearly horizontal excavation usually driven at right angles to a vein. Driven in waste rock to reach the next vein.
Double Jack- Eight-pound sledgehammer. The process of one miner swinging a hammer and hitting drill steel held by another miner to drill holes is called "double jacking."
Diamond Drill- A drill that uses a hollow, diamond-lined drill bit to obtain core samples from areas not accessible to sampling.
Drift- A nearly horizontal excavation driven along a vein, usually large enough for men and machinery to enter.
Drill Steel- Pieces of steel, usually seven-eights or one inch in diameter, used to drill into rock in preparation for blasting.

Drifting- The process of driving a tunnel is called drifting.
Drive- To drive a tunnel is to advance or extend the lengths of a tunnel.
Dump- A deposit of waste or ore derived from the interior of a mine. Usually disposed of or stored on the surface near the entrance of a mine.
Face- End of tunnel.
Fault- Crack or plane of weakness in the earth in which rock has slid past itself.
Float- Ore that has broken off from vein and moved downhill by gravity.
Flotation- Process used in concentration of sulfide ores. Metal minerals such as silver or gold will chemically coated in frothing tanks and then float off with the froth away from non-metal minerals and will then concentrate into thickening tanks.
Footwall- A mass of rock beneath a fault plane, vein, lode or bed of ore. Vein must be inclined from the vertical. The area below the inclined vein.
Fracture- Crack in the earth, usually considered to be smaller than a fault.
Free Gold- Gold in the native state, not in chemical combination.
Galena- Lead sulfide.
Giant Powder- An early form of dynamite used for blasting. Made by mixing nitroglycerin with sawdust.
Hanging Wall- The mass of rock above a fault plane, vein, lode, or bed of ore. The vein must be inclined from the vertical. The area above the inclined vein.
High-grade- Mineral in extremely rich concentrations. In the case of silver, usually visible to the naked eye.
Hoist- To bring people and supplies up a shaft or raise. Also, the actual machine used to hoist, usually a compressed air or electric-driven drum upon which a cable is wound and unwound.
Jackleg- Machine drill that pounds and turns drill steel for drilling holes. Pressure is exerted by a telescoping hydraulic leg that is attached to drill and pushes it toward the hole.
Lagging- Timber laid across caps or alongside of posts to prevent rock falls.
Lode- A fissure in a rock formation containing mineral. Synonymous with vein or ledge.
Low-grade- Ore that contains minimal values. Usually requires milling to recover the minerals.
Manway- Shaft or raise in which ladders have been installed on stulls set crossways in mined area.
Muck- Broken rock. The act of shoveling rock is called mucking.
Mucker- Machine used for shoveling up rock. Also, the laborer who shovels rocks or "muck."
Native Silver- Silver in its elemental state, usually combined with minor

amounts of gold or copper. Wire silver is native silver in the form of wires. The terms native and free silver are synonymous.
Ore- Mineral that can be mined at a profit.
Ore Pass- Hole used for transference of rock from one level to another by gravity.
Outcrop- The portion of a lode or vein that reaches the surface.
Pockets- Small pods or streaks of mineral within a vein structure.
Portal- The entrance to an adit or tunnel.
Post- Vertical support timber upon which a horizontal or angled cap is set for ground support.
Quartz- Silica Mineral.
Raise- An upward excavation driven to ready the next level above or any area above the working level.
Retort- A container in which gold-mercury amalgam is placed and heated to drive off and recover the volatile mercury, leaving the gold sponge behind. Mercury is condensed by a water jacket as it is vented from the retort through a pipe attached to the top of the container. The pipe is bent downward and the open end is kept in a stainless steel bucket to cold water.
Ribs- Sides of tunnel.
Rock Bolt- Six-foot long bolt driven into drilled hole and used with other rock bolts for ground support.
Round- Set of drilled holes that are loaded and blasted at end of tunnel face.
Salt- To artificially put silver/gold in a vein or sample for purposes of defrauding a potential buyer.
Sampling- Taking a small representative amount of mineral across the width of a vein, or from a pile of ore to determine it's value.
Sets- Posts and caps in succession and connected with lagging for purposes of ground support.
Shaft- A vertical or steeply inclined excavation used to reach lower portions of an ore body or mine.
Shear Zone- Zone along a fault or vein that contains multiple fractures.
Shot- The exploding of a drilled hole with blasting powder or dynamite.
Shoot- A lens of ore with greater vertical than horizontal dimensions. May be inclined within the plane of the vein or fault.
Sill- The floor of a tunnel.
Single Jack- One miner using a four-pound hammer to hit drill steel, also called single jacking.
Skip- Elevator used to raise ore in shafts or raises, also called a skipjack.
Slab- Rock that breaks off from the back or sides of a tunnel or drift.
Spiling- A process used for driving a tunnel through talus or broken rock. Steels are driven into the back of the tunnel at a low angle, caps and posts are then set beneath and the broken rock mucked out from below

them.
Streaks- Narrow veinlets of mineral within a vein. High-grade streaks would be bands of silver within a large band of lead or quartz.
Stope- An excavation from which ore is being extracted. If above an adit or tunnel, it is an overhand stope.
Stoper- A machine used to drill vertically.
Strike- To bear in the direction of a vein.
Structure- Faults, fractures, veins, etc. Any crack in the rock is called a structure.
Stull- Timber put between the hanging wall or footwall of a vein for support or installed in raises or shafts for making manways or ore passes.
Troy ounce- A method of weights used in measuring precious metals. A troy ounce is slightly heavier than an avoirdupois ounce.
Tunnel- A nearly horizontal (hole) passage that goes through a hill and may be open at both ends.
Vein- A fracture in the earth's crust that has been filled with minerals.
Windlass or whim A hand-cranked hoist designed to lift buckets of rock or ore from the bottom of a winze of shaft. If outside the shaft a horse whim hoisted the bucket or men.
Winze- A vertical or inclined opening or excavation driven to connect two levels of a mine. Sunk from the top down.

INDEX

Adams, Ashley D., 332
Adams, Fred, 208
Adams, Randolph, 94
Adamsville, 16, 92, 108, 198, 237
Adler, Dr., 200
Ainsworth, Walter, 285, 286, 288, 289
Alaska, 230, 231, 315
Allen, Jim, 115, 119
Almer, Red Jack, (also see "Elmer"), 347, 207
Anderson Brothers, 18, 19, 20, 270
Anderson, Bob, 19, 20, 270
Anderson, Dave, 19, 20, 270
Apache Kid, 124, 216, 217, 270, 276
Apache Leap, 4-6, 149, 135, 305, 338
Apache Tears, 6
Apache Trail, 201, 224, 231, 235, 238, 239, 276, 295
Apacheria, 1, 9
Apaches, 1-10, 13-15, 18-23, 31, 33, 40, 84, 100, 103, 107, 113, 149, 179, 217, 225-229, 271
Apex Mining Law, 249
Appel, Nathan B., 103
Arballo, B., 106
Arizona Stage Company, 187, 189, 191, 192
Arizona State University, 284, 315
Arizona Volunteers, 5, 6
Arnett Canyon, 134
Art, D.W., 190
Asarco Mining Company, 267, 269, 310
Avota, Jesus, 216
Bailey, Walter, 158
Bailey, William, 158
Bailey, Mary, 158
Baker, J. J., 118
Baldonado, Dolores, 103
Baldwin, M.A., 137
Bark Valley, 227, 239
Bark, James E., 104, 107, 235
Barkley, Augustus "Tex", 197
Barkley, William "Bill", 196
Barks Ranch, 240
Barnes, Richard, 144, 145
Barnes, Will C.148, 149
Barnett, Aaron, 103
Barney, Benjamin A., 55, 95
Barney, James M. (Col.), 6, 13, 27, 31, 32, 33, 55, 61, 91, 93, 99, 219, 283, 338, 341, 343, 347, 355
Baron of Arizona, 217, 218, 219, 347, 351, 353
Bassett, Jim, 89
Battalieu, Claude, 100
Bayless & Berkalew, 106
Beale, Ned, Lieutenant, 97, 98
Beeler, Frank, 168
Bellamy, Dad, 108
Bender Environmental, Inc., 333
Benson, Judge, 115, 139, 168
Benson, William H., 94, 172
Benson, William W., 134
Benton, F.E., 120
Berthoud, Stephen, 103
Bibliography, 351
Big Johnny Gulch, 270
Big Nose Kate, 167, 222
Big Red Ghost, 98
Bilk Shaft, 41, 53, 55, 248, 249
Binkley, John, 89
Black Jack, 83, 84, 100
Blair, Robert, 8
Blake Crusher, 30, 41, 55, 69, 84, 96
Blake, Eli Whitney, 41
Blake, Professor William P., 28, 41-45, 93, 94, 277, 279, 280, 351-353
Blaylock, Celia Ann "Mattie", 165-168
Blaylock, Sarah, Mrs., 167
Bley, W.J., 214
Block, Ben, 103
Bloody Tanks, 191, 270
Blue Point, 107, 239
Bluff Spring Canyon Trail, 228, 231, 239, 259
Bollinger, William, 267, 268
Booglebrand, Phineas T., 220, 221, 346
Booth, E.S., 50, 285
Bousche, Pete, (Deputy), 186
Bowen and Jones Saloon, 114, 121, 125, 149
Bowen, Elijah, 172, 179
Bowen, George, 173
Bowen, Robert, 50, 55, 90, 193, 169, 222, 233, 243, 244, 248, 285, 286, 339, 353
Bowen, Robert Phillip, 173 ,175, 176
Bowen, Ruth, 173,
Bowen, Stephen (Ben), 169, 172
Bowen, Susannah, 169, 172, 173
Bowen, Thomas Arthur, 173
Bowman, L., 191
Boyce-Thompson Arboretum, 9, 306
Brady, R.G., 106
Bramett, Jeff, 186, 187
Brash, John, 103
Bray, Hank, 110

Broerman, Ceslinger, 251
Broerman, F.F., 251
Brown & Wells, 107
Brown, Eliza Bowen, 172, 179
Brown, Harry, 179
Brown, Jesse H., 120
Brown, John N., 106
Brown, William Thomas (Bill), 21, 172
Bruch Electric Light Co., 141
Bruckner Cylinder Furnace, 61, 63, 69-71
Buchanan, Tom, 128, 164, 302, 303
Buckalew, O., 103, 149
Buffalo Mining and Smelting Company, 272
Buffalo Mountains, 318
Bull Dog Canyon, 239, 295
Business Directories and Census Reports, 150, 357
Butte, 5, 177, 257, 259, 260, 265, 297
Butterfield Overland Stage Company, 190
Bynum, Richard (Dick) Mark, 314, 316, 319, 320, 322
Caldwell, James, 103
California Column, 1, 2, 6
Camp Picket Post4, 9, 13, 14, 297, 298
Camp Supply, 12, 14, 20, 21, 55, 179, 297, 298, 300
Cananea Consolidated Copper Company, 273
Carlson, Jack, 8, 12, 25, 128, 217, 228, 239, 240, 241, 244, 246, 251, 252, 256, 257, 259, 260, 274, 298, 300, 301, 302, 307, 343, 350, 351, 354
Carpenter, Frank, 209, 210
Casanega, (Special Detective), 208
Castle Dome Mining and Smelting Company, 55, 274
Castle Rock, 238, 240
Cavaness, Alice Rowe, 104
Cavaness, Matt, 103, 105, 114, 158, 227, 348, 353
Caveness, Silas, 104, 105
Chamberlain, Lizzie, 107
Champion, Joe, 127, 150, 151, 257
Champion, W.S., 281
Charlebois, Alfred, 104
Cherry Creek, 16
Christy, George. D., 283, 286
Chuning, John, 223, 234, 235, 236, 347, 349
Cisneros, Antonio, 318, 350, 355
Clanton, Billy, 220, 221
Clanton, Ike, 220, 221
Clanton, Phin, 168, 220, 222, 264, 346, 353
Clark, J.E., 19
Clary, Alvin F, 298.
Clary, Tom, 298, 299, 304
Cleveland, Grover (President), 282
Clubfoot the Swamper, 228, 229, 230
Clum, John, 22, 265
Cochran Coke Ovens, 259, 260
Coke Ovens, 259, 260
Cole, George W., 106, 342, 351
Cole, Linda, 313
Conatser, Estee, 239, 348, 351
Connor, Jacob, 214
Consolidated Copper Company, 272, 273, 315
Copeland, Isaac, 19-30, 33, 280, 283
Corbusier, Walter, S. (Dr.), 21
Cottonwood Corral, 129, 194
Couper-Thwaite, E., 285
Cousin Jacks, 76, 77, 80
Cousin Jennies, 77
Cox, Quenton "Ted", 252-254
Craig, Dudley, 161, 304
Craig, Dudley, L., 303
Craig, Helen B., 299, 305, 350
Crawford, Chuck, 239, 240
Cribbing, 82, 358
Crisswell, Frank, 104
Crook, George (General), 14, 15, 19, 21, 275
crosscuts, 47, 77, 79, 81
Crowley, Con, 275, 276
Cummings, Mary K., 167
Czarnowski, F, 159
Daguerre, A., 103
Davis, Dr. H. H., 146
Davis, Greg, 8, 91, 97, 108, 109, 158, 170, 244, 256, 338, 340, 347, 353, 354, 355, 362
Davis, W.C., 107
Dead Man's Flat, 206
DeArnett, L., 106
Deen, Freddie Joseph, 310-337
Deen, Gary Todd, 308-337
Deen, Logan, 308-337
Deen, Mack Valentine, 308-312
Deen, Ronald Bedell, Jr., 308-337
Deen, Ronald Redell, Sr., 308-337
Deering, Joe, 89, 90, 124, 223, 233-237, 243, 354
Denver Exposition 1882, 47
Desert Well, 108
Deutch, William, 148
Devil's Canyon, 12, 100, 110, 164, 300-302, 304-306
Devine, Jon. J., 29
Devine, Laurie, 301
DeWalt, George, 251
Directory of Businesses, 110, 134-135, 137, 149

Dobbins, James C., 282
Dobie, Charlie, 123, 124
Doc Holiday, 167
Dodge, Caleb M., 272
Donnelly, F.O., 103, 127
Donnelly, Henry, 259
Doran, Andrew J. (Sheriff), 56, 91, 92, 207, 208, 209, 223.
Dorsey, Andy, 186, 187
double jacking, 79, 80, 358
Downe, Alex. B., 284
Downing, Major, 145
Drais, L.K., 151, 214
Dreamy Draw, 65, 51, 52, 53, 56, 57, 79, 81, 82, 317, 325, 326
drifts, 49
Dripping Springs, 214, 243
Dudley, Andy, 76
Duett, Helen, 136
Duran, Jose, 16
Dutchman's Cave, 243
Dyster Concentrating Table, 325
Eagle Mountains, 318
Earp, Mattie, 146, 165-170, 221, 345
Earp, Wyatt, 146, 165, 166, 167, 168, 169, 203, 220, 221, 222, 345, 351, 352
Egyptians, 80
El Groupo Mining Co., 237, 269, 310, 313
Elder, Kate, 167
Elliot, Bill, 84
Elliot, W.V., 214
Ellison, Slim, 136
Ellison, Susan, 136
Elmer, Red Jack (see also "Almer"), 207, 347
Ely, Sims, 230, 347
Emerick, George R., 103
Enyart, J., 89
Espinosa, Joe, 103
Evans, (U.S. Marshall), 209
Evans, George E., 214
Everybody's Drug Store, 110
Ewing, Thomas J., 301
Fales, Ned, 220, 201
Faylor & Parker's Saloon, 148
Feldman, Ron, 244, 252, 253, 254
Feldman, William, 306
Fenton, William, 103
Field, George L., 103
Finny, Samuel, 103
Fisher, Kate, 167
Float Cells, 324
Foley and Barrington, 120
Folsom, William, 19
Fook, Dang, 143
Fowle, John, 286, 289, 290
Foxworth-Galbraith, 292, 293
Frakes, Stella Lee, 260
Frakes, Violet, 260
Fraser, Jack, 50, 84, 105, 109, 110, 111, 125, 223, 235, 341, 355
Frisk, George, 103
Frue Vanners, 85, 61, 62, 63, 66, 70
Ft. McDowell, 7, 14, 16, 159, 223, 224, 226, 238, 239
Gabriel, Pete (Sheriff), 128, 140, 187, 198, 202, 203, 204, 205, 208, 210-217, 266
Gage, Jesse, 274
Gandara, Francisco A., 16
Gardiner, John J., 103
Gardner, Bob, 327
Gassler, Walt, 196, 344
Gayer, D.W. , 292
Gays, "Bat", 291, 292
Gee, Qui, 143
Gee, Sue, 143
Gibson, Sam, 274
Gilmer, Salisbury & Company Stage Line, 190
Girard, Henry, 186, 187
Gird, Dick, 34
Globe Ice Company, 137
Globe Ledge, 271
Glossary of Mining Terms, 358
Goldman, Leo, 105
Gonzales, Jose, 106
Gormley, Thomas, 235
Gradello, Juan, 7
Green, James, 235
Griffith, F.M., 29
Griffith, W.N., 186, 190
Grime, Cicero, 204, 205, 206, 281, 272
Grime, Lafayette, 204, 271
Grooves, Tom, 106
Guzman, Mike, Jr., 294, 295
Guzman, Mike, Sr., 294, 295
H. D. Wells & Company, 285, 357
Hackney, Aaron (Judge), 119
Hadji Ali, 97, 98
Hall, Andrew "Andy", 203, 204, 205, 271
Hall, L., 159
Hall, Maggie, 159
Hancock, John C., 103
Hanging Tree, 206
Happy Camp, 250, 255, 256
Happy Hollow Camp, 114
Harmon, Debbie, 311
Harony, Mary Katherine, 167

Harrington, W. (Special Deputy), 208
Harrison, Benjamin (President), 282
Hart, S., 259
Harte, Jules, 158
Harvey, (Doctor), 212, 213
Harvey, Walter, 145
Hastings, Town, 140, 149, 261, 262
Hawkins, Andrew, 302
Hawley, Curtis E., 144, 204, 205, 271
Hayden, Carl, 175
Hayden, Charles T., 119
Hayden, Don Carlos, 101
Hayden's Ferry, 3, 107, 180
Henderson, William, 274
Herman Mountain, 238
Herron, James, 129, 256, 292
Hewitt Station, 192, 250
Hi Jolly, 97
Hickey, N.M., 214
Hillebrand, A.W., 286, 288, 293
Historical Exploration and Treasures (H.E.A.T), 253, 254
Hoefer, Eugene, 94
Holiday, Doc, 167
Holland, Pat, 29
Holman, Wiley, 226, 261
Holmes, "Hunkydory", 16, 17, 18, 19, 197, 216, 270
Holmes, George "Brownie", 223, 237, 347, 348, 353
Holmes, Richard "Dick", 223, 232, 233, 236
Hoover, Herbert, 85
Horse Camp Basin, 238
Hoskins, Walter, 238
Hunt, George W.P. (Gov.), 135, 136, 150, 151
Irion, Elizabeth Craig, 161, 162, 163, 304
Irion, Robert, 164, 165, 302, 303, 304, 305, 344, 353
Irion, Robert A., 161, 163
Irish Buggies, 87
Irvine, Elizabeth, 161
Irvine, T., 19
Jenks, S.S., 103
Jinks, Jimmie, 253
John McCourt's Saloon, 116
Johnson, Dr.Walter F. , 291
Jones,.Charles H. (Dr), 33, 123, 282-286, 336
Kearns and Griffith, 190, 191
Keating, Jack, 183
Keating, John, 210, 212
Kelley, Tom, 262
Kellner Canyon, 190
Kellner, Ernest, 145, 305
Kelsey, Joe, 106
Kelvin, William Thompson, 268
Kennecott Copper Corporation, 268
Kennecott Mining Company, 309, 310
Kennedy, R.E., 137
Kenniard, Thomas (Dr), 89, 90, 129, 130, 168, 180, 235
Kerr, T. B. (Captain), 11
Kerr, Tom B., 16, 18, 198, 199
Kimball, William "Bill", 220
King, P., 19
King's Crown Mountain, 2, 9, 10, 24, 40, 42, 43, 50, 277, 298, 307, 317, 319
Kittle, Fred R., 118
Knight, Emma, 124
Knight, Emma Bray, 110
Knight, John, 76, 77, 110, 180
Knight, William George (Will), 110, 111
Knights of Pythias, 134, 152, 153
Kollenborn, Thomas, 107, 342, 346
La Barge Canyon, 238
Lamonica, Tom, 250
Lawhon, W.W., 286
Lincoln, Abraham (President), 17
Linkton, Henry, 259
Live Oak Development Company, 272
Lobb, Chester, 293, 295, 308, 317
Lobb, Dick, 292, 293, 294
Lobb, George, 76, 261, 262
Lockhart, Monty, 300, 305
Loffler's Saloon, 115
Logan, F., 259, 311, 320
Lombard, Joe, 127
Lomker, Henry, 284
Long, William, 19, 20, 22, 23, 26, 32
Lopez, V.R., 107
Lount, S.D., 136
Ludke's Store, 116, 120, 168
MacKenzie, Kenneth, 94
Macy, Arthur, 60, 91, 93, 94, 278-281
Maier, Frank, 113
Manlove, Mr., 159
Manuscripts and Thesis, 353
Marlowe, George, 104
Martinez, Apolonio, 103
Martinez, Francisco, 16
Marysville, 107, 108, 157, 158
Mason Valley, 13
Mason, Aaron, 41, 55, 68
Mason, Charles G., 10, 11, 13, 15, 20-31
Mason, Charles P., 125, 199, 207, 342, 346
Mason, Freeborn, 31, 91
Mason, Jim, 144

Masonic Order, 134
Masson, Francis C., 285
Matley, B., 284
Matthews, Jas., 214
Maxwell, 16
Mayer, Frank, 118, 150
Mazatzal Mountains, 16, 122
McCabe, John H., 284
McCoy, Jack, 186, 187, 201
McDonald, H.C., 189, 190
McGrew, John, 107
McGrew, Mike, 108, 138
McIntyre, Rev., 122
McMillen, Charles, 270
McNelly, Bill, 275
McNelly, Hal, 275
Meehan, John "Blackjack", 83, 84, 100
Melvin, Kate, 167
Merrill, W. Earl, 118, 342
Merrit, W.H., 148
Meyer, F., 102
Middleton, Eugene, 99, 100, 216
Middleton, Grace, 276, 292, 293, 295
Middleton, Lee, 275
Middleton, William, 3
Military Trail, 224, 239
Mill, Geo. L., 140
Miller, George L. 137, 139
Million Dollar Highway, 306
Miner, Tom, 16, 18
Mineral Hill Stage Line, 192
Miners Needle, 16, 225, 231, 238, 239, 240
Mitchell, H.F., 89
Monk, John Pipps, 231
Montgomery, H. B, 190
Moore, Mrs.Georgia,, 116, 341
Morgan, Patrick, 29
Morgan, William, 103
Morris, Eliza G., 178
Mounce, Henry, 206
Munson, 74, 271
Murbarger, Nell, 99, 341
Murray, Hank, 150, 201
New York Stock Exchange, 285
Newspapers, 30, 106, 201, 211, 356
Nick's Ranch, 12
Noble, H.H. 259, 277
Nolan, G.C., 272
Noriega, Francisco, 103
Noriega, Juan, 103
North Butte, 259
Notes, Letters, and Interviews, 354
O'Boyles, Bill, 122, 132
O'Boyles's Silver King Hotel, 132
Oates, Samuel, 89
Ochoa, J. M., 150, 151, 203
Odd Fellows, 134
Oehrlein, Elizabeth, 170, 172
OK Corral, 253
Old Tutaine, 180
Olson, Bob, 292
Pacific Chloridizing Furnace, 69
Padilla, Y., 103
Palace Saloon, 105
Palmer, C.F., 150, 213
Panknin, Ernest Albert, 230, 231
Papaianni, Michael, 166
Paul, Bob (Sheriff), 209
Peck, Ed, 16
Peebles, A.H., 103
Peralta Land Grant, 218, 219, 239
Peralta, Sofia, 218
Peralta-Reavis- James Addison, 217
Peralta Trail, 239, 240
Peralta Mine, 252, 253
Periodicals and Magazines, 355
Perkins, Fatty, 108, 138, 147
Phillips, Richard M., 91, 92
Phoenicians, 80
Phy, Jophesus "Joe", 213-216, 266
Picket Post, 2, 4, 5, 8, 9, 13, 14, 28, 29, 43, 66, 96, 106, 116, 134, 149, 190, 192, 296, 298
Picket Post Mountain, 9, 11, 12, 13, 14, 15, 103, 133, 146, 155, 192, 238, 240, 244, 246, 256, 292, 306, 338
Piggly Wiggly Store, 305
Pinal Baseball Club, 121
Pinal City, 4, 42, 54, 61, 81, 86, 96, 98, 101, 104, 114-153
Pinal City Brewery, 137
Pinal Consolidated Company, 259
Pinal Consolidated Mining Company, 259
Pinal Ranch, 12, 13, 99, 124, 159, 162, 164, 165, 298-305
Pinal Trail (Old), 96
Pine Canyon, 195
Pine Creek, 195
Pioneer Mining District, 25, 26, 114, 149, 257, 282, 253
Pioneer Pass, 191, 203
Pipps, "Dutch John", 223, 231-233, 347
Plan of Operations, 328, 330
Pomeroy, Frank, 118
Poncho Monroy, 107
Porter, Frank, 203
Porter, Wood, 200

Pottance, Dave, 106
Powell, J.L., 104, 204
Powers, Richard, 29
Professional Reports, 356
Pruett, Henry, 188
Pulaski Academy, 284
Pump Station, 21, 50, 117, 1749, 297
Purdue, Randolph, 104
Putnum, Marlene Ruby, 310
Quantrel, Susannah, 170
Quarter Circle U Ranch, 103, 196, 227, 240, 354
Queen Creek Canyon Tunnel, 12, 148
Queen Valley, 107, 192, 238
Quinlin, James, 102
raises, 49, 57, 79, 81, 358, 360
Ramboz Mining Camp, 197, 270
Randolph Canyon, 239, 243
Rapp, Charlie, 125, 200
Raymond, Rossiter, W., 25, 339
Reavis Ranch, 85, 110, 111, 159
Reavis, Elisha, 111, 159
Reavis, James Addison, 216
Red Jack Elmer (see Almer, Red Jack)
Red Wagon Brown, 189
Reeb, George, 103
Reese, Mrs. H. E. , 146, 147
Regan, Benjamin W., 13, 14, 19, 20, 22, 31, 32, 280
Renteria, J., 290
Reymert Mill Town, 245, 246
Reymert, J.D., 118, 134
Reynolds, Glen (Sheriff), 216
Rice, Mike, 214, 347
Riggs, Eunice Ann, 110
Ripsey Family, 107
Rivera, Christobal, 142
Riverside Stage Station, 207, 267, 309, 346
Riverside, Arizona, 153, 190, 191, 199, 209, 267, 269, 276, 296, 308, 310
Rogers Canyon, 28, 160, 243, 244, 250-253
Rogers District, 250
Rogers Trough Trailhead, 250
Rogers, James, 16
Rogers, M., 29
Rorebeck, Curtis G., 285
Rose, Ann Knight, 124, 342, 352, 111, 112, 116
Rose, Dan, 100, 342, 352
Roth, Lena, 171
Round Valley, 233
Rowes Crossing, 106
Rowland, Sheriff Billy, 215
Ruggles, Levi, Judge, 157
Russell Gulch, 191
Sabin, (Doctor), 212, 213
Safford, A. P. K. (Governor), 11, 16, 26, 33
Sam, Jim, 144
San Carlos (Camp), 14, 22, 120, 163
San Felice, Michael, 307
San Felice, Rusty, 319
San Felice, Tony (grandson), 246, 307
San Felice, Tony (son), 311
Sarrick, E.A., 255
Saviano, Ernie, 295
Scanlan, G.D., 208
Schieffelin, Ed, 34, 339
Schoose, Bob, 253, 254, 295
Schute, Gus, 274
Schuyler, Walter S. (Lt.), 21
Schwartz, Milton, 30, 338
Scofield, A.J., 284
Seiber, Al, 16, 21, 275, 276, 298, 350
Selby Works, 56, 61
Shaefer, F.A., 233
Shears, James, 188
Sherman Silver Purchase Act, 281, 282
Sherman, Barbara, 149, 344
Sherman, James E., 149, 344
Shill, Wright P., 119
Shoot-em up Dick, 144, 145
Shumway Case, 328, 330, 350
Siderits, Karl P., 334
Sieber, Al (see Seiber also), 16, 21, 274-276, 298, 349
Sierra Anchas, 16
Signal Mountain, 9
Signal Peak, 12, 13
Silver King & Florence Telegraph Company, 42
Silver King & Globe Stage Line, 186, 190
Silver King of Arizona Mining Company, 283-290, 349
Silver King Town, 28, 101, 110, 114, 115, 116, 118, 119, 120, 128, 132, 180, 297, 300, 304, 314
Single jacking, 79, 360
Sites, Candy, 315
Sites, George, 311, 314, 322, 336
Southern Pacific Railroad, 42, 68, 218
Southern Pacific Stage Line, 190
Spanish Trail, 239, 243
Spence, W.E., 186
Sprangler, D.B., 267
Square Sets, 82
Squaw Peak, 65, 362
St. Elmo's Saloon, 203, 205, 206, 207, 272

Stambach, Bob, 8, 256, 326
Stambaugh, Bill, 313
Stambaugh, Billie, 313
Stanfield, Johnny, 90
Starar, Andrew, 29,
Starar, Jacob, 29
Starar, Lewis, 29
Starr, C.H., 29, 214
Starrar, C. H. , 29
Steele, George, 119
Stern, E.J., 286
Stoneman Camp, 13, 298
Stoneman Grade, 9, 10, 11, 12, 14, 16, 20, 21, 23, 24, 26, 40, 43, 99, 102, 103, 179, 289, 297, 298, 300, 307
Stoneman Trail, 10, 99, 110, 117, 164, 296, 298, 300, 301, 304, 306, 314
Stoneman, George (General), 1, 4, 5, 6, 9, 10, 11, 14, 21, 50, 179, 298, 300
stopes, 49, 52, 56, 77, 79, 82, 263, 326
Sullivan, John, 4, 10, 11, 13, 14, 16, 18, 20, 21, 22, 23, 28, 29, 30, 33, 100
Summers, H.B., 270
Suter, Jake, 141, 150
Sutherland, W. H., 190
Swanson, James, 107, 343
Swift Trail, 305, 306, 313, 350
Swift, T. T., 305
Taylor, Joe, 88
Tempe Normal School, 284
Tenney, J.B., 296, 356
Tenny, J.B., 41
Texas & California Stage Line, 190
Thacker, (Detective), 208
The Miami Copper Company, 272, 274
Thomas, Hinton, 124
Thompson, Boyce William (Col.), 9, 263, 264, 306
Thompson, Gassy, 77, 78
Thompson, John H. (Sheriff), 206
Thompson, Mrs. Thomas, 160
Thrapp, Dan, 270, 298, 349
Thurston, Janet, 314
Tite, Harriet Bowen, 173
Tom Haley's Camp, 269
Tombstone, 34, 165-169, 220, 221, 265, 339
Tommyknockers, 77, 78
Tonto Creek, 16
Top of the World, 12
Tordillo Peak, 5
Tortilla Creek, 234, 238
Tortilla Mountains, 226, 238
Town of Barcelona, 268
Town of Florence, 2, 4, 30, 265
Town of Globe, 261, 270, 271
Town of Kearney, 190
Town of Miami, 270, 271, 272
Town of Sacaton, 6, 310
Town of Sonora, 268
Town of Superior, 261
Trap Canyon, 238
Tressler, Clyde, 300, 301, 302
Trevathan, Richard, 76
Trevethan, Philippa, 77
Trimble, Marshall, 23, 98, 341, 353
Truckee, California, 56
Trueman, William C., 94
Truman, William, 219
Tucker, Steven, 178, 179
Tucker, Velma Bowen, 170, 178, 179, 241, 345
Tunnel Saloon, 141, 210-214, 266
Tuttle, Joe, 208-210
Two Gun" Richard, 319, 320
United States Brewery, 137
University of Minnesota, 284
Up and Down Saloon, 272
Upper La Barge Canyon, 238
Vail, F. W. (Dr.), 203
Valentine, John, 124
Valenzuela, Frank, 145
Van Courtlandt, E.N., 94
Vekol Mountains, 7
Villa Verde Family, 259
Villa Verde, Peter, 259
W.J. Clemans Company, 109
Walker, Asa J., 107
Walker, Churga, 7
Walker, John D. (Captain), 5, 6, 210
Walker, Juana, 7
Wallace, Lois, 327
Wallach, Judge, 329
Waltz, Jacob, 7, 16, 17, 29, 108, 232, 236, 239, 243, 244
Wanna Whata Silver Mining Company, 136
Wardwell,.William L. (Dr.), 88, 89
Washoe & Co., 249
Weavers Needle, 225, 226, 241, 242, 246
Weeden, Thomas, 8
Weiser, Jacob, 8, 108, 338
Welch, M., 19
Welles, J. P., 282
Wells Fargo & Co., 33, 65, 67, 130, 137, 184, 186, 187, 188, 190 ,,192, 202, 204, 208, 265, 338
Wells, Frank D., 232
Western Union, 42

White Mountains, 819
Whitford, Chris, 107
Whitford, Henry, 107
Whitlow Canyon, 107, 192, 237
Whitlow Dam, 107
Whitlow Ferry, 107
Whitlow Ranch, 107
Whitlow Stage Station, 192
Whitlow, Charles, 105
Whitlow, Mary Elizabeth "Molly", 32
Whitlow, W., 107
Wide Awake Mining Company, 246
Wilburn, John, 275
Wildman, Perry, 100, 102, 115, 120, 122, 126, 127, 132, 139, 150, 183, 342
Williams, Lena Roth, 170
Williams, Robert, Mr., 27, 115 ,116, 117 132, 150, 171, 227
Williamson, Dan, 215, 275
Wilson, "Dump", 24
Windsor Consolidated Company, 65
winzes, 38, 49, 57, 77, 79, 317
Wolf, Rudy, 274
Wong Kong, 144
Wood Camp, 59, 75, 127, 128, 129
Woody, Clara T., 22
Worst, Clay, 249
Wratten, Carrie, 211
Wright, Bob, 181
Wright, Dan, 220, 221, 350
Wy, Wang, 143
Young, Otis E., 74, 338, 352
Zellweger cowboys, 108
Zenkendorf, Louis, 268

MINES

Anna Bell, 245, 257, 258
Australia, 254
Belcher, 148, 245
Belmont, 158
Big Dog #1, 317
Big Johnny, 108
Black Diamond, 245, 247
Bluebird, 270
Chloride King, 250, 251
Columbia, 245, 250, 257, 258, 259
Copper Depot, 269
Copper Gulch, 25
Copper Queen, 318
Daisy Dean, 197, 270
Dan & Mac, 275
El Medico Claim, 282, 293, 308, 334, 336
Emperor, 246, 247
Erie, 246
Eureka, 148, 245, 247, 248
Eureka Consolidated, 259
Famous McMillen, 270
Fortuna,, 318
Frankfort Mine, 249
Gem, 148, 149, 246, 261
Geo. Washington, 148
Gibson Mine, 274
Globe Ledge, 19, 20, 24, 32, 270, 272
Gold Leaf Ledge, 198
Hal & Al, 275
Hayes, 148
Hub and Irene Claims, 25, 262
Iron Mountain Ore Cart, 250
Little Molly, 91
Lost Coon, 275
Lost Dutchman, 7, 28, 90, 223, 231, 241, 242, 243, 252, 253, 274
Lost Soldiers, 235, 236, 242-244
Lucky Cuss, 34
Mac Morris, 203, 270
Mack Morris, 203, 270
Magma, 42, 85, 145, 149, 261, 263, 264, 271, 284, 304, 306, 312, 314
Manhattan, 250
Manhatten, 148, 251
Martinez Canyon, 256
Monarch, 250, 251
Monarch of the Sea, 247, 248
Monitor Camp, 267
Monroe, 269
Monroe Doctrine, 275
Munson's Chunk, 74, 271
Old Dominion, 90, 270, 272
Omya Marble, 12, 297
Orphan Boy, 245
Peachville , 245
Peralta Mine, 252, 253
Pike, 247
Pinal Chief, 245
Pinto Valley, 272, 273, 274
Pioneer Con, 253
Poor Man, 269
Ramboz, 197, 270
Ray, 210, 214, 217, 261, 267, 269, 310, 313, 314
Rescue, 270
Seventy-Six, 27, 28, 137
Silver Belle, 245, 257, 258, 259
Silver Brick Claim, 29
Silver Chariot, 148
Silver Chief, 244, 250, 251

Silver Queen, 11, 18, 19, 24, 25, 147, 148, 245, 261, 263, 264, 270, 285
Silverado, 318, 245
Silverlock, 247
Snap, 250, 251
Stonewall Jackson, 270
Supriser, 148
Uncle Billy, 255, 256
Vekol, 7, 8
Wana Whata, 148
Windsor, 65, 148, 244
Yo Tambian, 275

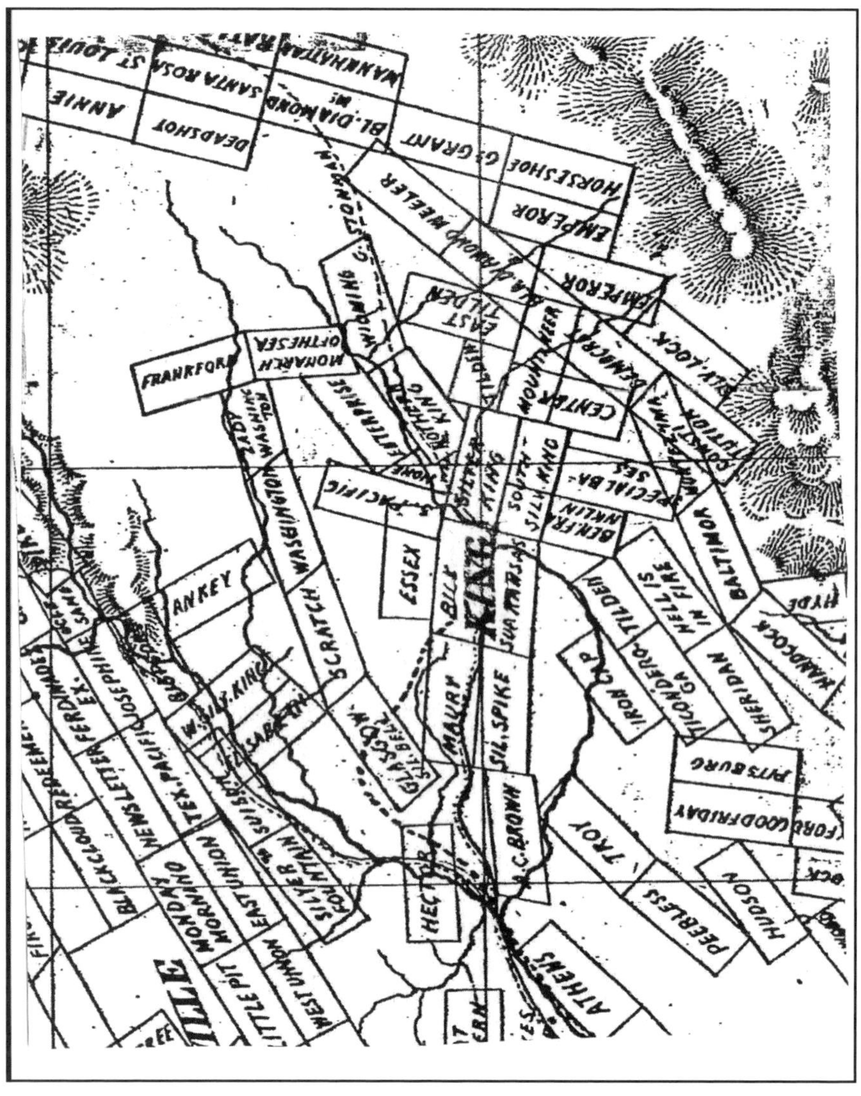

Mines near Silver King 1882 — University of Arizona Library, Tucson

About the Author

Originally from Beaver Falls, Pennsylvania, Jack San Felice retired as a Police Captain from a 30-year career in law enforcement in Prince George's County Maryland and Washington, D.C. Jack spent three years in the U.S. Army serving at Ft. Polk, Louisiana, Ft. Campbell, Kentucky and Ft. Belvoir, Virginia. He attended Geneva College in Beaver Falls, Pennsylvania, University of Maryland College Park, and the American University in Washington, D.C. where he received his B.S. and M.S. degrees. While in the Research and Development unit of the County Police, Jack honed his research skills under mentors Jim Burch and John Rhoads. Jack has been an avid hiker, horseback rider, and four-wheeler of the Superstition Mountains and Central Arizona for many years. For over twelve years he has studied, written, authored and lectured about history surrounding the Superstition Mountains, cowboy legends, Indian stories, mining and prospecting in Arizona Territory of the 1800s and beyond. Jack lives in Mesa, Arizona with his wife of 41 years, Wynne and near his adult children who followed Jack and his dreams out west to find their own. Jack and Wynne also have a son in Maryland and six grandchildren there, as well as three grandchildren in Arizona. He has been an instructor at Scottsdale Community College for over eleven years. He is the author of the *Lore of the Superstitions, Treasure Trails of the Superstitions, and Squaw Peak- A Hikers Guide.* (*Squaw Peak-A Hiker's Guide* was published prior to the name Squaw Peak being changed to Piestawa Peak.)

To my readers: I hope you have enjoyed reading about the history of the Silver King as much as I have enjoyed researching and reliving the experiences of the Arizona Territory pioneers.

Sincerely, Jack San Felice

Author decending into mine — *Greg Davis photo*